MATHEMATICS

Exploring Your World

AUTHORS

Grades K–2 Team
Herbert P. Ginsburg
Deborah B. Gustafson
Larry P. Leutzinger

Problem Solving Team
Lucille Croom
Gerald A. Goldin
Stephen Krulik
Henry O. Pollak
Jesse A. Rudnick
Dale G. Seymour

Grades 3–8 Team
Ruth I. Champagne
Carole E. Greenes
William D. McKillip
Lucy J. Orfan
Fernand J. Prevost
Bruce R. Vogeli
Marianne V. Weber

SILVER BURDETT & GINN

MORRISTOWN, NJ • NEEDHAM, MA
Atlanta, GA • Cincinnati, OH • Dallas, TX
Deerfield, IL • Menlo Park, CA

 ISBN 0-382-11844-8

Table of Contents

3 THEME Children in Action
Adding

4 THEME Third Graders: Busy at School
Subtracting

5 THEME Wonderful Things Around Us
Time • Graphing 138

6 THEME Fascinating Creatures
Multiplication Facts 0–5 170

7 THEME Mountain Adventure
Multiplication Facts 6–9 200

8 THEME People: Caring and Sharing
Division Facts 228

9 THEME Fun at Home
Fractions • Probability 264

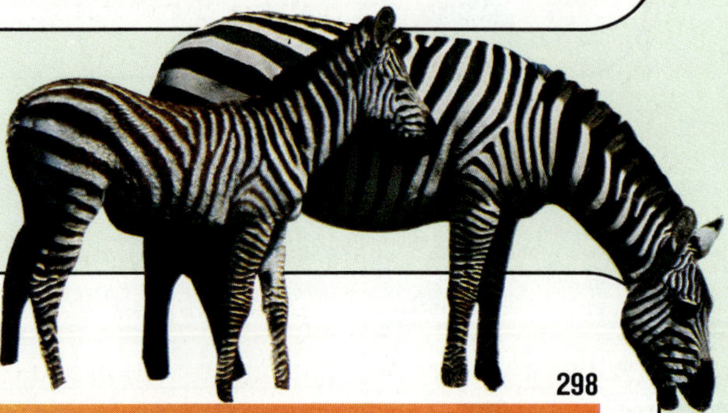

10 THEME The Great Outdoors
Understanding Measurement 298

GAS
PHONE
2.1 MI.

THEME Saving Our Planet

13 Exploring Multiplication and Division: One-Digit Numbers 394

CHAPTER RESOURCES

BOOK RESOURCES

GETTING STARTED

Come explore how mathematics plays an important role in your world.

MATHEMATICS

Exploring Your World

How is this chef using mathematics?

Why won't this seesaw balance?

What geometric shapes do you see?

Share Your Ideas Name other ways that mathematics is used in our world.

How will you learn mathematics?

This mathematics adventure will help you learn mathematics in many different ways.

Learn by Sharing Ideas with Others

Working together

What if all numbers in our world disappeared? List some problems we would face in a world without numbers.

In your group . . .

- Listen carefully.
- Agree or disagree pleasantly.
- Build on each others' ideas.
- Encourage everyone to share ideas.

Learn by Making Choices

CHOICES

Decision-making is part of mathematics. This symbol means you will have to make choices to solve problems.

Do you need an exact answer or an estimate? Which is the best way to find the answer?

- objects (like counters)
- paper and pencil
- calculator
- mental math

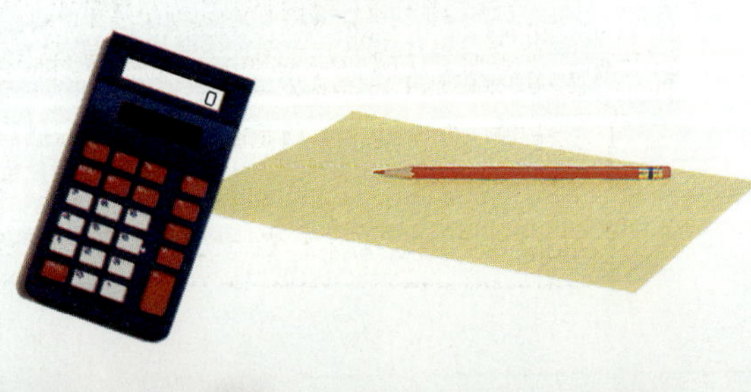

Learn by Solving Interesting Problems.

You are a problem solver! By learning to make choices, you will become a good problem solver.

THINK
EXPLORE
SOLVE
LOOK BACK

Look at each photograph. Choose a way to solve each problem. Explain your choices.

Buy Hugo a few toys. Will $10.00 be enough to pay for the toys?

Sew Hugo's rug. What other arrangements of squares can you make?

Hugo won't sit still. How can you find Hugo's height and weight?

Look ahead! You will learn new and exciting ways to do mathematics. You will discover that YOU are a mathematician.

SUMMING UP

1 Addition and Subtraction Facts

THEME People: Lending a Helping Hand

Sharing What You Know

Look at the pictures and tell the story. Think about the order of events. What happened first? What happened after the lion let the mouse go? What if the order of the events changed? Might the story have a different ending if the lion had been captured first? Why or why not?

Using Language

The order of events in a story tells what happened first, next, and so on. In mathematics, the **order property of addition** says that changing the order in which numbers are added does not change the sum. How is the order of events in a story like or different from the **order property of addition**?

Words to Know: doubles, sum, addend, order property, missing addend, difference, fact family

Be a Problem Solver

Look at the numbers. What pattern do you see?

1 9 2 8 3 7 4 6

Describe the pattern. Can you find more than one? **What if** you changed the order of the numbers? What new patterns can you make?

Create a new pattern of numbers. Describe your pattern.

GETTING STARTED

How are these facts alike?
5 + 5   3 + 3   1 + 1   8 + 8
Why do you think they are called doubles?

Mental Math Strategies: Addition Facts to 18

Third graders help the ranger plant seeds. They use mental math strategies to find out how many seeds they plant. Sue plants 8 seeds in one row and 7 in another row.

▸ You can use doubles.

Sue knows the double 7 + 7 = 14.
8 + 7 is 1 more.
So 8 + 7 = 15.

She also knows the double 8 + 8 = 16.
8 + 7 is 1 less.
So 8 + 7 = 15.

▸ You can make a sum of 10.

Al makes 10 by taking 2 from 7.
Now 8 + 2 = 10.
Then 10 + 5 = 15.

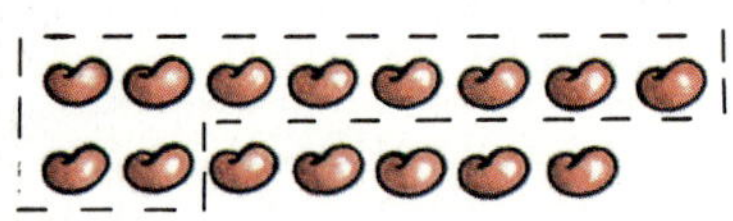

How could these strategies be used to find 5 + 4? 9 + 6? 8 + 4?

Check Your Understanding

Find each sum.

1. 7 + 7 **2.** 4 + 5 **3.** 3 + 2 **4.** 5 + 8 **5.** 6 + 6 **6.** 8 + 7 **7.** 9 + 4

Share Your Ideas Which strategies did you use for **1–4**? Why did you choose them?

Practice

Find each sum. Use mental math strategies to help you.

8. $4 + 4$	**9.** $2 + 1$	**10.** $5 + 5$	**11.** $7 + 8$	**12.** $9 + 6$	**13.** $3 + 3$	**14.** $6 + 5$
15. $2 + 3$	**16.** $6 + 7$	**17.** $7 + 5$	**18.** $8 + 8$	**19.** $5 + 7$	**20.** $8 + 5$	**21.** $2 + 2$
22. $6 + 9$	**23.** $1 + 2$	**24.** $9 + 9$	**25.** $6 + 8$	**26.** $5 + 9$	**27.** $7 + 9$	**28.** $4 + 3$

29. $5 + 6 = \square$	**30.** $8 + 3 = \square$	**31.** $1 + 1 = \square$	**32.** $9 + 7 = \square$
33. $9 + 8 = \square$	**34.** $5 + 4 = \square$	**35.** $4 + 7 = \square$	**36.** $0 + 0 = \square$
37. $\square = 2 + 9$	**38.** $\square = 3 + 4$	**39.** $\square = 8 + 9$	**40.** $\square = 7 + 6$

Follow each rule to complete.

Rule: Add 9.

	Input	Output
41.	6	
42.	8	
43.	3	
44.	7	

Rule: Double the input.

	Input	Output
45.	5	
46.	9	
47.	6	
48.	8	

Think and Apply

49. Write an addition fact for the fingers on 2 hands. the days in 2 weeks.

50. Fred planted 6 trees. Tom planted the same number of trees. How many trees did they plant?

What if you had to find the sum for 6 + 7? What different strategies could you use?

SUMMING UP

GETTING STARTED

Which would you rather have—a penny and a nickel or a nickel and a penny?

Addition Properties

Strategies and patterns help you to find sums.

If you know that 7 + 6 = 13, do you know what 6 + 7 is?

addend ↑ addend ↑ sum ↑

Are the sums the same?

Choose another basic fact. If you change the order of the addends, is the sum the same?

The Order Property

The order in which numbers are added does not change the sum.

7 + 6 = 6 + 7
13 = 13

Now add 0 + 5.

What happens when you add 0 and any number?
How does this pattern help you to know other addition facts?

The Zero Property

The sum of any number and 0 is that number.

0 + 5 = 5 and 5 + 0 = 5

Check Your Understanding

Add.

1.	3	2	**2.**	5	6	**3.**	9	0	**4.**	7	8	**5.**	0	7
	+2	+3		+6	+5		+0	+9		+8	+7		+7	+0

Share Your Ideas Draw a picture to show that
a. 3 + 5 = 5 + 3 **b.** 7 + 0 = 7 and 0 + 7 = 7.

Practice

Add.

6. $\begin{array}{r}2\\+1\\\hline\end{array}$ $\begin{array}{r}1\\+2\\\hline\end{array}$
7. $\begin{array}{r}8\\+2\\\hline\end{array}$ $\begin{array}{r}2\\+8\\\hline\end{array}$
8. $\begin{array}{r}3\\+0\\\hline\end{array}$ $\begin{array}{r}0\\+3\\\hline\end{array}$
9. $\begin{array}{r}6\\+3\\\hline\end{array}$ $\begin{array}{r}3\\+6\\\hline\end{array}$
10. $\begin{array}{r}7\\+2\\\hline\end{array}$ $\begin{array}{r}2\\+7\\\hline\end{array}$
11. $\begin{array}{r}5\\+9\\\hline\end{array}$ $\begin{array}{r}9\\+5\\\hline\end{array}$
12. $\begin{array}{r}8\\+6\\\hline\end{array}$ $\begin{array}{r}6\\+8\\\hline\end{array}$
13. $\begin{array}{r}0\\+5\\\hline\end{array}$ $\begin{array}{r}5\\+0\\\hline\end{array}$
14. $\begin{array}{r}2\\+5\\\hline\end{array}$ $\begin{array}{r}5\\+2\\\hline\end{array}$
15. $\begin{array}{r}4\\+1\\\hline\end{array}$ $\begin{array}{r}1\\+4\\\hline\end{array}$
16. $\begin{array}{r}3\\+9\\\hline\end{array}$ $\begin{array}{r}9\\+3\\\hline\end{array}$
17. $\begin{array}{r}6\\+4\\\hline\end{array}$ $\begin{array}{r}4\\+6\\\hline\end{array}$

18. $8 + 1 = \square$ $1 + 8 = \square$
19. $7 + 9 = \square$ $9 + 7 = \square$
20. $4 + 8 = \square$ $8 + 4 = \square$
21. $0 + 6 = \square$ $6 + 0 = \square$
22. $4 + 3 = \square$ $3 + 4 = \square$
23. $6 + 2 = \square$ $2 + 6 = \square$
24. $0 + 4 = \square$ $4 + 0 = \square$
25. $5 + 8 = \square$ $8 + 5 = \square$
26. $0 + 1 = \square$ $1 + 0 = \square$
27. $7 + 5 = \square$ $5 + 7 = \square$
28. $9 + 8 = \square$ $8 + 9 = \square$
29. $7 + 3 = \square$ $3 + 7 = \square$

Find each missing number.

30. $5 + 3 = \square + 5$
31. $4 + 7 = 7 + \square$
32. $1 + 6 = \square + 1$
33. $6 + 3 = \square + 7$
34. $1 + 9 = 8 + \square$
35. $7 + \square = 3 + 4$

Think and Apply

DATA 36. Find things that are in groups of 1–9. Choose 3 groups. Write an addition story for each. Use a strategy to solve each.

37. Val sold 2 tulip bulbs and 3 crocus bulbs. Ben sold 3 tulip bulbs and 2 crocus bulbs. How many bulbs did each person sell?

Visual Thinking

Write an addition fact for each domino. Then mentally turn the domino the other way and write another fact.

38. 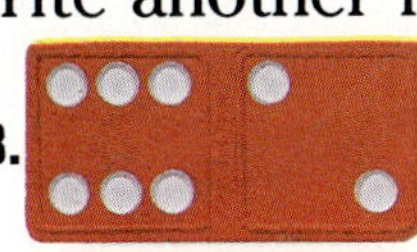

39. 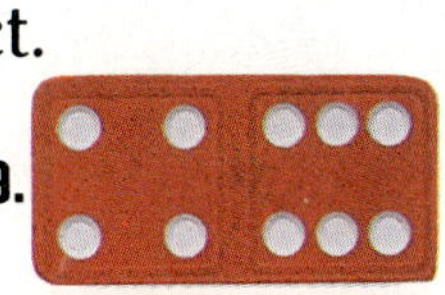

40. 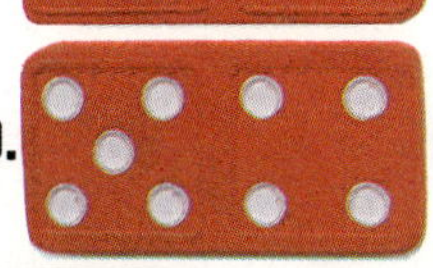

41. 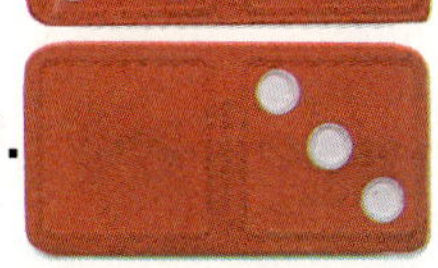

SUMMING UP

Explain the order property and the zero property in your own words.

GETTING STARTED

Which addition facts have a sum of ten?

Adding More Than Two Numbers

How many patients at Children's Hospital will get a balloon?

4 + 5 + 2 = □

We can group the addends in different ways.

(4 + 5) + 2 = □ → **9 + 2 = 11**

4 + (5 + 2) = □ → **4 + 7 = 11**

▸ The way in which numbers are grouped does not change the sum.

We can add down or up.

$$\begin{array}{r} 6 \\ 3 \\ +5 \\ \hline \end{array} \quad \begin{array}{r} 9 \\ +5 \\ \hline 14 \end{array} \downarrow \qquad \begin{array}{r} 6 \\ 3 \\ +5 \\ \hline \end{array} \quad \begin{array}{r} 6 \\ +8 \\ \hline 14 \end{array} \uparrow$$

We can also look for a sum of ten.

$$\begin{array}{r} 3 \\ 5 \\ +7 \\ \hline \end{array} \quad \begin{array}{r} 10 \\ +\ 5 \\ \hline 15 \end{array}$$

Numbers are grouped to make adding easier.

Check Your Understanding

Add.

1. 3 + 6 + 2 = □ **2.** 5 + 2 + 5 = □ **3.** 4 + 5 + 3 = □

4. $\begin{array}{r} 6 \\ 3 \\ +0 \\ \hline \end{array}$ **5.** $\begin{array}{r} 3 \\ 4 \\ +5 \\ \hline \end{array}$ **6.** $\begin{array}{r} 2 \\ 2 \\ +7 \\ \hline \end{array}$ **7.** $\begin{array}{r} 3 \\ 2 \\ +6 \\ \hline \end{array}$ **8.** $\begin{array}{r} 8 \\ 2 \\ +2 \\ \hline \end{array}$ **9.** $\begin{array}{r} 6 \\ 3 \\ +4 \\ \hline \end{array}$

Share Your Ideas Explain how you added in **2**, **5**, and **8**.

Practice

Add.

10. $\begin{array}{r} 4 \\ 3 \\ +2 \\ \hline \end{array}$ 11. $\begin{array}{r} 4 \\ 2 \\ +7 \\ \hline \end{array}$ 12. $\begin{array}{r} 3 \\ 1 \\ +6 \\ \hline \end{array}$ 13. $\begin{array}{r} 6 \\ 0 \\ +2 \\ \hline \end{array}$

14. $\begin{array}{r} 5 \\ 3 \\ +6 \\ \hline \end{array}$ 15. $\begin{array}{r} 7 \\ 2 \\ +4 \\ \hline \end{array}$ 16. $\begin{array}{r} 6 \\ 2 \\ +4 \\ \hline \end{array}$ 17. $\begin{array}{r} 2 \\ 6 \\ +3 \\ \hline \end{array}$

18. $\begin{array}{r} 1 \\ 4 \\ 3 \\ +5 \\ \hline \end{array}$ 19. $\begin{array}{r} 2 \\ 1 \\ 1 \\ +8 \\ \hline \end{array}$ 20. $\begin{array}{r} 5 \\ 1 \\ 2 \\ +7 \\ \hline \end{array}$ 21. $\begin{array}{r} 9 \\ 0 \\ 6 \\ +3 \\ \hline \end{array}$

Choose the correct sum.

22. 4 + 3 + 6 = ☐
a. 10 b. 13
c. 9 d. 7

23. 5 + 4 + 5 = ☐
a. 9 b. 15
c. 14 d. 10

Think and Apply

Use the numbers [4] [1] [3] [5] [2]. **Find the sum of**

24. the numbers greater than 2.

25. the numbers less than 4.

Solve.

26. Bill has a total of 14 balloons to give out. He gave out 5, 3, and 4 balloons. Does he have any balloons left? Explain.

27. Sue has 16 balloons to give to three children. She gives 7 balloons to the first child and 2 balloons to the second child. How many balloons will the third child get?

SUMMING UP

Explain how you can add 6 + 2 + 4 in different ways.

Mixed Review

1. $\begin{array}{r} 5 \\ +5 \\ \hline \end{array}$ 2. $\begin{array}{r} 8 \\ +3 \\ \hline \end{array}$

3. $\begin{array}{r} 0 \\ +9 \\ \hline \end{array}$ 4. $\begin{array}{r} 4 \\ +6 \\ \hline \end{array}$

5. $\begin{array}{r} 3 \\ +3 \\ \hline \end{array}$ 6. $\begin{array}{r} 7 \\ +4 \\ \hline \end{array}$

7. $\begin{array}{r} 1 \\ +5 \\ \hline \end{array}$ 8. $\begin{array}{r} 5 \\ +8 \\ \hline \end{array}$

9. $\begin{array}{r} 9 \\ +9 \\ \hline \end{array}$ 10. $\begin{array}{r} 8 \\ +7 \\ \hline \end{array}$

11. $\begin{array}{r} 6 \\ +5 \\ \hline \end{array}$ 12. $\begin{array}{r} 9 \\ +7 \\ \hline \end{array}$

13. 7 + 2 = ☐

14. 4 + 8 = ☐

15. 9 + 3 = ☐

16. 6 + 7 = ☐

17. 5 + 3 = ☐

18. 4 + 1 = ☐

19. 3 + 3 = ☐

20. 4 + 5 = ☐

21. 3 + 1 = ☐

Midchapter Review

Find each sum. pages 2–5

1. $\begin{array}{r}2\\+3\\\hline\end{array}$ 2. $\begin{array}{r}4\\+4\\\hline\end{array}$ 3. $\begin{array}{r}8\\+0\\\hline\end{array}$ 4. $\begin{array}{r}1\\+5\\\hline\end{array}$ 5. $\begin{array}{r}9\\+5\\\hline\end{array}$ 6. $\begin{array}{r}7\\+6\\\hline\end{array}$

7. $\begin{array}{r}6\\+8\\\hline\end{array}$ 8. $\begin{array}{r}8\\+9\\\hline\end{array}$ 9. $\begin{array}{r}9\\+4\\\hline\end{array}$ 10. $\begin{array}{r}4\\+8\\\hline\end{array}$ 11. $\begin{array}{r}6\\+6\\\hline\end{array}$ 12. $\begin{array}{r}0\\+9\\\hline\end{array}$

13. $6 + 4 = \square$ 14. $7 + 9 = \square$ 15. $\square = 5 + 8$

Add. pages 6–7

16. $\begin{array}{r}4\\3\\+6\\\hline\end{array}$ 17. $\begin{array}{r}7\\0\\+9\\\hline\end{array}$ 18. $\begin{array}{r}5\\2\\+7\\\hline\end{array}$ 19. $\begin{array}{r}2\\5\\+4\\\hline\end{array}$ 20. $\begin{array}{r}6\\3\\+6\\\hline\end{array}$ 21. $\begin{array}{r}1\\7\\1\\+8\\\hline\end{array}$

22. $8 + 1 + 6 = \square$ 23. $7 + 2 + 7 = \square$ 24. $\square = 3 + 1 + 8$

Choose the correct word to complete each sentence.

25. The answer in addition is called the _____.
26. Each number that is added is called an _____.
27. Facts in which both addends are the same are called _____.

Words To Know
doubles addend sum

Solve.

28. Jody helped at the hospital 4 days one week and the same number of days the next week. How many days is that in all?
29. Look at Jake's schedule. How many hours was Jake scheduled to help out at the hospital?

Exploring Problem Solving

THINK
EXPLORE
SOLVE
LOOK BACK

How Many Different Lengths?

This rope can be used to measure different lengths.

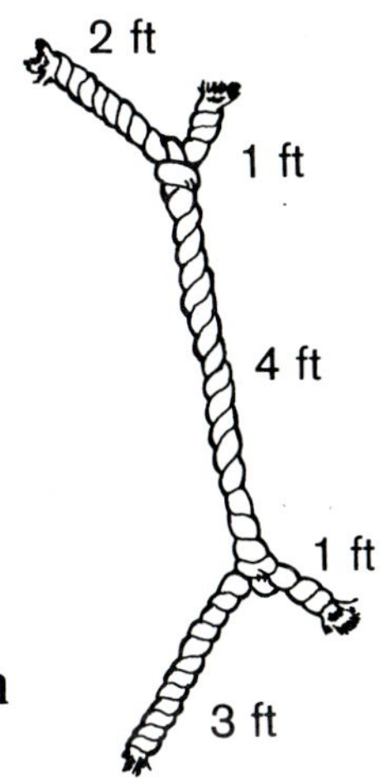

Thinking Critically

What lengths can you measure with it?

When you do problem solving lessons like this, work in a small group. Keep a written record of your work so you can share your thinking.

Analyzing and Making Decisions

1. What lengths can you measure if you use one part of the rope? two parts of the rope? three parts of the rope?
2. Can you use four parts of the rope? Explain.
3. List all the lengths you can measure, in order, from the shortest to the longest.

Look Back Design your own measuring rope. Find all the lengths you can measure with it.

Using Facts from Pictures and Text

Harry and his mother take the next ticket. How many numbers will be called before their number is called?

This four-step plan can help you solve problems.

Solving the Problem

Think What is the question? What are the facts? Look at the picture. What number was called last? What number will Harry and his mother take?

Explore How can you solve the problem? Will counting help?

Solve How many numbers will be called before their number is called?

Look Back Does your answer make sense? **What if** some students got 8 as an answer? How might they have gotten that answer? Why is it not correct?

Share Your Ideas

1. Look at the number of people in the picture. Why might the number of people in front of the counter be different from the number of tickets?
2. The rope problem on page 9 and this problem used drawings. How might seeing a drawing help you solve a problem?

THINK
EXPLORE
SOLVE
LOOK BACK

Practice

CHOICES

Solve. Use a calculator where appropriate.
Use the picture at the right to answer 3 and 4.

3. Erica wants to put a stamp on each letter. Can she? Explain.
4. **What if** Erica buys 5 more stamps? Will she have enough stamps to put 1 on each letter? Explain.
5. Carl wants to have 12 fish. He has 6 fish. How many more should he buy?
6. The Tigers scored 7 runs. The Royals scored 6 runs. Who won the baseball game?
7. Look at the beach scene below. Do you think it is 5:00 P.M. at the beach? Explain.
8. Vanessa saw 8 cans of dog food in the cupboard. She used 2 cans to feed her dogs. How many cans of dog food are in the cupboard now?
9. Harry drank 4 glasses of milk on Monday and 3 on Tuesday. How many glasses of milk did he drink altogether?
10. Cheryl read four books last week. She read the same number this week. How many did she read in the two weeks?

Create Your Own

Draw a picture and create a problem to go with it.

GETTING STARTED

How is tying and untying shoelaces like adding and subtracting numbers?

Relating Addition and Subtraction

Jose helps Carlos study. Jose has drawn 8 word cards. If there were 15 cards in the bag, how many are left?

Subtract to find how many are left.

15 − 8 = □

You can find the difference by finding the missing addend!

Think
8 + □ = 15
8 + 7 = 15
↑
missing addend

$$\begin{array}{r} 15 \\ -\ 8 \\ \hline 7 \end{array}$$ ← difference

There are 7 cards left.

What if 3 word cards are drawn from a bag of 12? How many cards are left? What addition fact could you use to solve this problem?

Check Your Understanding

Find each missing addend. Then write each difference.

1. $\begin{array}{r} 3 \\ +\square \\ \hline 6 \end{array}$ $\begin{array}{r} 6 \\ -3 \\ \hline \end{array}$ **2.** $\begin{array}{r} 4 \\ +\square \\ \hline 4 \end{array}$ $\begin{array}{r} 4 \\ -4 \\ \hline \end{array}$ **3.** $\begin{array}{r} 7 \\ +\square \\ \hline 10 \end{array}$ $\begin{array}{r} 10 \\ -\ 7 \\ \hline \end{array}$ **4.** $\begin{array}{r} 5 \\ +\square \\ \hline 9 \end{array}$ $\begin{array}{r} 9 \\ -5 \\ \hline \end{array}$

5. 2 + □ = 4 **6.** 0 + □ = 5 **7.** 6 + □ = 7 **8.** 8 + □ = 16

4 − 2 = □ 5 − 0 = □ 7 − 6 = □ 16 − 8 = □

Share Your Ideas What addition fact could you use to solve 11 − 8 = □?

Practice

Find each missing addend. Then write each difference.

9. $\begin{array}{r} 9 \\ +\square \\ \hline 17 \end{array}$ $\begin{array}{r} 17 \\ -\ 9 \\ \hline \end{array}$ **10.** $\begin{array}{r} 3 \\ +\square \\ \hline 7 \end{array}$ $\begin{array}{r} 7 \\ -3 \\ \hline \end{array}$ **11.** $\begin{array}{r} 5 \\ +\square \\ \hline 12 \end{array}$ $\begin{array}{r} 12 \\ -\ 5 \\ \hline \end{array}$ **12.** $\begin{array}{r} 2 \\ +\square \\ \hline 10 \end{array}$ $\begin{array}{r} 10 \\ -\ 2 \\ \hline \end{array}$

13. $\begin{array}{r} 5 \\ +\square \\ \hline 13 \end{array}$ $\begin{array}{r} 13 \\ -5 \\ \hline \end{array}$ **14.** $\begin{array}{r} 7 \\ +\square \\ \hline 11 \end{array}$ $\begin{array}{r} 11 \\ -\ 7 \\ \hline \end{array}$ **15.** $\begin{array}{r} 7 \\ +\square \\ \hline 7 \end{array}$ $\begin{array}{r} 7 \\ -7 \\ \hline \end{array}$ **16.** $\begin{array}{r} 1 \\ +\square \\ \hline 9 \end{array}$ $\begin{array}{r} 9 \\ -1 \\ \hline \end{array}$

17. $\begin{array}{r} 6 \\ +\square \\ \hline 15 \end{array}$ $\begin{array}{r} 15 \\ -\ 6 \\ \hline \end{array}$ **18.** $\begin{array}{r} 0 \\ +\square \\ \hline 9 \end{array}$ $\begin{array}{r} 9 \\ -0 \\ \hline \end{array}$ **19.** $\begin{array}{r} 4 \\ +\square \\ \hline 10 \end{array}$ $\begin{array}{r} 10 \\ -\ 4 \\ \hline \end{array}$ **20.** $\begin{array}{r} 9 \\ +\square \\ \hline 13 \end{array}$ $\begin{array}{r} 13 \\ -\ 9 \\ \hline \end{array}$

21. $9 + \square = 18$ $18 - 9 = \square$ **22.** $6 + \square = 9$ $9 - 6 = \square$ **23.** $8 + \square = 13$ $13 - 8 = \square$ **24.** $2 + \square = 8$ $8 - 2 = \square$

25. $3 + \square = 11$ $11 - 3 = \square$ **26.** $6 + \square = 12$ $12 - 6 = \square$ **27.** $1 + \square = 10$ $10 - 1 = \square$ **28.** $4 + \square = 13$ $13 - 4 = \square$

Complete each flowchart. Then tell what you discovered.

29. Input 9 → Add 6 → Subtract 6 → Output ?

30. Input 16 → Subtract 7 → Add 7 → Output ?

31. Make a flowchart of your own.

Think and Apply

32. Sara had 12 pencils. April borrowed some. Now, Sara has 9 pencils. How many pencils did April borrow?

33. Carl gave Rob 4 pens on Monday. On Friday he gave Rob two more pens. How many pens did Carl give to Rob?

Draw a picture to explain how $5 + 4 = 9$ and $9 - 4 = 5$ are related.

SUMMING UP

GETTING STARTED

How are adding and subtracting related?

Subtraction Strategies

A team of 16 people is needed to help clean the river. Seven people have volunteered. How many more people are needed?

Subtract to find how many more are needed.

$16 - 7 = \square$

Remember Every subtraction fact has a related addition fact.

$7 + \square = 16$
$7 + 9 = 16$

Nine more people are needed.

Patterns can help you with subtraction facts.

$\begin{array}{r} 9 \\ -9 \\ \hline 0 \end{array}$ $\begin{array}{r} 1 \\ -1 \\ \hline 0 \end{array}$ $\begin{array}{r} 3 \\ -3 \\ \hline 0 \end{array}$ $\begin{array}{r} 6 \\ -6 \\ \hline 0 \end{array}$

When you subtract a number from itself, the difference is 0.

$\begin{array}{r} 5 \\ -0 \\ \hline 5 \end{array}$ $\begin{array}{r} 3 \\ -0 \\ \hline 3 \end{array}$ $\begin{array}{r} 8 \\ -0 \\ \hline 8 \end{array}$ $\begin{array}{r} 1 \\ -0 \\ \hline 1 \end{array}$

When you subtract 0 from a number, the difference is that number.

Check Your Understanding

Find each difference.

1. $\begin{array}{r} 5 \\ -2 \\ \hline \end{array}$ **2.** $\begin{array}{r} 6 \\ -0 \\ \hline \end{array}$ **3.** $\begin{array}{r} 11 \\ -\ 5 \\ \hline \end{array}$ **4.** $\begin{array}{r} 15 \\ -\ 7 \\ \hline \end{array}$ **5.** $\begin{array}{r} 6 \\ -4 \\ \hline \end{array}$ **6.** $\begin{array}{r} 5 \\ -5 \\ \hline \end{array}$ **7.** $\begin{array}{r} 10 \\ -\ 3 \\ \hline \end{array}$

Share Your Ideas Look back at **1–7**. When did you use addition facts? When did you use patterns?

Practice

Find each difference.

8. $8 - 8$ **9.** $9 - 8$ **10.** $4 - 0$ **11.** $2 - 2$ **12.** $8 - 5$ **13.** $13 - 6$ **14.** $0 - 0$

15. $8 - 3$ **16.** $6 - 5$ **17.** $7 - 5$ **18.** $1 - 1$ **19.** $7 - 0$ **20.** $5 - 1$ **21.** $8 - 2$

22. $10 - 8$ **23.** $9 - 6$ **24.** $7 - 1$ **25.** $9 - 0$ **26.** $9 - 3$ **27.** $8 - 4$ **28.** $7 - 3$

Choose the correct difference.

29. $9 - 7 = \square$
a. 7 **b.** 3
c. 2 **d.** 6

30. $2 - 0 = \square$
a. 2 **b.** 4
c. 1 **d.** 0

31. $\square = 5 - 3$
a. 8 **b.** 6
c. 9 **d.** 2

32. $\square = 1 - 1$
a. 1 **b.** 0
c. 2 **d.** 6

Think and Apply

33. John needs 12 rakes and 3 shovels for his volunteers. He has 4 rakes. How many more rakes does John need?

34. Ed mows the front yard in 1 hour. Lisa mows the back yard in 1 hour. If they start mowing together at noon when will they finish?

Visual Thinking

Complete the pattern. Tell what comes next.

35.

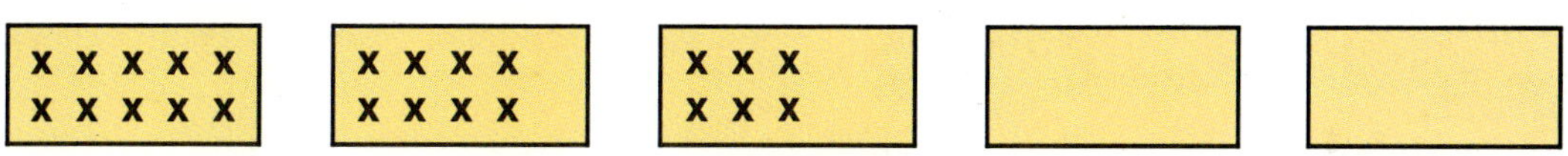

36.

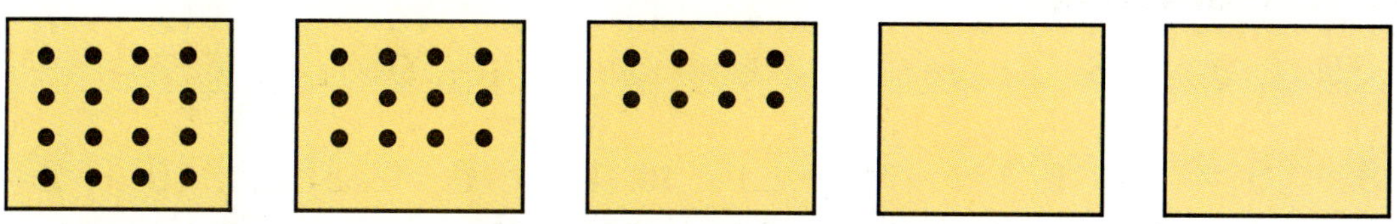

Look back at **8–11**. What strategies and patterns did you use to find each difference?

SUMMING UP

GETTING STARTED Suppose one box of dog biscuits has more biscuits than another. What are some ways you can find out how many more?

Subtraction Facts to 18

Jay likes animals. On Saturdays he helps the veterinarian.

There are 11 rawhide bones and 3 dogs. How many more bones are there than dogs?

Subtract to find out how many more are in one group than in another group.

$$\begin{array}{r} 11 \\ -\ 3 \\ \hline 8 \end{array}$$

$11 - 3 = 8$

There are 8 more bones than dogs.

Check Your Understanding

Subtract.

1. $\begin{array}{r} 10 \\ -\ 9 \\ \hline \end{array}$ **2.** $\begin{array}{r} 14 \\ -\ 6 \\ \hline \end{array}$ **3.** $\begin{array}{r} 17 \\ -\ 8 \\ \hline \end{array}$ **4.** $\begin{array}{r} 11 \\ -\ 2 \\ \hline \end{array}$ **5.** $\begin{array}{r} 15 \\ -\ 9 \\ \hline \end{array}$ **6.** $\begin{array}{r} 12 \\ -\ 7 \\ \hline \end{array}$ **7.** $\begin{array}{r} 14 \\ -\ 8 \\ \hline \end{array}$

8. $15 - 8 = \square$ **9.** $11 - 9 = \square$ **10.** $11 - 6 = \square$ **11.** $10 - 2 = \square$

Share Your Ideas Does $13 - 4 = 9$ show how many more biscuits there are than dogs? Explain.

Practice

Subtract.

12. $\begin{array}{r} 4 \\ -1 \\ \hline \end{array}$ 13. $\begin{array}{r} 12 \\ -3 \\ \hline \end{array}$ 14. $\begin{array}{r} 7 \\ -7 \\ \hline \end{array}$ 15. $\begin{array}{r} 10 \\ -6 \\ \hline \end{array}$ 16. $\begin{array}{r} 14 \\ -9 \\ \hline \end{array}$ 17. $\begin{array}{r} 6 \\ -0 \\ \hline \end{array}$ 18. $\begin{array}{r} 10 \\ -5 \\ \hline \end{array}$

19. $\begin{array}{r} 12 \\ -5 \\ \hline \end{array}$ 20. $\begin{array}{r} 7 \\ -6 \\ \hline \end{array}$ 21. $\begin{array}{r} 16 \\ -9 \\ \hline \end{array}$ 22. $\begin{array}{r} 11 \\ -8 \\ \hline \end{array}$ 23. $\begin{array}{r} 15 \\ -6 \\ \hline \end{array}$ 24. $\begin{array}{r} 14 \\ -7 \\ \hline \end{array}$ 25. $\begin{array}{r} 6 \\ -2 \\ \hline \end{array}$

26. $\begin{array}{r} 7 \\ -4 \\ \hline \end{array}$ 27. $\begin{array}{r} 10 \\ -7 \\ \hline \end{array}$ 28. $\begin{array}{r} 8 \\ -6 \\ \hline \end{array}$ 29. $\begin{array}{r} 12 \\ -9 \\ \hline \end{array}$ 30. $\begin{array}{r} 7 \\ -2 \\ \hline \end{array}$ 31. $\begin{array}{r} 9 \\ -3 \\ \hline \end{array}$ 32. $\begin{array}{r} 15 \\ -7 \\ \hline \end{array}$

33. $8 - 1 = \square$ 34. $2 - 0 = \square$ 35. $10 - 4 = \square$ 36. $14 - 5 = \square$

37. $16 - 8 = \square$ 38. $13 - 5 = \square$ 39. $12 - 4 = \square$ 40. $11 - 4 = \square$

Decide which has the greater difference.

41. $11 - 3$ or $11 - 9$ 42. $13 - 6$ or $12 - 0$ 43. $14 - 5$ or $15 - 9$

44. $16 - 7$ or $13 - 7$ 45. $12 - 6$ or $15 - 6$ 46. $16 - 9$ or $18 - 9$

Think and Apply

47. Dr. Barry treated 13 dogs. Six went home. How many were still at the hospital?

48. One day 12 dogs and 8 cats were treated. How many more dogs than cats were treated?

Logical Thinking

Follow the chain. Write the answer.

49. $0 + 3 + 5 - 5 - 3 = \bigcirc$

50. $0 + 4 + 2 - 4 - 2 = \bigcirc$

51. How could you find the answer without using paper and pencil?

Draw a picture for **47** and **48**. Then tell how you solved each problem.

SUMMING UP

GETTING STARTED

How are 5 + 7 and 7 + 5 alike?

Fact Families

Some facts use the same numbers. Together these facts make up a fact family. Here is the fact family for 5, 7, and 12.

5 + 7 = 12

7 + 5 = 12

12 − 5 = 7

12 − 7 = 5

Look at the addition and subtraction facts above. Tell in your own words how these facts are related.

Check Your Understanding

Complete each fact family.

1. $\begin{array}{r}6\\+9\\\hline\end{array}$ $\begin{array}{r}9\\+6\\\hline\end{array}$ $\begin{array}{r}15\\-\ 6\\\hline\end{array}$ $\begin{array}{r}15\\-\ 9\\\hline\end{array}$

2. $\begin{array}{r}4\\+9\\\hline\end{array}$ $\begin{array}{r}9\\+4\\\hline\end{array}$ $\begin{array}{r}13\\-\ 4\\\hline\end{array}$ $\begin{array}{r}13\\-\ 9\\\hline\end{array}$

Share Your Ideas Explain how to write a fact family for 7, 9, and 16.

Practice

Complete each fact family.

3. 7 + 6 = ☐
   6 + 7 = ☐
   13 − 7 = ☐
   13 − 6 = ☐

4. 3 + 6 = ☐
   6 + 3 = ☐
   9 − 3 = ☐
   9 − 6 = ☐

5. 5 + 0 = ☐
   0 + 5 = ☐
   5 − 5 = ☐
   5 − 0 = ☐

6. 9 + 8 = ☐
   8 + 9 = ☐
   17 − 9 = ☐
   17 − 8 = ☐

7. $\begin{array}{r}8\\+5\\\hline\end{array}$ $\begin{array}{r}5\\+8\\\hline\end{array}$ $\begin{array}{r}13\\-8\\\hline\end{array}$ $\begin{array}{r}13\\-5\\\hline\end{array}$

8. $\begin{array}{r}2\\+9\\\hline\end{array}$ $\begin{array}{r}9\\+2\\\hline\end{array}$ $\begin{array}{r}11\\-2\\\hline\end{array}$ $\begin{array}{r}11\\-9\\\hline\end{array}$

9. $\begin{array}{r}9\\+7\\\hline\end{array}$ $\begin{array}{r}7\\+9\\\hline\end{array}$ $\begin{array}{r}16\\-7\\\hline\end{array}$ $\begin{array}{r}16\\-9\\\hline\end{array}$

10. $\begin{array}{r}9\\+9\\\hline\end{array}$ $\begin{array}{r}18\\-9\\\hline\end{array}$

11. $\begin{array}{r}8\\+8\\\hline\end{array}$ $\begin{array}{r}16\\-8\\\hline\end{array}$

Think and Apply

Write a fact family for each.

12. 4, 7, 11
13. 5, 9, 14
14. 0, 9, 9
15. 5, 5, 10
16. How did you know what facts to use for **14**?
17. **Look back** at page 18. Write a fact family using the number of students in the picture.

Mixed Review

1. $\begin{array}{r}2\\+7\\\hline\end{array}$
2. $\begin{array}{r}8\\-4\\\hline\end{array}$
3. $\begin{array}{r}11\\-7\\\hline\end{array}$
4. $\begin{array}{r}8\\+5\\\hline\end{array}$
5. $\begin{array}{r}13\\-9\\\hline\end{array}$
6. $\begin{array}{r}9\\+6\\\hline\end{array}$
7. $\begin{array}{r}12\\-4\\\hline\end{array}$
8. $\begin{array}{r}7\\+7\\\hline\end{array}$
9. $\begin{array}{r}14\\-8\\\hline\end{array}$
10. $\begin{array}{r}0\\+9\\\hline\end{array}$
11. $\begin{array}{r}4\\1\\+7\\\hline\end{array}$
12. $\begin{array}{r}5\\3\\+2\\\hline\end{array}$
13. $\begin{array}{r}4\\2\\+5\\\hline\end{array}$
14. $\begin{array}{r}6\\3\\+3\\\hline\end{array}$

Complete each fact family.

15. 5 + 4 = ☐
    4 + 5 = ☐
    9 − 5 = ☐
    9 − 4 = ☐

16. 8 + 7 = ☐
    7 + 8 = ☐
    15 − 8 = ☐
    15 − 7 = ☐

Explain why the fact family 6 + 6 = 12, 12 − 6 = 6 has only two facts.

SUMMING UP

Using Problem Solving

Odd and Even Numbers

All of Whitney's socks are white, so it is easy to put them into pairs. Sometimes all the socks can be put into pairs. Sometimes they cannot.

Working together

Materials: 30 counters

A. Take a handful of counters and count them.

B. Make a guess as to whether you can pair all the counters or whether there will be a counter left over. Then try it.

C. Try **B** again with these groups of counters.

17 counters	15 counters
12 counters	19 counters
14 counters	16 counters

For which groups of counters could you make all pairs? For which groups did you have one counter left over?

- **Even numbers** can be shown by objects that can be put into groups of two.
- **Odd numbers** can be shown by objects that cannot be put into groups of two. There is always one object left over.

Sharing Your Ideas

1. What even numbers did you find?
2. What odd numbers did you find?
3. How can you tell if a number is even or odd? How would knowing this help Whitney?

THINK
EXPLORE
SOLVE
LOOK BACK

Extending Your Thinking

4. Use counters to show these additions.

$6 + 4 = \square$ $2 + 8 = \square$
$8 + 6 = \square$ $2 + 4 = \square$

Are the sums of these even numbers even or odd numbers? Look at the set of 8 counters and the set of 6 counters. Can the counters in each set be paired? When you combine the sets to make a set of 14, can all the counters be paired?

5. Use your counters to show these additions.

$3 + 5 = \square$ $9 + 1 = \square$
$7 + 3 = \square$ $5 + 9 = \square$

Are the sums of these odd numbers odd or even numbers? Look at the sets of counters for the addition $7 + 3$. Can you pair all of the counters in a set of 7 counters? in a set of 3 counters? in a set of 10 counters?

6. **What if** you add an even number and an odd number? Show these additions with counters.

$2 + 1 = \square$ $4 + 3 = \square$ $5 + 6 = \square$

What are the sums of these odd and even numbers? Are they even or odd numbers? Look at the counters for the addition sentence $5 + 6 = 11$. Can you pair the 5 counters? the 6 counters? the 11 counters?

Summing Up

7. Complete these sentences. Explain your answer.

a. even + even = ______ b. odd + odd = ______
c. odd + even = ______

8. What happens when you subtract? Show each of these subtractions with counters.

$6 - 4 = \square$ $9 - 3 = \square$ $10 - 3 = \square$

$8 - 2 = \square$ $7 - 5 = \square$ $9 - 6 = \square$

9. Complete these sentences. Explain your answer.

a. even − even = ______ b. odd − odd = ______
c. even − odd = ______ d. odd − even = ______

Chapter Review

Add. pages 2–7

1. $\begin{array}{r} 3 \\ +4 \\ \hline \end{array}$ 2. $\begin{array}{r} 7 \\ +7 \\ \hline \end{array}$ 3. $\begin{array}{r} 7 \\ +8 \\ \hline \end{array}$ 4. $\begin{array}{r} 4 \\ +0 \\ \hline \end{array}$ 5. $\begin{array}{r} 5 \\ +8 \\ \hline \end{array}$ 6. $\begin{array}{r} 9 \\ +1 \\ \hline \end{array}$

7. $6 + 3 = \square$
 $3 + 6 = \square$

8. $0 + 5 = \square$
 $5 + 0 = \square$

9. $7 + 5 = \square$
 $5 + 7 = \square$

10. $\square = 4 + 8$
 $\square = 8 + 4$

11. $3 + 5 + 2 = \square$ 12. $3 + 2 + 6 = \square$ 13. $\square = 2 + 3 + 2 + 3$

Write each missing addend. Then subtract. pages 12–13

14. $\begin{array}{r} 8 \\ +\square \\ \hline 17 \end{array}$ $\begin{array}{r} 17 \\ -8 \\ \hline \end{array}$

15. $\begin{array}{r} 9 \\ +\square \\ \hline 9 \end{array}$ $\begin{array}{r} 9 \\ -9 \\ \hline \end{array}$

16. $\begin{array}{r} 3 \\ +\square \\ \hline 12 \end{array}$ $\begin{array}{r} 12 \\ -3 \\ \hline \end{array}$

Subtract. pages 14–17

17. $\begin{array}{r} 11 \\ -\ 6 \\ \hline \end{array}$ 18. $\begin{array}{r} 12 \\ -\ 4 \\ \hline \end{array}$ 19. $\begin{array}{r} 7 \\ -0 \\ \hline \end{array}$ 20. $\begin{array}{r} 9 \\ -8 \\ \hline \end{array}$ 21. $\begin{array}{r} 5 \\ -5 \\ \hline \end{array}$ 22. $\begin{array}{r} 16 \\ -\ 7 \\ \hline \end{array}$

23. $17 - 9 = \square$ 24. $14 - 5 = \square$ 25. $11 - 2 = \square$

26. $\square = 9 - 5$ 27. $\square = 7 - 7$ 28. $\square = 15 - 6$

Write a fact family for each. pages 18–19

29. 8, 9, 17 30. 3, 9, 12 31. 3, 7, 10 32. 4, 9, 13

Choose the correct words to complete each sentence.

Words To Know
missing addend difference fact family

33. The answer in subtraction is the ______.
34. Facts that use the same numbers belong to a ______.
35. To solve $6 + \square = 9$ you find the ______.

Solve. pages 9–11

36. How many roses and tulips are there in all?
37. How many more roses than tulips are there?

Chapter Test

Add.

1. $\begin{array}{r} 6 \\ +7 \\ \hline \end{array}$ 2. $\begin{array}{r} 5 \\ +8 \\ \hline \end{array}$ 3. $\begin{array}{r} 9 \\ +9 \\ \hline \end{array}$ 4. $\begin{array}{r} 3 \\ +2 \\ \hline \end{array}$ 5. $\begin{array}{r} 1 \\ +4 \\ \hline \end{array}$ 6. $\begin{array}{r} 8 \\ +0 \\ \hline \end{array}$

7. $3 + 8 = \square$ 8. $4 + 6 = \square$ 9. $\square = 5 + 7$

$8 + 3 = \square$ $6 + 4 = \square$ $\square = 7 + 5$

10. $7 + 2 + 1 = \square$ 11. $6 + 2 + 4 = \square$ 12. $\square = 2 + 4 + 0 + 4$

Write each missing addend. Then subtract.

13. $\begin{array}{r} 2 \\ +\square \\ \hline 11 \end{array}$ $\begin{array}{r} 11 \\ -\ 2 \\ \hline \end{array}$ 14. $\begin{array}{r} 8 \\ +\square \\ \hline 16 \end{array}$ $\begin{array}{r} 16 \\ -\ 8 \\ \hline \end{array}$ 15. $\begin{array}{r} 7 \\ +\square \\ \hline 7 \end{array}$ $\begin{array}{r} 7 \\ -7 \\ \hline \end{array}$

Subtract.

16. $\begin{array}{r} 9 \\ -0 \\ \hline \end{array}$ 17. $\begin{array}{r} 11 \\ -\ 3 \\ \hline \end{array}$ 18. $\begin{array}{r} 7 \\ -7 \\ \hline \end{array}$ 19. $\begin{array}{r} 13 \\ -\ 4 \\ \hline \end{array}$ 20. $\begin{array}{r} 17 \\ -\ 8 \\ \hline \end{array}$ 21. $\begin{array}{r} 10 \\ -\ 3 \\ \hline \end{array}$

22. $9 - 4 = \square$ 23. $12 - 5 = \square$ 24. $14 - 9 = \square$

25. $\square = 15 - 6$ 26. $\square = 6 - 5$ 27. $\square = 8 - 8$

Write a fact family for each.

28. 1, 6, 7 29. 0, 8, 8 30. 5, 9, 14 31. 3, 8, 11

Solve.

32. How many rawhide bones and dog biscuits are there?

33. How many more bones than biscuits are there?

THINK Al collected 3 books, Ellen collected 5 books, and Sue collected 4 books. They gave out 7 books at the children's hospital. How many books were left?

EXTENSION

Graphing Sums

Materials: 2 number cubes, Workmat 1

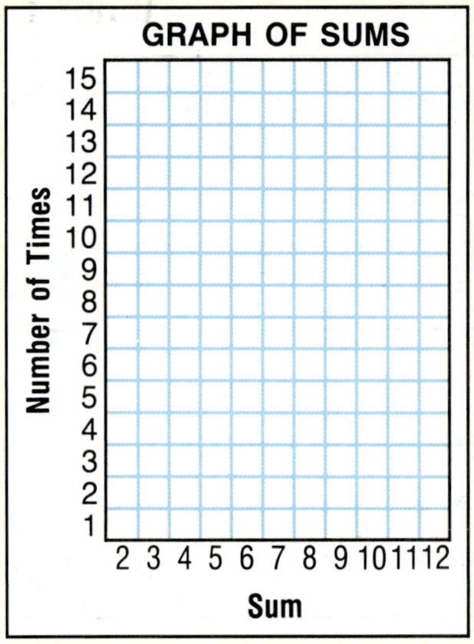

Working Together

Predict which sum will come up most often when you toss 2 number cubes.

Now try this experiment to find out.

A. Toss the cubes 25 times.

B. Each time, find the sum of the numbers that land face up.

C. Start at the bottom of the graph. For each toss, shade one square to show the sum.

1. List all the ways you can get a sum of 7. of 2.
2. Which sum came up most often? least often? Explain.
3. Write about what your graph shows. Then talk about it with a classmate. Do you have the same results? Why or why not?

Maintaining Skills

Choose the correct answer. Write A, B, C, or D.

1. $\begin{array}{r} 9 \\ +5 \\ \hline \end{array}$

   A 13 **C** 16
   B 14 **D** not given

2. Find the missing number.
   $\square + 7 = 7 + 6$

   A 6 **C** 5
   B 7 **D** not given

3. $3 + 4 + 7 = \square$

   A 10 **C** 13
   B 14 **D** not given

4. Find the missing addend.
   $7 + \square = 15$

   A 8 **C** 7
   B 6 **D** not given

5. $\begin{array}{r} 12 \\ -\ 8 \\ \hline \end{array}$

   A 0 **C** 5
   B 4 **D** not given

6. $13 - 4 = \square$

   A 7 **C** 6
   B 8 **D** not given

7. $18 - 9 = \square$

   A 10 **C** 8
   B 9 **D** not given

8. $\begin{array}{r} 15 \\ -\ 7 \\ \hline \end{array}$

   A 2 **C** 8
   B 9 **D** not given

9. What is another fact in this fact family? $11 - 7 = 4$

   A $4 + 7 = 10$ **C** $11 + 7 = 18$
   B $11 - 4 = 7$ **D** not given

10. What is another fact in this fact family? $14 - 6 = 8$

   A $6 + 8 = 14$ **C** $14 - 6 = 7$
   B $8 + 8 = 16$ **D** not given

Solve. Use the picture.

11. How many books are there altogether?

   A 7 **C** 6
   B 10 **D** not given

12. How many are not poetry books?

   A 3 **C** 7
   B 6 **D** not given

2 Building Number Sense: Place Value, Money

THEME The World of Imagination

Sharing What You Know

Help Zorp from Zorno. He wants to learn about humans and their world. Think about how you would explain money to someone from another world. Talk about money, how it is used, and how you can count it.

Using Language

Zorp has four fingers. We sometimes call fingers **digits.** People used their digits to count. Now, **digits** are also the numbers 0 to 9. We use these digits for counting numbers. Use your fingers. Show Zorp how you count pennies, nickels, and dimes. How could Zorp count the coins?

Words to Know: place value, digit, ordinal numbers, rounded numbers, greater than, less than

Be a Problem Solver

Zorp brings money back from Earth to Zorno. He gives his friend Zorta one piece of United States money. Zorp keeps ten pieces of the money. Now Zorp and Zorta each have the same amount of money. What pieces could each one have?

Design your own coins for use on Zorno. Decide how much each coin is worth.

GETTING STARTED

Name some things that come in groups of ten; in groups of one hundred.

Hundreds, Tens, and Ones

"Look Zorp," Katie said. "These blocks show a number. "We use the digits 0, 1, 2, 3, 4, 5, 6, 7, 8, and 9 to write numbers. A place value chart shows the value of each digit in a number."

hundreds	tens	ones
2	3	6

The value of the digit 2 is 200. What is the value of the digit 3? the digit 6?

2 hundreds + **3** tens + **6** ones

200 + **30** + **6** = **236**

number **236**
word name two hundred thirty-six

Check Your Understanding

Write each number. Then read it.

1. 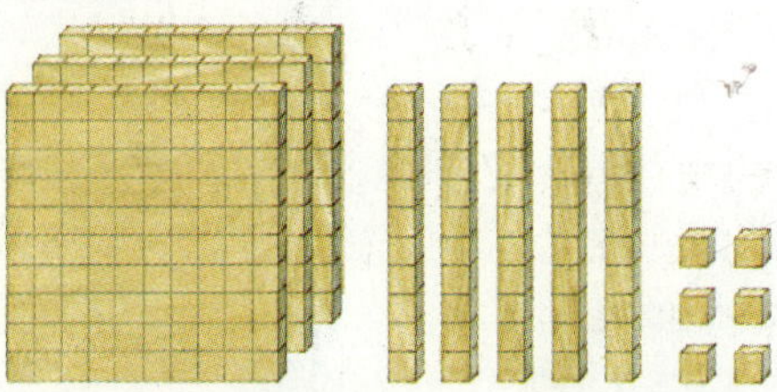

2. 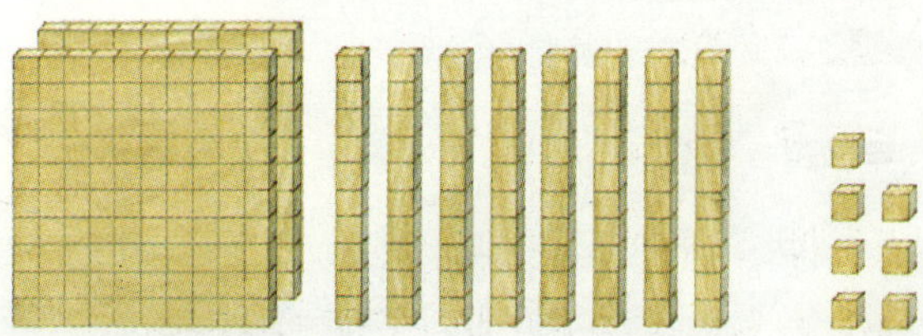

3. 2 tens 5 ones

4. 8 hundreds 9 tens 3 ones

5. 70 + 5

6. 900 + 10 + 0

Share Your Ideas Show Zork how you could use base-ten blocks to show 409.

Practice

Write each number.

7.

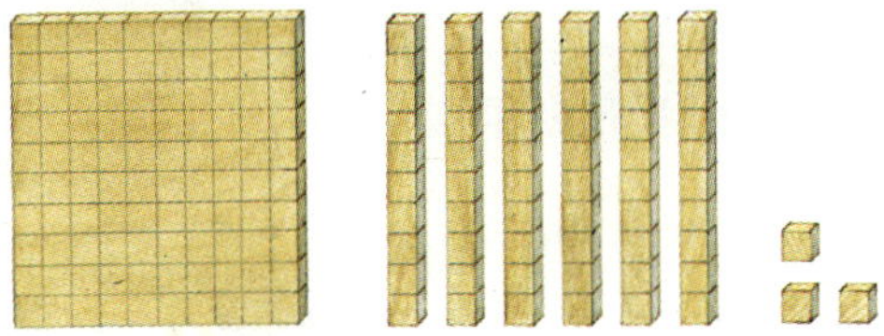

8.

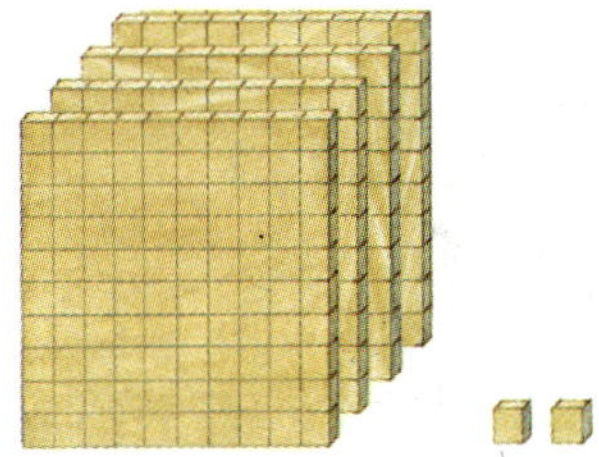

9. 80 + 6

10. 300 + 40 + 2

11. 7 tens 1 one

12. 4 hundreds 9 tens 5 ones

13. six hundreds six ones

Write the word name for each number.

14. 54

15. 73

16. 109

17. 684

18. 4,070

Give the value of the digit 5.

19. 45

20. 57

21. 251

22. 572

23. 500

Think and Apply

24. Use the digits [4][6][3].
Write the least 3-digit number.

Display each number on a calculator. Then write the number.

Example: 7 tens Press [7] [0].
Write 70.

25. 3 tens

26. 9 hundreds

27. 8 tens 9 ones

28. 6 hundreds 3 tens 2 ones

29. 5 hundreds 0 tens 3 ones

Logical Thinking

30. What number am I? My ones digit is 5. My tens digit is 4 more than my ones digit. My hundreds digit is 2 less than my tens digit.

31. Create your own puzzle. Have a classmate try it.

SUMMING UP

What number is 10 more than 210? 100 more than 307? Tell how you know.

GETTING STARTED

It is quicker to count to 100 by ones or by tens? Try it if you wish.

Counting and Order

Ben told Zork to imagine counting all the stars in the sky.

First, he counted by ones: 1, 2, 3, 4, 5, . . .

Then, he counted by tens: 10, 20, 30, 40, 50, . . .

Next, he counted by hundreds: 100, 200, 300, 400, 500, . . .

Imagine counting numbers by ones, tens, or hundreds. Would you ever reach a last number? Explain.

A **number line** helps you to show the order of numbers.

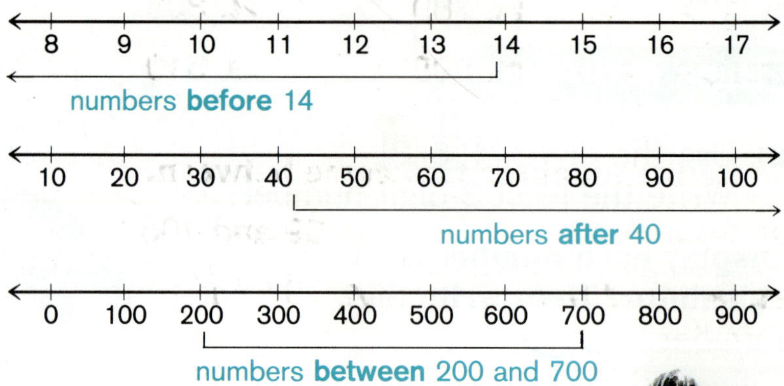

Check Your Understanding

1. Count by ones: 20, 21, 22, ___, ___, ___, ___
2. Count by tens: 50, 60, 70, ___, ___, ___, ___
3. Count by hundreds: 400, 500, 600, ___, ___, ___, ___

Share Your Ideas How do the digits change when you count by ones from 92 to 108? by tens from 430 to 530? by hundreds from 391 to 891?

Practice

Use the pattern to find the missing numbers.

4. Count by ones: 67, 68, 69, ___, ___, ___, ___
5. Count by tens: 120, 130, 140, ___, ___, ___, ___
6. Count by hundreds: 250, 350, 450, ___, ___, ___, ___
7. 20, ___, 40, ___, ___
8. 211, 212, ___, ___, ___
9. 93, 92, ___, ___, ___
10. 45, 55, ___, ___, ___
11. 12, ___, 32, ___, ___
12. 200, ___, ___, 500, ___
13. 365, 465, ___, ___, ___
14. ___, ___, ___, 700, 800
15. 430, 420, ___, ___, ___
16. 1,300, 1,200, ___, ___, ___

Write the number that comes just before and just after.

17. 82
18. 9
19. 17
20. 41
21. 63
22. 122
23. 789
24. 303
25. 299
26. 400
27. 609
28. 898
29. 510
30. 741
31. 901

Write the numbers that come between.

32. 38 and 42
33. 99 and 103
34. 60 and 64
35. 541 and 545
36. 797 and 801
37. 324 and 328

Think and Apply

38. Twelve people are in line for the Space Fair. If two are admitted, how many people are still waiting?

39. Your turn is number 62. Your friend's turn is number 71. How many people are between you and your friend?

SUMMING UP

Draw a number line marked in hundreds from 0 to 900. Show the numbers **before** 300, **after** 600, **between** 300 and 600.

GETTING STARTED

In what different ways are the words "first," "second," and "third" used in ball games?

Ordinals

Each year Rita dreams of running in the New York Marathon. Look at her list of winners. Who finished first? Who finished twenty-first?

▸ Numbers like first and twenty-first are **ordinal numbers.** They show order or position.

Read 22nd as, "twenty-second." Try counting from 23rd to 29th by ordinal numbers. Now count from 30th to 40th.

Check Your Understanding

In what position did each runner finish?

1. Judy **2.** Tony

3. Bob **4.** Marsha

Write the word name for each.

5. 23rd **6.** 36th **7.** 59th **8.** 75th

Share Your Ideas Many ordinal numbers are formed by adding *th.* Name some ordinal numbers that are not formed this way.

Practice

Use the list on page 32 to answer 9–20. In what position did each runner finish?

9. Ivy
10. Louis
11. David
12. Mu Lan
13. Ryan
14. Doris
15. Tomás
16. Joe

Name the position of the runner who finished.

17. two behind the third runner
18. one before the tenth runner
19. three behind the fifteenth runner
20. one before the twenty-first runner

Write the word name for each.

21. 5th
22. 16th
23. 24th
24. 33rd
25. 40th
26. 62nd
27. 85th
28. 200th

Think and Apply

29. Lisa is fifty-third. Ti is fifty-fifth. Name the position of the person between them.

30. There are 99 runners. Jay is next to last. Name the position of the person in front of Jay.

Name the position of each shaded square. First, start from the left. Then, start from the right.

31. 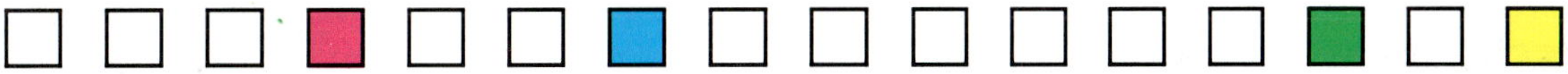

32. 

Write the word names for the numbers that come between 45th and 55th.

SUMMING UP

GETTING STARTED

Is the number of students in your class closer to 10, 20, 30, 40, or 50? How can you tell?

Rounding to the Nearest Ten

Mr. Samuel wants to know about how many students signed up to help make the Imaginary Time Machine. Adam said there were about 20. How did Adam know?

We **round** numbers to tell **about** how many instead of **exactly** how many.

Round 21 to the nearest ten.

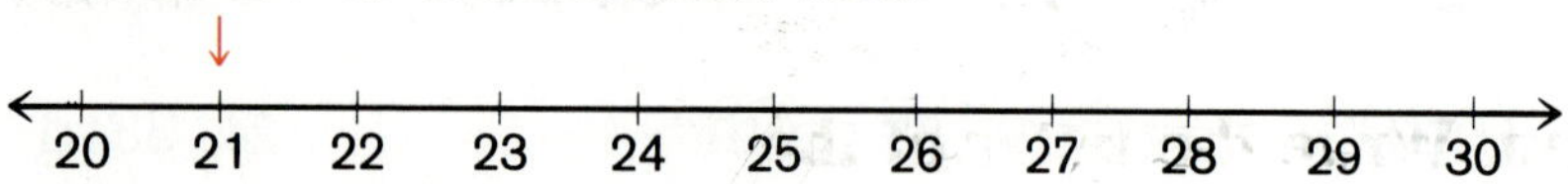

21 is between 20 and 30.
Which ten is closer? 21 is closer to 20.
21 rounded to the nearest ten is 20.
So about 20 children signed up.

Round 75 to the nearest ten.

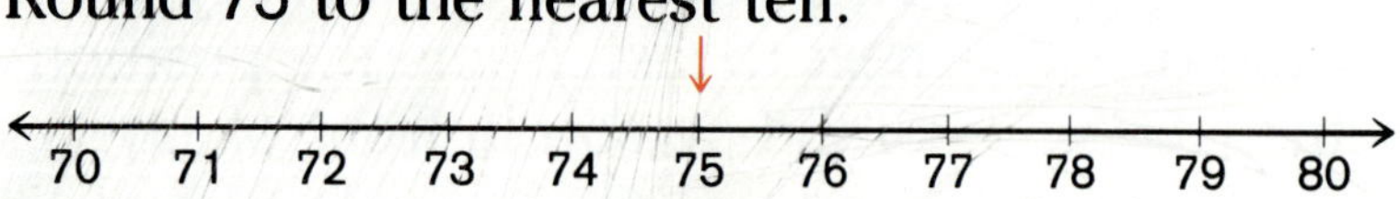

75 is between 70 and 80.
When a number is halfway between, round to the greater number.
80 > 70.
So 75 rounded to the nearest ten is 80.

Check Your Understanding

Round to the nearest ten. Use the number lines above.

1. 28 **2.** 25 **3.** 23 **4.** 71 **5.** 79 **6.** 76

Share Your Ideas Name some whole numbers that round to 80 when rounded to the nearest ten.

Practice

Round to the nearest ten. Use the number line to help you.

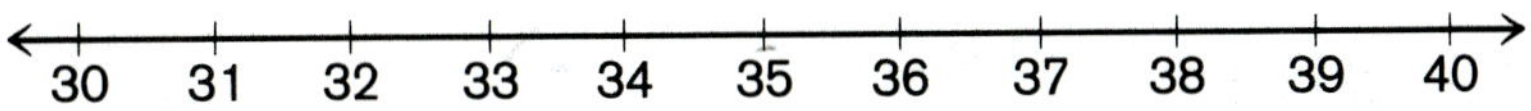

7. 33 **8.** 35 **9.** 38 **10.** 32 **11.** 36 **12.** 34

Round to the nearest ten. You may draw a number line to help you.

13. 63 **14.** 67 **15.** 62 **16.** 69 **17.** 65 **18.** 66

19. 45 **20.** 92 **21.** 31 **22.** 18 **23.** 59 **24.** 52

25. 71 **26.** 89 **27.** 7 **28.** 77 **29.** 145 **30.** 186

Round to the nearest ten. Write the letter of the correct answer.

31. 58
a. 70 **b.** 50
c. 60 **d.** 40

32. 71
a. 80 **b.** 70
c. 60 **d.** 50

33. 55
a. 50 **b.** 30
c. 40 **d.** 60

Think and Apply

34. Sara checked the supplies for the Time Machine. She rounded the number for each supply to the nearest ten. What did she write?
a. 84 metal bolts
b. 35 long head nails
c. 46 pieces of plywood
d. 12 quarts of paint

35. Name the greatest whole number that when rounded to the nearest ten is 90.

Look back at **13–18**. Which numbers did you round to 70? Explain.

SUMMING UP

GETTING STARTED

How many pages are in your math book? Is the number closer to 100, 200, 300, 400, or 500? How can you tell?

Rounding to the Nearest Hundred

About how many of each item are needed to make the Time Machine?

cardboard rectangles	108
foil paper squares	150
paper fasteners	181

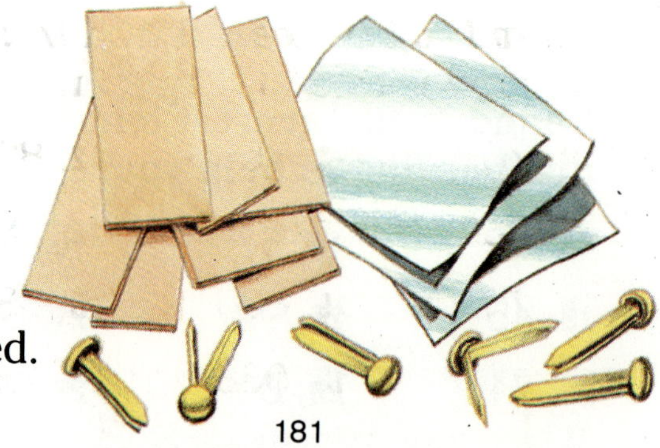

Round each number to the nearest hundred.

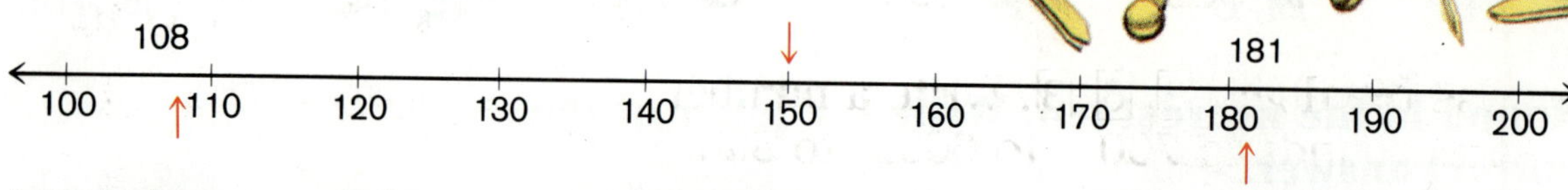

108 is between 100 and 200.
108 is closer to 100.

108 rounded to the nearest hundred is 100.

About 100 cardboard rectangles are needed.

150 is halfway between 100 and 200. When a number is halfway between, round to the greater number.

150 rounded to the nearest hundred is 200.

About 200 foil paper squares are needed.

Which hundred is 181 closer to?
181 is closer to 200.

181 rounded to the nearest hundred is 200.

About 200 paper fasteners are needed.

Check Your Understanding

Round to the nearest hundred.

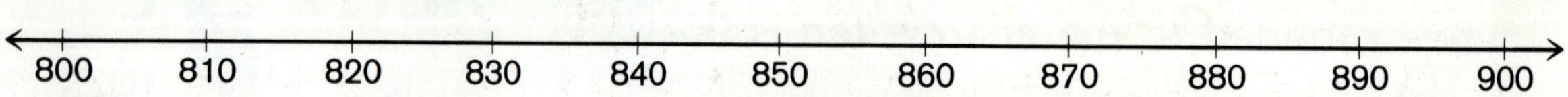

1. 876 **2.** 802 **3.** 850 **4.** 881 **5.** 845

Share Your Ideas Name the greatest whole number that when rounded to the nearest hundred is 800.

Practice

Round to the nearest hundred.

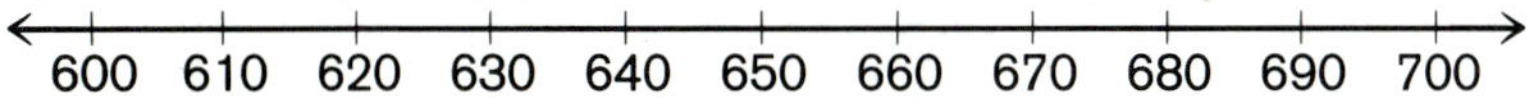

6. 647 **7.** 675 **8.** 620 **9.** 688

Round to the nearest hundred. You may draw a number line to help you.

10. 731	**11.** 456	**12.** 839	**13.** 503
14. 350	**15.** 651	**16.** 186	**17.** 209
18. 755	**19.** 899	**20.** 331	**21.** 695
22. 150	**13.** 938	**24.** 1,650	**25.** 1,286

26. Use the digits [8] [6] [3]. Write a number that rounds to 700. to 600. to 900.

27. Use the digits [4] [2] [9]. Write the number that rounds to 900. to 300. to 500.

Think and Apply

28. Jack is studying last year's Space Fair attendance. He wants to know about how many people to expect this year. If Jack rounds each number to the nearest hundred, what numbers does he get?

Name the least number that when rounded to the nearest hundred is 800. Explain.

Mixed Review

1. $6 + 6$

2. $8 - 5$

3. $18 - 9$

4. $0 + 5$

5. $14 - 6$

6. $7 + 1$

7. $9 + \square = 14$
$14 - 9 = \square$

8. $7 + \square = 10$
$10 - 7 = \square$

9. $8 + \square = 12$
$12 - 8 = \square$

10. $8 + \square = 9$
$9 - 8 = \square$

Complete each fact family.

11. $7 + 9 = \square$
$9 + 7 = \square$
$16 - 9 = \square$
$16 - 7 = \square$

12. $6 + 9 = \square$
$9 + 6 = \square$
$15 - 6 = \square$
$15 - 9 = \square$

SUMMING UP

Activity

Exploring Thousands

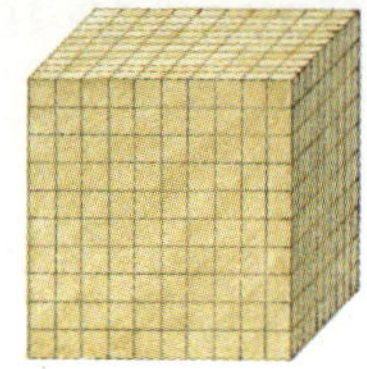

Show Zorp how you can use different blocks to show 1,000.

Working together

Materials: base-ten blocks, Workmat 2

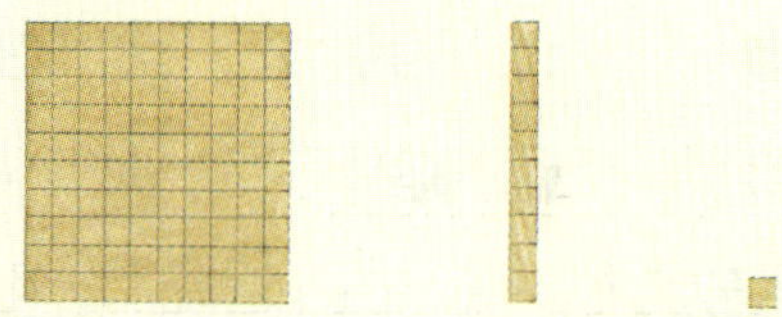

Show 1,000 in different ways.
Record your findings on Workmat 2.

A. Use only hundreds blocks.

B. Use tens blocks to make a hundreds block. How many tens blocks would you need to show 1,000?

C. Use ones blocks to make a tens block. How many ones blocks would you need to show 1,000?

D. Use different combinations of base-ten blocks to show 1,000.

	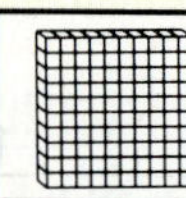	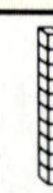	
A.			
B.			
C.			
D.			

Sharing Your Results

Look back at the numbers you recorded.

1. You can show 1,000 with how many hundreds blocks? tens blocks? ones blocks?

2. Which combinations of blocks did you use for **D**? Explain why each combination shows 1,000.

Extending the Activity

You can use a place value table to show thousands.

Work in a group. Complete the table. Record the blocks you could use to show each number.

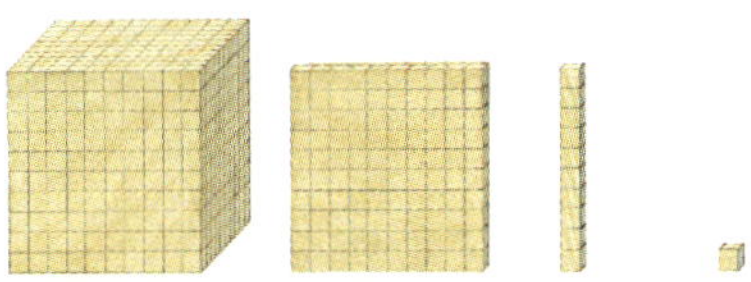

		thousands	hundreds	tens	ones
3.	1,487	1	4	8	7
4.	2,000			0	0
5.	2,346				
6.	3,589				19
7.	7,842		18		
8.	8,231	7		2	
9.	9,043	8			

10. **Look back** at **3–9**. Can you use fewer blocks to show any of the numbers? Explain.

11. What are the fewest blocks that show
a. 2,164? **b.** 5,379? **c.** 9,999?

Summing Up

12. How many third graders are in your school? Imagine 1,000 third graders. Would they fit in the principal's office? in the school gym?

13. How would you describe 1,356 using base-ten blocks? Explain.

GETTING STARTED

In one minute can you count to 1,000 by hundreds? by tens? by ones? Try it.

Place Value to Thousands

Jose imagined climbing Cadillac Mountain with Zorp. They had 1,532 feet to go.

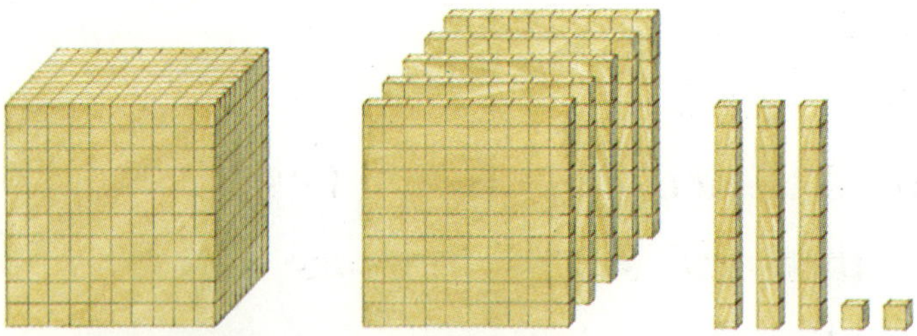

thousands	hundreds	tens	ones
1,	5	3	2

The value of the digit 1 is 1,000. What is the value of the digit 5? the digit 3? the digit 2?

1 thousand + **5** hundreds + **3** tens + **2** ones
1,000 + **500** + **30** + **2** = **1,532**

number **1,532**

A comma between the thousands and hundreds makes the number easier to read.

word name one thousand, five hundred thirty-two

Check Your Understanding

Write each number.

1. 9 thousands 8 hundreds 3 tens 2 ones
2. 6,000 + 700 + 10 + 3
3. six thousand, one hundred two

Write the value of the digit 7.

4. 7,539
5. 4,078
6. 1,702
7. 517
8. 7,000

Share Your Ideas How are the numbers 3,456 and 6,543 alike? How are they different?

Practice

Write each number.

9. 5 thousands 6 hundreds 4 tens 2 ones
10. 8 thousands 2 hundreds 0 tens 5 ones
11. 3 thousands 8 hundreds 1 ten 1 one
12. 2,000 + 900 + 60 + 4
13. 5,000 + 700 + 50 + 6
14. 6,000 + 900 + 2
15. nine thousand, one hundred fifty-six
16. two thousand, six hundred nine
17. eight thousand, eighty-one

Write the value of the digit 4.

18. 841 19. 1,764 20. 4,532 21. 9,419 22. 8,040

Write the word name for each.

23. 850 24. 3,205 25. 9,874 26. 6,200 27. 7,001

Think and Apply

28. **Look back** at 17. How did you know how many digits to write?

Display each on a calculator. Then write the number.

29. 6 hundreds 4 tens 2 ones
30. nine hundred eight
31. 8 thousands 7 hundreds 5 tens 3 ones
32. five thousand, eighty

Common Error

33. This is how one thousand, three hundred two was written: 1,32. ← incorrect
Explain how to write the number correctly.

SUMMING UP

Which number is 100 less than 3,405? 1,000 less? 1,000 more?

Midchapter Review

Write the number. pages 28–29, 40–41

1. 9 tens 2 ones
2. 6 hundreds 4 tens 0 ones
3. 700 + 60 + 3
4. 5,000 + 400 + 30 + 1

Write the value of the digit 3. pages 28–29, 40–41

5. 135
6. 603
7. 3,040
8. 4,362
9. 3,999

Write the missing numbers. pages 30–31

10. 220 230 _____ _____
11. 4,600 4,700 _____ _____ _____

Round to the nearest ten. pages 34–35

12. 23
13. 55
14. 78
15. 31
16. 17

Round to the nearest hundred. pages 36–37

17. 826
18. 452
19. 790
20. 317
21. 506

Choose the correct word to complete each sentence.

22. _____ _____ are numbers we use to show order.
23. _____ _____ are numbers we use to tell about how many.

Words to Know
Rounded numbers Ordinal numbers

Solve.

24. Sara dreamed she came in seventh in the marathon. Put these in order to show the order of the people before Sara.

third	fifth	second
first	fourth	sixth

25. Jason spent $28 on play tickets. Which of these describes what he spent?
    a. more than $30
    b. less than $20
    c. about $30
    d. about $20

Exploring Problem Solving

THINK
EXPLORE
SOLVE
LOOK BACK

How Many Pictures?

The students at West School are drawing pictures to display at the city hall. Pictures are needed for only one wall that is 100 feet long. One picture will not be placed above another.

Thinking Critically

How many pictures should they bring to put on the 100-foot wall?
Work in a small group. You may wish to use drawing paper to help solve this problem.

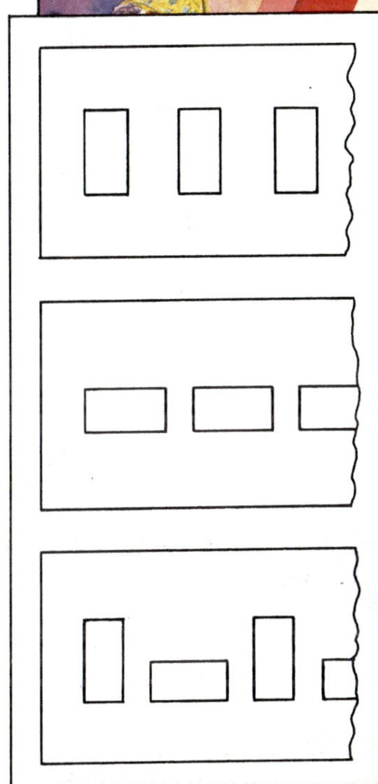

Analyzing and Making Decisions

1. Would it help you if you saw a sample of the drawings? Explain.
2. How many ways can you hang a picture? **What if** the picture is longer than it is wide? Which way will take more room?
3. Would an experiment help you? Explain. Must you find a wall that is 100 feet long? Explain. Can you use a shorter wall? Explain.
4. How many pictures can you put in one row along the wall? Explain why you chose that number.

Look Back What could you do if you did not have enough pictures? What could you do if you had too many? Would you rather take more pictures than you think you needed or the exact number?

Problem Solving Strategies

Experiment

What if you covered your hand with pennies? How many pennies would you need? How much is your hand worth?

Sometimes the best way to solve a problem is to experiment. You can act out the problem to find the answer.

Solving the Problem

Work in a small group. You will need to use either real or play pennies.

Think What is the question?

Explore How many pennies do you think you will need? Would you rather try to cover your hand or a drawing of your hand? Explain. What would be an easy way to draw your hand? How can you cover it with pennies?

Solve How many pennies fit on your hand? How much is your hand worth?

Look Back Did you cover your whole hand? Did the pennies cover more than your hand? Explain.

Share Your Ideas

1. Did you draw your hand with your fingers together or with space between your fingers? Does it make a difference? How do you know?
2. **What if** you used nickels or dimes? How would that change the amount of money?

Practice

Solve. Use a calculator where appropriate.

CHOICES
3. Fred and Alex measured a desk. Fred used four hand spans to measure the length. Alex used 3 hand spans. Whose hand span is longer?

4. Choose one pencil. Estimate how many pencils of this length will fit end to end across your desk. Experiment to check.

5. Pick a pencil of a different length. Estimate how many pencils of this length will fit across your desk. Experiment to check.

6. Tom wants to make 16 equal sections on his paper. What is the fewest number of folds that are needed?

Create Your Own

Write a problem for which you must estimate a length, and then experiment.

Mixed Strategy Review

7. Ms. Martinez works at the bank. She leaves her house and stops at the post office on the way to work. How far does she drive?

8. She stops at both the post office and the store on her way to work. How far is it if she goes the shorter way?

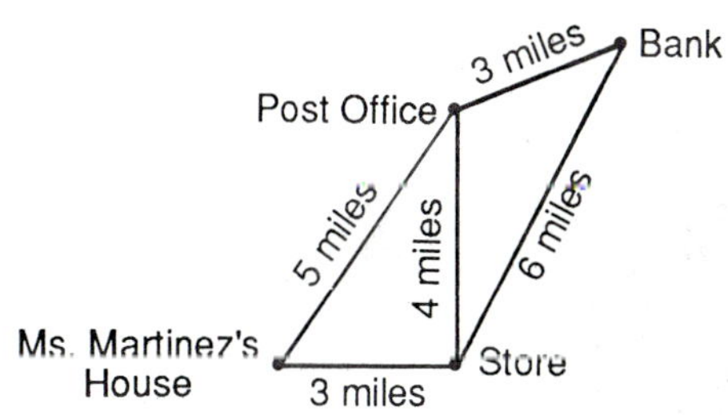

9. Look at the picture of the tissue carton. Ms. Gardner buys 2 boxes of tissues. Mr. Bruce buys 1 box of tissues. How many boxes of tissues are left in the carton?

10. Rewrite **9** so that there are only 4 boxes of tissues left.

GETTING STARTED

Would you rather have 87 cents or 78 cents? Why?

Comparing and Ordering Numbers

Mike and Zorp pretend they rode bicycles in competition. Mike rode 925 miles. Zorp rode 914 miles. Who rode more miles?

▸ To compare numbers, start at the left. Compare the digits in each place.

hundreds	tens	ones
9	2	5
9	1	4

↓	↓	
same	2 > 1	

\> means is greater than
< means is less than

So, **925** > **914** and **914** < **925.**

Mike rode more miles.

▸ To order numbers from least to greatest, compare them two at a time.

Order **986**, **931**, and **1,021**.

Compare **986** and **931**.
931 < **986**

Compare **986** and **1,021**.
Hundreds are less than thousands.
986 < **1,021**

So, **931** < **986** < **1,021**

The numbers in order are 931, 986, 1,021.

Check Your Understanding

Compare. Use > or < for ●.

1. 39 ● 35 **2.** 720 ● 750 **3.** 1,001 ● 1,100 **4.** 508 ● 5,518

Write the numbers in order from least to greatest.

5. 490 670 231 875 **6.** 9,618 905 9,541

Share Your Ideas Explain your answer for **3.**

Practice

Compare. Use > or < for ●.

7. 89 ● 85
8. 78 ● 87
9. 20 ● 200
10. 72 ● 27
11. 380 ● 354
12. 520 ● 550
13. 4,508 ● 4,518
14. 7,302 ● 7,202
15. 6,911 ● 918
16. 9,996 ● 9,991
17. 5,400 ● 540
18. 11,000 ● 1,100

Write the numbers in order from least to greatest.

19.	967	976	694	975
20.	5,473	5,074	5,862	5,194
21.	7,801	3,806	7,849	7,124
22.	930	8,001	9,010	938

Choose the numbers that are ordered from greatest to least.

23.	a. 82	280	28	480	24.	a. 1,017	1,524	1,701	1,961
	b. 725	516	901	317		b. 8,802	8,082	8,280	6,802
	c. 416	414	409	361		c. 5,839	5,700	5,200	5,522
	d. 815	199	591	223		d. 7,363	6,523	732	462

Think and Apply

25. **What if** you want to travel by bicycle to these cities? Which city is farthest? nearest?

Houston	1,686 miles
Austin	1,837 miles
Dallas	1,561 miles

Visual Thinking

These shapes are not in any order. Describe a way to order them.

26.

Look back at **19.** Where would you put 965? 697? 692?

SUMMING UP

GETTING STARTED

Numbers can describe great amounts or long distances. Give examples.

Ten Thousands and Hundred Thousands

Imagine flying around the earth at the Equator. You would fly about 24,859 miles! 24,859 can be shown on a place value table.

thousands		ones		
tens	ones	hundreds	tens	ones
2	4,	8	5	9

The value of the digit 2 is 20,000. What is the value of the digit 4?

number **24,859**
word name twenty-four thousand, eight hundred fifty-nine

To reach the moon you would fly about 238,751 miles!

thousands			ones		
hundreds	tens	ones	hundreds	tens	ones
2	3	8,	7	5	1

The value of the digit 2 is 200,000. What is the value of the digit 3? the digit 8?

number **238,751**
word name two hundred thirty-eight thousand, seven hundred fifty-one

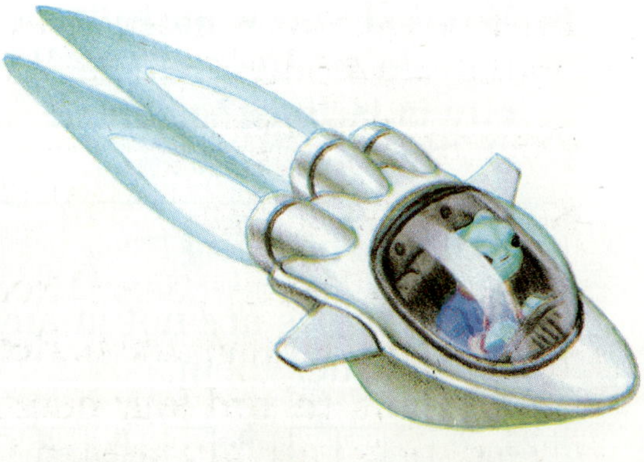

Check Your Understanding

Write each number.

1. 172 thousands 3 hundreds 1 ten 6 ones
2. 30,000 + 9,000 + 400 + 20 + 5
3. sixty-five thousand, one hundred fifteen

Write the value of the digit 8.

4. 837,204
5. 33,280
6. 18,591
7. 505,862

Share Your Ideas Is 65,490 the same number as 650,490? Explain.

Practice

Write each number or word name.

8. 21 thousands 3 hundreds 2 tens 8 ones
9. 60 thousands 5 hundreds 9 tens 9 ones
10. 115 thousands 7 hundreds 5 ones
11. 956 thousands 6 hundreds 3 tens 7 ones
12. 60,000 + 700 + 80 + 2
13. 892,000 + 500 + 70 + 4
14. fifty-one thousand, six hundred forty-five
15. nine hundred two thousand, sixty-nine
16. 570,924 17. 75,185 18. 1,102,030

Write the value of the digit 4.

19. 94,016 20. 458,721 21. 346,819

Give the next number in the pattern.

22. 79,700 79,800 79,900 ____.
23. 520,000 510,000 500,000 ____.

Think and Apply

Display each on a calculator. Then remove only one digit without changing the other digits.

Example: 654321

Remove the 2. Press: − 2 0.

24. 987654 Remove the 8. Remove the 9.
25. 456789 Remove the 6. Remove the 7.
26. DATA Find and record four numbers that name great amounts in a newspaper. Compare with a classmate. Which number was greatest? least?

You have just flown 15,999 miles into space. Name the number that is ten miles before and after this number.

SUMMING UP

Mixed Review

1. $15 - 8$
2. $4 + 9$
3. $13 - 7$
4. $7 + 9$
5. $8 + 4$
6. $6 - 0$
7. $7 + 2 + 4$
8. $8 + 0 + 5$
9. $6 + 3 + 6$
10. $2 + 5 + 4$
11. $\square + 6 = 12$
12. $8 - \square = 0$
13. $5 + 4 = \square$
14. $16 - 9 = \square$
15. $\square - 9 = 7$
16. $6 + \square = 9$
17. $6 + 3 + 4 = \square$
18. $5 + 3 + 4 = \square$
19. $4 + 4 + 5 = \square$
20. $6 + 0 + 9 = \square$
21. $8 + 1 + 7 = \square$

GETTING STARTED

Pretend you have a penny, a dime, a quarter, and a nickel. Put the coins in order from greatest to least value.

Counting Coins

Emile wants to buy the toy car. The car is on sale for 76¢.

▸ To count coins, start with the coin of greatest value. End with the coin of least value.

penny = 1¢

nickel = 5¢

dime = 10¢

quarter = 25¢

half dollar = 50¢

Emile counted this way.

25, 50, 60, 70, 75, 76 cents (76¢)

The value of these coins is 76 cents.

Check Your Understanding

Count to find each value.

1. 

2. 

Share Your Ideas Describe what you would do to count a handful of coins.

Practice

Count to find each value.

3. 

4. 

5. 

6. 

7. 

8. 

Tell how many coins the change machine gives.

	Input	Output
9.	1 nickel	☐ pennies
10.	1 dime	☐ pennies
11.	1 quarter	☐ pennies

	Input	Output
12.	1 dime	☐ nickels
13.	1 quarter	☐ nickels
14.	1 half dollar	☐ quarters

Think and Apply

15. Use 3 different combinations of coins to show 75 cents. Which way uses the fewest coins? Which way uses the most coins?

Visual Thinking

Continue the pattern.

16.  ____ ____ ____ ____

Order some coins from the greatest to the least value. Count them. Now count them in a different order. Which way was easier?

SUMMING UP

Activity

Exploring Working With Coins

Would Zorp and Lee have enough money to buy a globe for 85¢?

To find out, show their coins. Count how much money they have together. Do they have enough? How much do they have left?

Working together

Materials: 5 objects; tags labeled 98¢, 82¢, 77¢, 65¢, and 39¢; play coins in a bag; Workmat 3

Record your work.

My Amount	Partner's Amount	Total Value	We Spent	We Have Left

A. Put a tag on each object.

B. Each partner takes 4 coins and counts them.

C. Combine the money. Find the total value.

D. Decide if you can buy another object. If not, take 2 more coins each. Count out the coins you need to buy the object. (Exchange, if necessary.)

E. How much money is left? Return all coins to the bag.

F. Repeat **B** through **E** until all the objects are bought.

Sharing Your Results

1. How did you decide if you could buy an object?

2. How did you count the money you had left?

3. What is the easiest way to count coins? Why?

Extending the Activity

Work with a partner. Record your work.

Use the items below for 4–7.

4. You have 5 dimes, 3 nickels, and 2 pennies. Can you buy the space doll?

5. You have 3 quarters. Which items could you buy?

6. You have a quarter and 4 dimes. You add 5 nickels. Which items could you buy?

7. You have a half dollar and 3 dimes. How much money will be left if you buy the space helmet?

Summing Up

8. **What if** you have 99¢? Show four different ways you can spend it on the above items.

GETTING STARTED

You have a five-dollar bill, a ten-dollar bill, and a one-dollar bill. Put them in order from the greatest value to the least value.

Working with Coins and Bills

Pam is showing Zorp how to count the money she saved.

> To count money, start with the coin or bill of greatest value. End with the coin or bill of least value.

10, 15, 16 dollars, and 25, 35, 40 cents

Pam has sixteen dollars and forty cents or $16.40.

> Use the dollar sign ($) in front of the dollars. Use the decimal point (.) in front of the cents.

1 dollar = 100¢ = $1.00

5 dollars = $5.00

10 dollars = $10.00

Check Your Understanding

Write each value.

1. 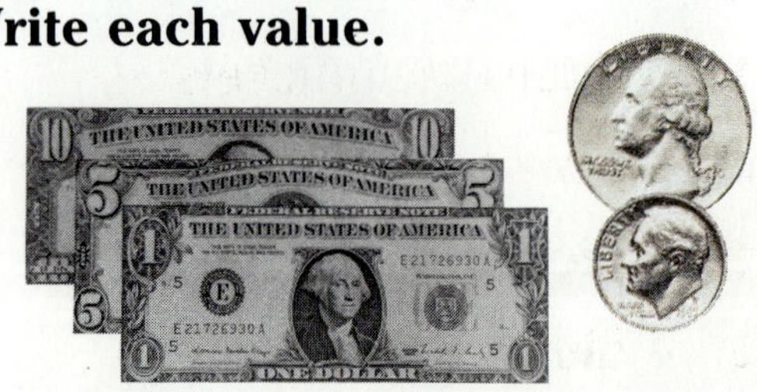

2. 

Share Your Ideas Find three different ways to make $16.26. Which way uses the fewest coins and bills?

Practice

Write each value.

3. 

4. 

5. 4 one-dollar bills, 9 pennies
6. 2 five-dollar bills, 6 dimes
7. 7 ten-dollar bills, 8 nickels
8. 3 ten-dollar bills, 29 pennies
9. 3 five-dollar bills, 3 quarters
10. 6 five-dollar bills, 8 nickels
11. 4 ten-dollar bills, 2 one-dollar bills, 3 quarters, 6 pennies
12. 8 five-dollar bills, 4 dimes, 2 quarters
13. 5 ten-dollar bills, 2 one-dollar bills, 1 five-dollar bill, 19 pennies
14. 6 pennies, 3 nickels, 1 quarter, 7 five-dollar bills
15. 2 nickels, 5 dimes, 1 ten-dollar bill, 10 one-dollar bills
16. 5 quarters, 15 pennies, 1 dime, 9 five-dollar bills

Name the fewest bills and coins needed to make each amount.

17. $4.16 18. $8.57 19. $122.00 20. $150.89

Think and Apply

21. Name 4 coins that have a total value of $.50.
22. Name 6 bills that have a total value of $40.
23. Name 2 bills and 6 coins that have a total value of $10.50.

Logical Thinking

Which would you rather have? Explain why.

24. 6 dimes or 6 nickels
25. 3 quarters or 8 dimes
26. 5 nickels or 20 pennies
27. 4 dimes or 50 pennies

SUMMING UP

What if Zorp has $2.85? What possible combinations of bills and coins could he have?

Using Problem Solving

Counting Change

Ian helps his parents by making change at the snack stand. One day someone bought carrots for $.37 and gave Ian $1.00. He made change in this way.

He counted from the cost. "37 cents for the carrots."

$.38 $.39 $.40 $.50 $.75 $1.00

He ended with the amount given. The change was 63 cents or $.63.

Working together

Materials: play coins or real pennies, nickels, dimes, and quarters

Use the coins and make change the way that Ian might do it.

Customer's bill	Customer gives Ian	Coins given as change
A. $.25	$1	3 quarters
B. $.42	$1	________
C. $.34	$1	________

Customer's bill	Customer gives Ian	Coins given as change
D. $.19	$1	________
E. $.55	$1	________
F. $.72	$1	________

Sharing Your Ideas

1. Tell how you would make change when someone pays for something with a dollar.

2. **What if** someone bought an apple for $.49 and paid for it with a $5 bill? Tell how to make change.

THINK
EXPLORE
SOLVE
LOOK BACK

Extending Your Thinking

3. It is near closing time and Ian does not have much change. He needs to keep some of each type of coin. Then he can give each customer change. He has these coins left.

4 quarters   8 dimes   10 nickels   15 pennies

Give each customer change. Record your work. Use real or play coins if you wish.

Customer's bill	Customer gives Ian	Change needed	Coins given as change
$.45	$1	______	______
$.53	$1	______	______
$.64	$1	______	______
$.26	$1	______	______
$.89	$1	______	______

4. Make up a customer's bill for less than $1. Have a partner find at least 3 different ways to make change.

Summing Up

5. Were you able to make change for all the customers in **3**? Did you have one of each coin left? If so, why? If not, why not?

6. Did you find at least 3 ways to give your partner change in **4**? Explain.

Chapter Review

Write the value of the digit 6. pages 28–29, 40–41, 48–49

1. 163 **2.** 4,633 **3.** 16,052 **4.** 699,000 **5.** 465,425

Write the missing numbers. pages 30–31

6. 480 490 ____ ____

7. 3,200 ____ ____ ____ 3,600

Round to the nearest ten. pages 34–35

8. 31 **9.** 67 **10.** 79 **11.** 45 **12.** 22 **13.** 53

Round to the nearest hundred. pages 36–37

14. 352 **15.** 526 **16.** 680 **17.** 267 **18.** 409

Write the numbers in order from least to greatest. pages 46–47

19. 652 530 660 606

20. 5,280 5,820 5,291 5,288

Find the value. pages 50–55

21.

22.

Choose the correct words to complete each sentence.

23. The symbol $>$ means ____.
The symbol $<$ means ____.

Words To Know
is less than is greater than

Solve. pages 43–45

24. **What if** you have 3 dimes and 2 nickels in one pocket? In another pocket you have 1 nickel and 4 pennies. How much do you have in all?

25. First guess. Then experiment. How many times can you write the numbers 1-10 across your paper?

Chapter Test

Write the value of the digit 9.

1. 968 2. 7,639 3. 29,350 4. 490,222 5. 987,430

Write the missing numbers.

6. 1,250 1,350 _____ _____ 7. 5,925 5,935 _____ _____

Round to the nearest ten.

8. 27 9. 76 10. 41 11. 85 12. 49 13. 93

Round to the nearest hundred.

14. 731 15. 174 16. 829 17. 308 18. 451 19. 790

Write the numbers in order from least to greatest.

20. 782 700 718 800 21. 1,670 1,600 1,699 1,598

Find the value.

22.

23.

Solve.

24. Seth needs $1.85. He has 3 quarters, 5 dimes, and 2 nickels. Does he have the amount he needs?

25. First guess. Then experiment. How many letters of the alphabet can you write in a row across your paper?

THINK What is the least whole number possible that rounds to 5,000? the greatest whole number possible?

EXTENSION

Building Numbers

What if you use these digits? 1 9 5

How many different numbers can you build? Which one is the greatest number? How do you know?

Materials: a set of cards numbered 0-9, a paper bag, Workmat 4

Building Numbers

Round 1 ___ , ___ ___ ___
Round 2 ___ , ___ ___ ___
Round 3 ___ , ___ ___ ___
Round 4 ___ ___ , ___ ___ ___
Round 5 ___ ___ , ___ ___ ___

Working Together

Try to build the greatest number.

1. Place the cards in the bag.
2. One player draws numbers from the bag.
3. As each number is drawn, each player writes it in any space next to Round 1. Then the card is placed face down on the table.
4. Play continues until enough numbers have been drawn to complete Round 1.
5. Now read and compare the numbers built by each player.
6. The player with the greatest number scores a point.
7. Continue playing. Use Round 2, then Round 3, and so on.

Create your own new ways to play.

Some ideas are:

- The player with the least number scores a point.
- The player with the number closest to 50,000 scores a point.

Maintaining Skills

Choose the correct answer. Write A, B, C, or D.

1. $\begin{array}{r} 7 \\ +3 \\ \hline \end{array}$

   A 8 **C** 10

   B 11 **D** not given

2. 9 + 9 = ☐

   A 17 **C** 18

   B 16 **D** not given

3. Find the missing addend.
   6 + ☐ = 11

   A 6 **C** 7

   B 5 **D** not given

4. 5 + 6 + 3 = ☐

   A 13 **C** 15

   B 12 **D** not given

5. 11 − 6 = ☐

   A 3 **C** 5

   B 6 **D** not given

6. 17 − 8 = ☐

   A 8 **C** 7

   B 9 **D** not given

7. What is the value of the digit 4 in 346?

   A 40 **C** 4

   B 400 **D** not given

8. What is the ordinal name for the number after 16?

   A seventh **C** seventeenth

   B seventeen **D** not given

9. Round 68 to the nearest ten.

   A 60 **C** 80

   B 70 **D** not given

10. Choose the number for 7,000 + 500 + 40 + 3.

    A 7,543 **C** 7,354

    B 7,453 **D** not given

Solve. Use the picture.

11. How many sea creatures are there altogether?

    A 11 **C** 12

    B 13 **D** not given

12. How many more starfish are there than sea horses?

    A 11 **C** 5

    B 10 **D** not given

3 Adding

THEME Children in Action

Sharing What You Know

Do you see your favorite activity in the pictures on this page? Discuss activities you enjoy. Which ones are fun to do alone? Which ones are more fun to do in a group? Talk about how you use addition in some of the games you play with your friends.

Using Language

Many games call for the players to develop a plan. A soccer team might begin the game with one plan. Later, the team may need to **regroup** and use a new plan. For example, a player might be moved to a new position. In mathematics, you sometimes need to **regroup** 10 ones as 1 ten. How are moving players on a team and regrouping in mathematics alike? How are they different?

Words to Know: sum, estimate, regroup, front digits, decimal point

Be a Problem Solver

You and a partner are playing a game. At each turn, a player can put 1 or 2 sticks in the circle. The person who puts the sixth stick in the circle wins. There are 2 sticks already in the circle. It is your turn. What will you do?

Write about your next move in the game.

Activity

Exploring Addition

Make a zoo game for addition. Play the game to collect as many cards as possible.

Working together

Materials: ten 3 x 5 cards, base-ten blocks

A. Draw an animal on each card. Then write one of these numbers on each card: 65, 39, 16, 43, 61, 28, 57, 72, 35, 84.

B. Place the cards face down on the table. Take turns drawing a pair of cards.

C. Show each number with base-ten blocks.

D. Combine the blocks to show the sum. Use the fewest blocks possible. Exchange if necessary.

E. If the sum of a pair is 100, keep the cards. Record each number and the sum. Then take another turn. If the sum is not 100 return the cards face down. Continue playing until all the cards have been paired to make 100.

Sharing Your Results

1. How did you know when to exchange blocks?
2. How did showing the sum with the fewest blocks make it easier to record the sum?
3. Which pairs of cards made the sum 100?

Extending the Activity

Now make a safari game like the first game. The object of this game is to collect the most safari cards.

Work with a partner.

Materials: ten 3 x 5 cards, base-ten blocks

4. Make a set of 10 safari cards. Then write one of these numbers on each card: 115, 184, 58, 16, 123, 151, 77, 142, 85, 49.
5. Take turns drawing pairs of cards. Show each number with base-ten blocks.
6. Combine ones blocks, tens blocks, hundreds blocks. Show the sum using the fewest blocks. Exchange when necessary. Record your work this way.

hundreds	tens	ones
1	1	
1	8	4
	5	8
2	4	2

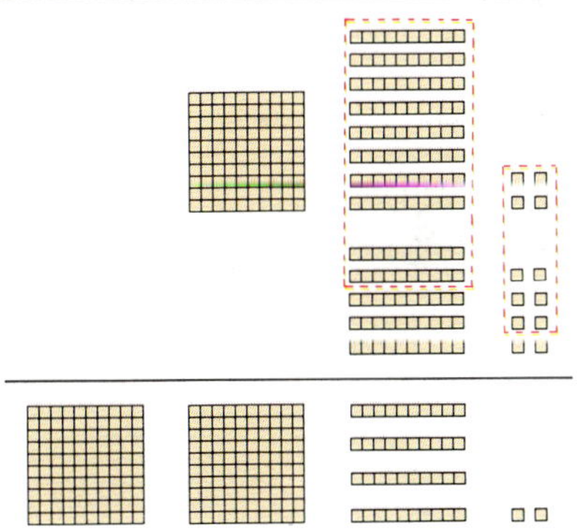

7. If the sum is 200, keep the cards and take another turn. If not, return the cards face down.

Summing Up

8. **Look back** at your records. How can you tell when you regrouped the ones? the tens?
9. **What if** you wanted to invent another game for a sum of 500? What are some numbers you could use for a set of ten cards? Try the game.

GETTING STARTED

Find the pairs that name the same number.

5 tens 3 ones	8 tens 12 ones	4 tens 13 ones
9 tens 2 ones	6 tens 3 ones	5 tens 13 ones

Adding Two-Digit Numbers

Jan ran 32 yards in one event and 46 yards in another event. How many yards did she run?

32 + 46 = □

Step 1
Add ones.

tens	ones
3	2
+4	6
	8

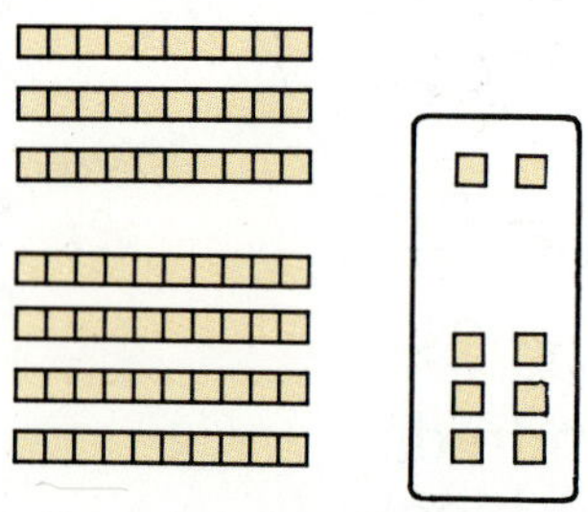

Step 2
Add tens.

tens	ones
3	2
+4	6
7	8

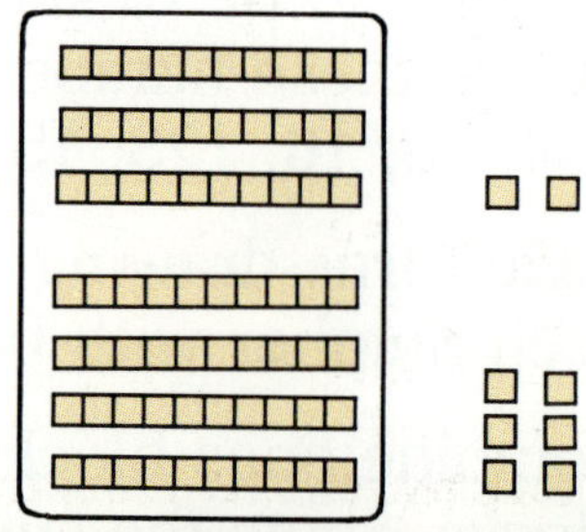

Jan ran 78 yards in all.

Find 26 + 35.

Line up the ones and tens digits.

Step 1
Add ones.
Regroup.

tens	ones
1	
2	6
+3	5
	1

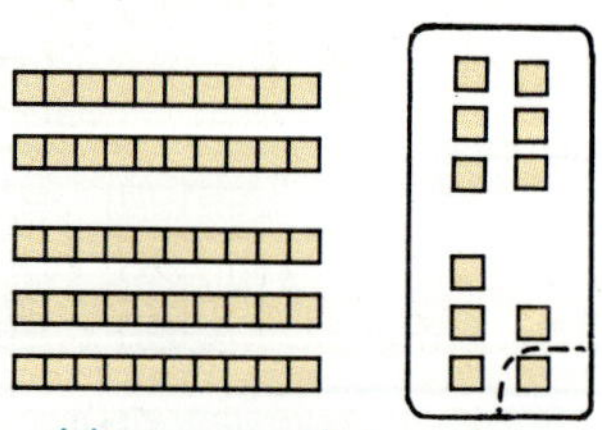

11 ones = 1 ten 1 one

Step 2
Add tens.

tens	ones
1	
2	6
+3	5
6	1

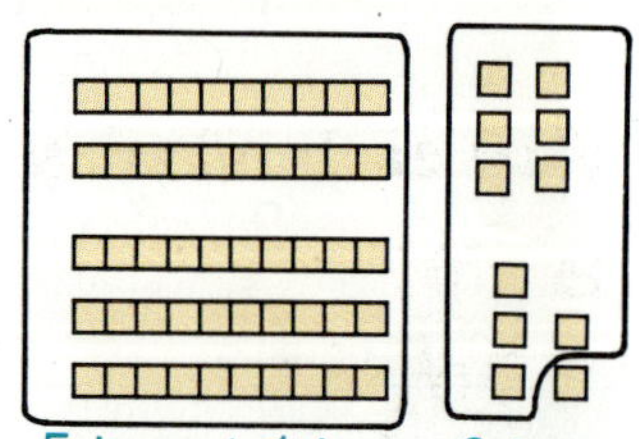

5 tens + 1 ten = 6 tens

Check Your Understanding

Add.

1. 15 + 22 = □ **2.** 13 + 36 = □ **3.** 53 + 8 = □

Share Your Ideas Explain what could happen if you did not line up the ones and tens digits.

Practice

Add.

4.

tens	ones
3	5
+6	0

5.

tens	ones
4	1
+3	6

6.

tens	ones
6	8
+	2

7.

tens	ones
1	7
+5	1

8.

tens	ones
5	5
+2	7

9. $47 + 12$
10. $37 + 8$
11. $76 + 12$
12. $46 + 36$
13. $26 + 53$
14. $7 + 64$

15. $70 + 19$
16. $9 + 47$
17. $56 + 12$
18. $68 + 24$
19. $45 + 24$
20. $29 + 33$

21. 14 + 74 = □
22. 13 + 38 = □
23. 16 + 57 = □
24. □ = 33 + 18
25. □ = 31 + 28
26. □ = 47 + 15

Add across and down.

27.

50	19	
35	62	

28.

39	45	
57	23	

Find each missing number.

29.
```
  67
+ 2□
  □1
```

30.
```
  39
+ 5□
  □2
```

31.
```
  6□
+ □8
  92
```

32.
```
  8□
+ □9
  99
```

Think and Apply

33. Third graders need 50 points to win the event. They have 35 points. If they score 15 more points, will they win? Explain.

34. Fourth graders scored 20 points in each of two races. How many points did they score in all?

Look back at **20**. What happens if you add the tens and then the ones?

SUMMING UP

GETTING STARTED

Explain what it means to regroup in addition.

Regrouping Twice

Ray and Ed are making a path to a clubhouse. So far Ray put down 56 stones and Ed put down 78 stones. How many stones did they put down?

56 + 78 = □ Add to find how many in all.

Step 1
Add ones.
Regroup.

t	o
1	
5	6
+7	8
	4

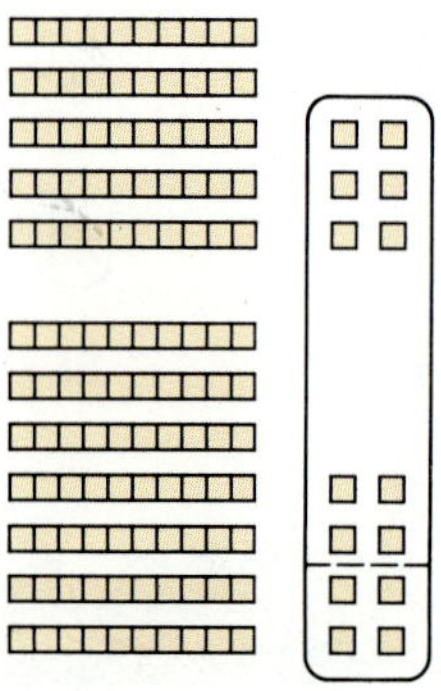

14 ones = 1 ten 4 ones

Step 2
Add tens.
Regroup.

h	t	o
	1	
	5	6
+	7	8
1	3	4

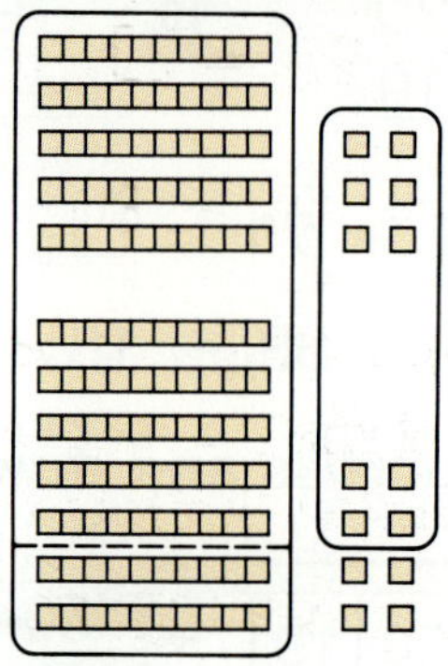

13 tens = 1 hundred 3 tens

Check by adding up.

$$\begin{array}{r} {}^{1}\\ 56 \\ +78 \\ \hline 134 \end{array} \uparrow$$

Why are the sums the same?

They put down 134 stones.

Check Your Understanding

Add. For 1–6 check by adding up.

1. $\begin{array}{r} 49 \\ +51 \\ \hline \end{array}$
2. $\begin{array}{r} 87 \\ +75 \\ \hline \end{array}$
3. $\begin{array}{r} 99 \\ +\ 9 \\ \hline \end{array}$
4. $\begin{array}{r} 74 \\ +38 \\ \hline \end{array}$
5. $\begin{array}{r} 85 \\ +97 \\ \hline \end{array}$
6. $\begin{array}{r} 78 \\ +46 \\ \hline \end{array}$

7. 54 + 66 = □
8. 59 + 94 = □
9. 8 + 97 = □

Share Your Ideas How can you tell by looking at the addends when to regroup ones? tens?

Practice

Add. For 10–27 check by adding up.

10. 79 + 53
11. 64 + 71
12. 52 + 59
13. 38 + 68
14. 73 + 82
15. 28 + 53
16. 41 + 79
17. 65 + 67
18. 98 + 54
19. 46 + 37
20. 94 + 65
21. 82 + 49
22. 87 + 29
23. 46 + 73
24. 68 + 49
25. 39 + 84
26. 86 + 75
27. 59 + 89
28. 5 + 95 = □
29. 69 + 85 = □
30. 62 + 48 = □
31. □ = 18 + 86
32. □ = 28 + 83
33. □ = 49 + 51

Compare. Use >, <, or = for ●. Choose mental math, paper and pencil, or a calculator. Explain your choices.

CHOICES

34. 16 + 92 ● 92 + 16
35. 96 + 50 ● 96 + 40
36. 83 + 0 ● 0 + 93
37. 85 + 43 ● 75 + 53

Think and Apply

38. When would you regroup in 95 + 27?

39. Ray used 40 stones. Ed used double that number. How many stones did they use in all?

40. Ed found 34 stones on Monday and 78 on Tuesday. Ray found 47 on Monday and 69 on Tuesday. Who found more stones?

Make up an addition problem that needs regrouping twice. How will you choose the numbers?

SUMMING UP

GETTING STARTED

Pattern, pattern, who sees the pattern?
13, 23, 33, 43, 53, . . .
16, 26, 36, 46, 56, . . .

Mental Math: Strategies for Addition

Barbara likes to add mentally whenever she can.

You can add other numbers mentally by thinking of basic facts.

Basic Fact $6 + 8 = 14$

$16 + 8 = 24$ Think $14 + 10$ more

$26 + 8 = 34$ $14 + 20$ more

$36 + 8 = 44$ $14 + 30$ more

What do you notice about the number of ones in each sum? the number of tens in each sum?

Knowing basic facts and patterns can help you add three numbers mentally.

$6 + 7 + 5 = \square$ → $13 + 5 = 18$

$9 + 7 + 6 = \square$ → $16 + 6 = 22$

Check Your Understanding

Use mental math. Find each sum.

1. $9 + 7$ $19 + 7$ $29 + 7$ $39 + 7$ $49 + 7$

2. $3 + 3$ $13 + 3$ $23 + 3$ $33 + 3$ $43 + 3$

3. $7 + 8 + 9 = \square$

4. $6 + 8 + 7 = \square$

5. $3 + 6 + 9 = \square$

Share Your Ideas Look back at **3–5**. Explain how you found each sum.

Practice

Use mental math. Find each sum.

6. 4 + 8 = □
14 + 8 = □
24 + 8 = □
34 + 8 = □
44 + 8 = □

7. 7 + 5 = □
17 + 5 = □
27 + 5 = □
37 + 5 = □
47 + 5 = □

8. $\begin{array}{r}8\\+5\\\hline\end{array}$ $\begin{array}{r}18\\+\ 5\\\hline\end{array}$ $\begin{array}{r}28\\+\ 5\\\hline\end{array}$ $\begin{array}{r}38\\+\ 5\\\hline\end{array}$

9. $\begin{array}{r}4\\+2\\\hline\end{array}$ $\begin{array}{r}14\\+\ 2\\\hline\end{array}$ $\begin{array}{r}24\\+\ 2\\\hline\end{array}$ $\begin{array}{r}34\\+\ 2\\\hline\end{array}$

10. $\begin{array}{r}8\\3\\+2\\\hline\end{array}$ **11.** $\begin{array}{r}5\\7\\+8\\\hline\end{array}$ **12.** $\begin{array}{r}8\\7\\+8\\\hline\end{array}$ **13.** $\begin{array}{r}9\\5\\+7\\\hline\end{array}$

14. $\begin{array}{r}7\\8\\+9\\\hline\end{array}$ **15.** $\begin{array}{r}9\\9\\+7\\\hline\end{array}$ **16.** $\begin{array}{r}8\\9\\+6\\\hline\end{array}$ **17.** $\begin{array}{r}8\\8\\+9\\\hline\end{array}$

Think and Apply

18. Arrange the numbers 1–9 so that the sum along each side of the triangle is the same.

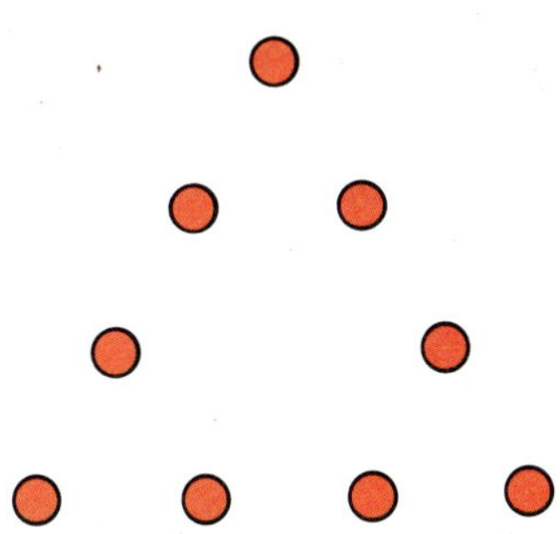

Explain how you can use a basic fact to find 46 + 7 = □.

SUMMING UP

Mixed Review

1. $\begin{array}{r}4\\+7\\\hline\end{array}$ **2.** $\begin{array}{r}9\\+8\\\hline\end{array}$

3. $\begin{array}{r}16\\-\ 8\\\hline\end{array}$ **4.** $\begin{array}{r}13\\-\ 7\\\hline\end{array}$

5. $\begin{array}{r}4\\2\\+5\\\hline\end{array}$ **6.** $\begin{array}{r}9\\0\\+6\\\hline\end{array}$

7. $\begin{array}{r}4\\3\\2\\+7\\\hline\end{array}$ **8.** $\begin{array}{r}9\\0\\8\\+1\\\hline\end{array}$

9. 9 + □ = 16

10. 7 + □ = 13

11. □ − 8 = 2

12. □ − 9 = 3

13. 3 + 4 + 5 = □

14. 6 + 2 + □ = 15

15. 8 + □ + 3 = 12

16. □ + 4 + 4 = 8

17. 5 + 3 + □ = 13

18. 6 + □ + 6 = 18

GETTING STARTED

There are two paths around the flower bed. One is 5 + 4 + 5 steps long. The other is 6 + 2 + 7 steps. Which path is shorter?

Adding More Than Two Numbers

The zookeeper allows 110 children to visit the zoo each morning. So far 29 children are there. **What if** a busload of 38 children and a busload of 35 children come? Will the zookeeper allow all of the children to visit?

29 + 38 + 35 = □

Step 1 Add ones. Regroup.	Step 2 Add tens. Regroup
$\begin{array}{r} {}^{2}\\ 29\\ 38\\ +35\\ \hline 2 \end{array}$ 17 + 5 = 22 ones = 2 tens 2 ones	$\begin{array}{r} {}^{2}\\ 29\\ 38\\ +35\\ \hline 102 \end{array}$

Check by adding up.

$\begin{array}{r} {}^{2}\\ 29\\ 38\\ +35\\ \hline 102 \end{array}$ $\begin{array}{r} 9\\ +13\\ \hline 22 \end{array}$

The zookeeper will allow all of the children to visit.

Check Your Understanding

Write each sum. For 1–6 check by adding up.

1.	2.	3.	4.	5.	6.
16	24	29	26	16	35
15	48	7	98	47	68
+23	+22	+43	+19	+31	+21

7. 6 + 49 + 39 = □   8. 96 + 4 + 79 = □   9. 36 + 67 + 20 = □

Share Your Ideas Look back at 1–5. When did you use mental math? Explain.

Practice

Write each sum. For 10–21 check by adding up.

10. 23 + 14 + 34	**11.** 25 + 28 + 27	**12.** 6 + 48 + 39	**13.** 29 + 54 + 18	**14.** 46 + 16 + 17	**15.** 62 + 36 + 20
16. 99 + 31 + 22	**17.** 73 + 45 + 89	**18.** 47 + 52 + 6	**19.** 98 + 79 + 62	**20.** 19 + 58 + 62 + 51	**21.** 47 + 22 + 94 + 51

22. 17 + 11 + 18 = □

23. 25 + 62 + 75 = □

24. □ = 48 + 16 + 2

25. □ = 53 + 50 + 68 + 9

CHOICES

Solve each. Choose paper and pencil or a calculator for 26–28. Explain your choices.

26. Find 33 + 47 + 15 = □.

27. Change the order of the addends. Find the sum.

28. Change the order again. Find the sum.

29. Write a sentence telling what you discovered.

Think and Apply

30. Predict which path to the elks is longer. Now find out.

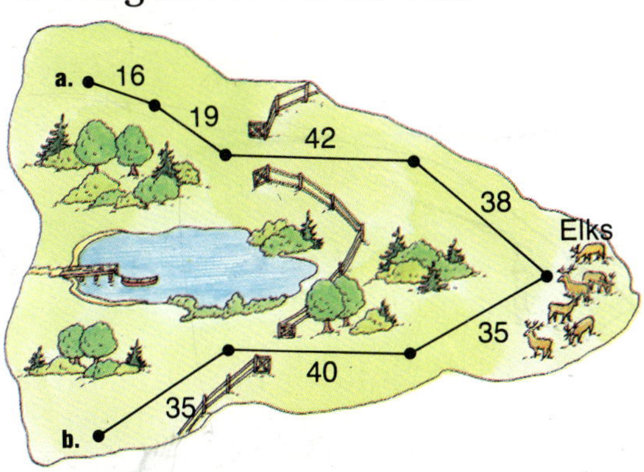

Logical Thinking

Complete each pattern.

31. 1, 2, 4, 7, 11, ___, ___, ___.

32. 0, 8, 15, 21, 26, ___, ___, ___.

33. 0, 15, 30, 45, ___, ___, ___.

SUMMING UP

How can you use what you know to find the sum of 5 addends?
Try 15 + 32 + 17 + 91 + 8 = □.

Midchapter Review

Add. pages 64–73

1. $\begin{array}{r} 24 \\ +56 \\ \hline \end{array}$
2. $\begin{array}{r} 30 \\ +18 \\ \hline \end{array}$
3. $\begin{array}{r} 51 \\ +52 \\ \hline \end{array}$
4. $\begin{array}{r} 5 \\ +95 \\ \hline \end{array}$
5. $\begin{array}{r} 48 \\ +73 \\ \hline \end{array}$
6. $\begin{array}{r} 46 \\ +46 \\ \hline \end{array}$
7. $\begin{array}{r} 39 \\ +84 \\ \hline \end{array}$
8. $\begin{array}{r} 73 \\ +62 \\ \hline \end{array}$
9. $\begin{array}{r} 89 \\ +59 \\ \hline \end{array}$
10. $\begin{array}{r} 75 \\ +86 \\ \hline \end{array}$
11. $\begin{array}{r} 47 \\ +\ 8 \\ \hline \end{array}$
12. $\begin{array}{r} 78 \\ +94 \\ \hline \end{array}$
13. $\begin{array}{r} 15 \\ +69 \\ \hline \end{array}$
14. $\begin{array}{r} 40 \\ +83 \\ \hline \end{array}$
15. $\begin{array}{r} 72 \\ +58 \\ \hline \end{array}$

16. 93 + 9 = ☐
17. 62 + 48 = ☐
18. 74 + 61 = ☐
19. 57 + 88 = ☐
20. 94 + 63 = ☐
21. 6 + 24 = ☐
22. ☐ = 7 + 54
23. ☐ = 14 + 47
24. ☐ = 19 + 87

25. $\begin{array}{r} 54 \\ 13 \\ +40 \\ \hline \end{array}$
26. $\begin{array}{r} 23 \\ 61 \\ +\ 8 \\ \hline \end{array}$
27. $\begin{array}{r} 28 \\ 10 \\ +16 \\ \hline \end{array}$
28. $\begin{array}{r} 66 \\ 19 \\ +\ 8 \\ \hline \end{array}$
29. $\begin{array}{r} 77 \\ 38 \\ +25 \\ \hline \end{array}$

30. 67 + 5 + 13 = ☐
31. 41 + 26 + 93 = ☐
32. 39 + 20 + 48 = ☐
33. 68 + 19 + 4 = ☐
34. ☐ = 9 + 11 + 27
35. ☐ = 52 + 8 + 21

36. Explain what it means to "regroup" in this example.

$\begin{array}{r} 36 \\ +27 \\ \hline \end{array}$

Solve.

37. Jason collected 152 baseball cards. Steve collected 195 cards. How many cards did they collect in all?

38. Hillside School has 401 boys, 356 girls, and 29 teachers involved in sports. How many are involved in sports?

Exploring Problem Solving

THINK
EXPLORE
SOLVE
LOOK BACK

Who Should Win?

The Blue and the Green teams are playing Race to Eleven. These are the rules.

- The first team picks either 2 or 3.
- The second team adds 2 or 3 to that number.
- The teams take turns adding 2 or 3 to the total.
- The team that reaches 11 or more first wins.

Thinking Critically

The Green Team has just added 3, and the total is 5. What number should the Blue Team add to 5? Your class can form teams, and play the game as you answer the questions.

Analyzing and Making Decisions

1. Try the game a few times.
2. **What if** the total is 8? Pretend that it is the Green Team's turn. What number should they add to 8? Explain.
3. **What if** the total is 7? What number should the Green Team add to 7? Who should win if they add that number? Explain.
4. Now try 5. It is the Blue Team's turn. What number should they add to 5? Explain.

Look Back When playing the game, would you rather play first or second? Explain.

Problem Solving Strategies

What's Extra?

At the Amusement Park, a big sign shows how long it takes to wait in line from that point to take each ride. Anna and Rose want to ride both the Twirling Saucers and the Upside-Down Rocket. They have to leave in 35 minutes. Will they have time for both rides? Use the sign to help you.

Sometimes a problem has too much information. You have to be sure to use only the information you need.

Solving the Problem

Think What is the question? What do Anna and Rose want to do?

Explore Which two rides on the sign should Anna and Rose look at? How long will those rides take? Are there facts that you do not need? Explain.

Solve How long will the two rides take altogether? Do Anna and Rose have enough time for both rides?

Look Back Did you use the right facts to find the answer? If you did, you know that Anna and Rose had to decide something. What did they have to decide?

Share Your Ideas

1. Many problems in real life have too much information. How is grocery shopping like a problem with too much information?

Create Your Own

Write a problem that has too much information. Have someone solve it.

THINK
EXPLORE
SOLVE
LOOK BACK

Practice

Solve. Use a calculator where appropriate.

CHOICES

2. You rode on the Whip. Janet rode on the Ferris Wheel. Luis rode on the Cable Car. How many tickets did you and Janet use?
3. Marlene rode the Carousel twice. Yin rode the Carousel three times. How many tickets did Yin use?
4. Marlene bought 12 tickets. Susan bought 15. They both rode on the Ferris Wheel, the Cable Car, and the Whip. Can Marlene visit the Mystery House? Explain.
5. On the trip to the Amusement Park the third grade bus had 27 students from Ms. Anthony's class and 31 students from Mrs. Berry's class. The fourth grade bus had 29 students from Ms. Carter's class and 24 students from Mr. Denny's class. How many students were on the third grade bus?

Mixed Strategy Review

6. Ms. Stevens told the class to do the 28 problems on page 74 and the first 15 problems on page 75. How many problems in all should they solve?

Use the information in the sign to solve 7 and 8.

Mr. and Mrs. Franklin and their children—ages 7, 9, and 13 all want to go on a train ride.

7. How much will it cost for all the family?
8. **What if** there are 26 people in the car already? Can all of the Franklin family ride this time? Explain.

Activity

Exploring Estimation

During the event *Hands Across America,* people held hands all the way across the United States! Before the event, planners estimated that it would take over 5 million people to reach across the country.

Estimate how many students it will take to reach across your classroom.

Working together

A. Compare the different guesses made by the class.

B. Choose one guess to try out.

C. Test one guess by lining up that number of students. Make another guess if you wish.

D. Find the actual number of students it takes to reach across the room.

Sharing Your Results

1. Explain why some guesses were different from others.

2. Did seeing students lined up help you make a better guess? Explain.

Extending the Activity

Work in a small group.

Record your work.

3. Trace your hand to make a hand print. Cut it out.
4. Estimate how many hand prints will fit across your desk. Now find the actual number.
5. Next estimate the number of hand prints
   - across the chalkboard.
   - across the teacher's desk.
   - across a large map.
   - across some other object.
6. Now trace your footprint and cut it out.
7. Estimate how many footprints will reach across the chalkboard. Then find the actual number.
8. Now estimate the number of footprints
   - along a wall of your classroom.
   - from your classroom to the front door.

Summing Up

9. How do your estimates compare with those of your classmates? Why might they be different?
10. Think about the actual number of hand prints across your desk and footprints across the chalkboard. How did knowing these numbers help you to make the other estimates?
11. How does knowing the actual number of units in one length help you estimate the number of units in another length?

GETTING STARTED

Estimate the number of words on this page. Compare estimates. Discuss how you made your estimate.

Estimating Sums

Ben earned 65 coupons for reading books last week. He earned 72 coupons this week. Does he have enough for a gift?

You can estimate to find out. One way is to add the values of the **front digits**. Use zeros for the other places.

$$\begin{array}{r} 65 \\ +\ 72 \\ \hline 130 \end{array}$$

estimated sum

Ben has enough coupons for the poster.

What if Sue earned 164 coupons and then 150? Which gift would she get?

$$\begin{array}{r} 164 \\ +\ 150 \\ \hline 200 \end{array}$$

estimated sum

Is this estimate good enough for this situation? Explain.

Check Your Understanding

Estimate each sum.

1. 34 + 28
2. 83 + 61
3. 71 + 92
4. 210 + 530
5. 249 + 710
6. 495 + 373
7. 30 + 64 = □
8. 95 + 33 = □
9. 27 + 54 = □

Sharing Your Ideas Look back at 1–4. Is your estimate greater than or less than the actual sum? Explain.

Practice

Estimate each sum.

10. 62 +43	**11.** 51 +24	**12.** 75 +85	**13.** 67 +82	**14.** 31 +95	**15.** 24 +68
16. 132 +243	**17.** 507 +436	**18.** 301 +420	**19.** 627 +382	**20.** 289 +654	**21.** 513 129
22. 296 +213	**23.** 347 +256	**24.** 460 +316	**25.** 466 +935	**26.** 462 102 +329	**27.** 104 689 +726

28. 70 + 63 = ☐ **29.** 54 + 37 = ☐ **30.** 92 + 18 = ☐

31. 81 + 35 = ☐ **32.** 60 + 93 = ☐ **33.** 37 + 95 = ☐

34. 138 + 452 = ☐ **35.** 439 + 920 + 140 = ☐

Estimate each sum.
Choose the correct answer.

36. 25 + 76 = ☐
a. 91 **b.** 180 **c.** 90 **d.** 80

37. 482 + 273 = ☐
a. 500 **b.** 600 **c.** 650 **d.** 800

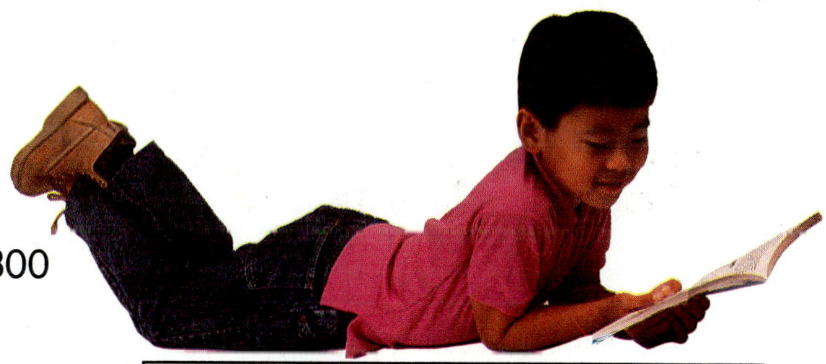

Think and Apply

38. Use the numbers in **15** to write a word problem. Is estimation a good way to solve your problem? Explain.

39. Tomas earned 113 coupons last week and 208 coupons this week. Estimate how many coupons he earned in all.

Visual Thinking

Which tower has the most blocks? How can you tell the number of blocks that you cannot see?

40. 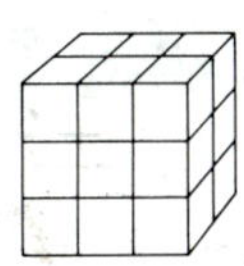

41. 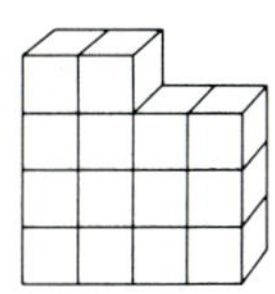

42. 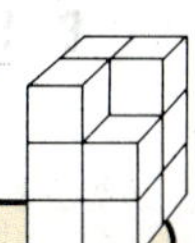

43. 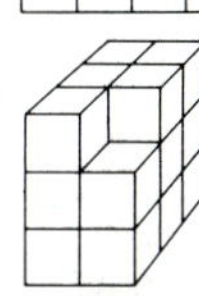

SUMMING UP

Is this statement *always, sometimes,* or *never* true? Explain. When you estimate using front digits the actual sum is less than the estimate.

GETTING STARTED

What is the value of each digit?

135 642 709 How did you know?

Adding Three-Digit Numbers

In 1989 children in Bernardsville, New Jersey, planned their own playground. How many volunteers were needed?

172 + 134 = □

Step 1 Add ones.	**Step 2** Add tens. Regroup.	**Step 3** Add hundreds.
$\begin{array}{r} 172 \\ +134 \\ \hline 6 \end{array}$	$\begin{array}{r} {}^{1} \\ 172 \\ +134 \\ \hline 06 \end{array}$	$\begin{array}{r} {}^{1} \\ 172 \\ +134 \\ \hline 306 \end{array}$

There were 306 volunteers needed.

More Examples

a. No regrouping.

$$\begin{array}{r} 156 \\ +243 \\ \hline 399 \end{array}$$

b. Regroup ones.

$$\begin{array}{r} {}^{1} \\ 248 \\ +735 \\ \hline 983 \end{array}$$

c. Regroup hundreds.

$$\begin{array}{r} {}^{1} \\ 746 \\ +\ 952 \\ \hline 1{,}698 \end{array}$$

16 hundreds = 1 thousand 6 hundreds

In which place will you need to regroup 487 + 812?

Check Your Understanding

Add. Check your work.

1. $\begin{array}{r} 201 \\ +736 \\ \hline \end{array}$ **2.** $\begin{array}{r} 116 \\ +\ 45 \\ \hline \end{array}$ **3.** $\begin{array}{r} 326 \\ +873 \\ \hline \end{array}$ **4.** $\begin{array}{r} 689 \\ +121 \\ \hline \end{array}$ **5.** $\begin{array}{r} 371 \\ +155 \\ \hline \end{array}$

Share Your Ideas Look back at **3–5**. How did you know when to regroup?

Practice

Add. Check your work.

6. 368 + 122
7. 166 + 253
8. 232 + 845
9. 717 + 46
10. 183 + 173
11. 541 + 56
12. 134 + 618
13. 97 + 492
14. 844 + 923
15. 213 + 766
16. 548 + 327
17. 693 + 232
18. 502 + 160 + 31
19. 325 + 246 + 300
20. 242 + 365 + 152
21. 231 + 10 + 152 + 395
22. 243 + 621 + 100 + 925
23. 308 + 253 + 425 + 113

24. 731 + 25 = ☐
25. 656 + 229 = ☐
26. 901 + 683 + 95 = ☐
27. ☐ = 634 + 75
28. ☐ = 249 + 38
29. ☐ = 182 + 21 + 376

Compare. Use >, <, or = for each ⬬. Choose mental math, paper and pencil, or a calculator. Explain your choices.

CHOICES

30. 723 + 20 ⬬ 20 + 732
31. 596 + 500 ⬬ 500 + 569
32. 793 + 621 ⬬ 456 + 698
33. 406 + 891 ⬬ 650 + 647

Think and Apply

34. The castle needed 265 screws. The hut needed 112 screws. How many screws were needed altogether?

35. The tire bridge used 72 feet of chain. The suspension bridge used 260 feet of chain. How many feet of chain were needed for both bridges?

SUMMING UP

How is adding three-digit numbers like adding two-digit numbers? How is it different?

GETTING STARTED

Use estimation to find the two numbers with the greatest sum.

746   351   882   270   965

Regrouping More Than Once

Tim put 363 pieces on the town's mosaic. Lil added 579 pieces. How many pieces did they place in all?

363 + 579 = □

Step 1 Add ones. Regroup.	**Step 2** Add tens. Regroup.	**Step 3** Add hundreds.
1 363 +579 2 12 ones = 1 ten 2 ones	1 1 363 +579 42 14 tens = 1 hundred 4 tens	1 1 363 +579 942

Tim and Lil placed 942 pieces in all.

Find 639 + 891.

Step 1 Add ones. Regroup.	**Step 2** Add tens. Regroup.	**Step 3** Add hundreds. Regroup.
1 639 +891 0 10 ones = 1 ten 0 ones	1 1 639 +891 30 13 tens = 1 hundred 3 tens	1 1 639 +891 1,530 15 hundreds = 1 thousand 5 hundreds

Explain how you could add 363 + 99.

Check Your Understanding

Add. Check your work.

1. 147 + 785 **2.** 814 + 378 **3.** 796 + 351 **4.** 627 + 849 **5.** 535 + 467

Share Your Ideas Explain how you regrouped in **5**.

Practice

Add. Check your work.

6. $\begin{array}{r} 639 \\ +284 \\ \hline \end{array}$ 7. $\begin{array}{r} 576 \\ +825 \\ \hline \end{array}$ 8. $\begin{array}{r} 487 \\ +599 \\ \hline \end{array}$ 9. $\begin{array}{r} 860 \\ +129 \\ \hline \end{array}$ 10. $\begin{array}{r} 724 \\ +997 \\ \hline \end{array}$ 11. $\begin{array}{r} 496 \\ +326 \\ \hline \end{array}$

12. $\begin{array}{r} 370 \\ +619 \\ \hline \end{array}$ 13. $\begin{array}{r} 993 \\ +\ 72 \\ \hline \end{array}$ 14. $\begin{array}{r} 685 \\ +976 \\ \hline \end{array}$ 15. $\begin{array}{r} 508 \\ +927 \\ \hline \end{array}$ 16. $\begin{array}{r} 148 \\ +874 \\ \hline \end{array}$ 17. $\begin{array}{r} 812 \\ +493 \\ \hline \end{array}$

18. $\begin{array}{r} 467 \\ 589 \\ +835 \\ \hline \end{array}$ 19. $\begin{array}{r} 61 \\ 654 \\ +707 \\ \hline \end{array}$ 20. $\begin{array}{r} 378 \\ 396 \\ +469 \\ \hline \end{array}$ 21. $\begin{array}{r} 793 \\ 847 \\ +654 \\ \hline \end{array}$ 22. $\begin{array}{r} 421 \\ 368 \\ 37 \\ +101 \\ \hline \end{array}$ 23. $\begin{array}{r} 895 \\ 239 \\ 546 \\ +764 \\ \hline \end{array}$

24. $979 + 42 = \square$ 25. $625 + 743 = \square$ 26. $287 + 50 + 941 = \square$

27. $\square = 839 + 71$ 28. $\square = 345 + 777$ 29. $\square = 656 + 901 + 34$

Follow each rule to complete.

Rule: Add 435.

	Input	Output
30.	297	
31.	968	
32.	641	
33.	787	

Rule: Add 654.

	Input	Output
34.	932	
35.	690	
36.	178	
37.	356	

Rule: Add 199.

	Input	Output
38.	495	
39.	99	
40.	962	
41.	834	

Think and Apply

42. Tim and Lil together placed 942 pieces on the mosaic. John added 130 pieces. Now how many pieces are on the mosaic?

43. The mosaic pieces come in boxes of 1,000. Dina placed about 580 pieces on the mosaic. She wants to place about 500 more pieces on it. Will one box be enough?

Make up an example in which you need to regroup in the ones and tens places. Give it to a classmate to solve.

SUMMING UP

GETTING STARTED

What happens when you add one more to 999? What happens when you add one more to 9,999?

Adding Four-Digit Numbers

In a bikeathon, members of one group rode 2,352 laps. Members of another group rode 3,469 laps. How many laps did they ride altogether? **2,352 + 3,469 = □**

Knowing how to add two- and three-digit numbers helps you to add numbers with many digits.

Step 1 Add ones. Regroup.	**Step 2** Add tens. Regroup.	**Step 3** Add hundreds.	**Step 4** Add thousands.
$\begin{array}{r} {}^{1} \\ 2,352 \\ +3,469 \\ \hline 1 \end{array}$	$\begin{array}{r} {}^{1\,1} \\ 2,352 \\ +3,469 \\ \hline 21 \end{array}$	$\begin{array}{r} {}^{1\,1} \\ 2,352 \\ +3,469 \\ \hline 821 \end{array}$	$\begin{array}{r} {}^{1\,1} \\ 2,352 \\ +3,469 \\ \hline 5,821 \end{array}$

They rode 5,821 laps altogether.

More Examples

a. $\begin{array}{r} 6,153 \\ +3,620 \\ \hline 9,773 \end{array}$ **b.** $\begin{array}{r} {}^{1} \\ 5,464 \\ +2,912 \\ \hline 8,376 \end{array}$ **c.** $\begin{array}{r} {}^{1\,1\,1} \\ 9,675 \\ +8,635 \\ \hline 18,310 \end{array}$

Look back at **c.** Explain the regrouping.

Check Your Understanding

Add. Check your work.

1. $\begin{array}{r} 4,105 \\ +5,831 \\ \hline \end{array}$ **2.** $\begin{array}{r} 8,275 \\ +\quad 956 \\ \hline \end{array}$ **3.** $\begin{array}{r} 2,859 \\ +5,104 \\ \hline \end{array}$ **4.** $\begin{array}{r} 3,462 \\ +7,539 \\ \hline \end{array}$ **5.** $\begin{array}{r} 6,976 \\ +\quad 92 \\ \hline \end{array}$

Share Your Ideas How is adding four-digit numbers like adding two- and three-digit numbers?

Practice

Add. Check your work.

6. 392 + 7,045
7. 8,152 + 4,788
8. 7,847 + 6,823
9. 6,369 + 5,735
10. 4,836 + 3,053
11. 3,408 + 2,597
12. 41,887 + 26,022
13. 139,814 + 650,173
14. 2,723 + 6,569 + 406 + 9,735

15. 3,945 + 4,498 = □
16. 9,975 + 576 = □
17. 7,592 + 8,434 = □
18. 7,355 + 875 = □
19. □ = 4,865 + 8,789
20. □ = 149 + 6,017

Use the numbers 9,945 | 9,493 | 9,944 | 9,496 **to find the sums for 21–25.**

Choose mental math, paper and pencil, or a calculator. Explain your choices.

CHOICES

21. the greatest and the least numbers.
22. the two greatest numbers.
23. the even numbers.
24. the odd numbers.
25. the numbers between 9,400 and 10,000.

Think and Apply

26. The Blazers need to ride 2,500 laps to set a record for the bikeathon. By darkness they rode 2,050 laps. Did they set the bikeathon record yet?

How many digits could the sum of each have?
2 four-digit numbers   2 five-digit numbers
Explain your answers.

SUMMING UP

Mixed Review

1. 7 + 9
2. 15 − 8
3. 13 − 6
4. 9 + 9
5. 2 + 7 + 2
6. 3 + 2 + 6
7. 2 + 1 + 8 + 1
8. 4 + 4 + 2 + 4

Compare. Use >, <, or = for ⬬.

9. 568 ⬬ 564
10. 1,723 ⬬ 3,723
11. 86,425 ⬬ 68,245
12. 95,602 ⬬ 95,601
13. 2 + 4 ⬬ 6 + 1
14. 13 − 7 ⬬ 18 − 5
15. 5 + 3 ⬬ 2 + 9

Give the value of the digit 2.

16. 342
17. 2,976
18. 5,429
19. 9,243
20. 265,074
21. 28,370

GETTING STARTED

You have 2 ten-dollar bills, 5 one-dollar bills, 6 dimes, and 3 pennies. How much money do you have?

Adding Money

Sally has $25.00 to spend at the craft show. Does she have enough to buy the backpack and wallet?

We add amounts of money the way we add other numbers.

Line up the decimal points. Then add.

$$\begin{array}{r} {}^{1} \\ \$12.75 \\ +\ 11.50 \\ \hline \$24.25 \end{array}$$

Write the dollar sign and decimal point in the answer.

The total cost is $24.25. Since $25.00 > $24.25, Sally has enough to buy the backpack and wallet.

Could Sally have bought the mirror and the cat pillow?

$$\begin{array}{r} {}^{1\ 1} \\ \$\ 9.99 \\ +\ 13.99 \\ \hline \$23.98 \end{array}$$

$25.00 > $23.98

Sally could have bought the mirror and the pillow.

Check Your Understanding

Add. Check your work.

1. $\begin{array}{r} \$5.67 \\ +\ 1.19 \\ \hline \end{array}$ **2.** $\begin{array}{r} \$34.50 \\ +\ 83.74 \\ \hline \end{array}$ **3.** $\begin{array}{r} \$73.09 \\ +\ 5.80 \\ \hline \end{array}$ **4.** $\begin{array}{r} \$42.85 \\ +\ 68.19 \\ \hline \end{array}$

5. $8.27 + $.46 = ☐ **6.** $53.95 + $86.94 = ☐

Share Your Ideas How is adding money the same as adding other numbers? How is it different?

Practice

Add. Check your work.

7. $\begin{array}{r} \$8.50 \\ +\ 1.25 \\ \hline \end{array}$
8. $\begin{array}{r} \$\ \ .87 \\ +\ 39.02 \\ \hline \end{array}$
9. $\begin{array}{r} \$67.55 \\ +\ 23.18 \\ \hline \end{array}$
10. $\begin{array}{r} \$84.49 \\ +\ 15.31 \\ \hline \end{array}$
11. $\begin{array}{r} \$59.56 \\ +\ 36.96 \\ \hline \end{array}$
12. $\begin{array}{r} \$45.82 \\ +\ 10.43 \\ \hline \end{array}$
13. $\begin{array}{r} \$87.63 \\ +\ 3.75 \\ \hline \end{array}$
14. $\begin{array}{r} \$73.09 \\ +\ 18.96 \\ \hline \end{array}$
15. $\begin{array}{r} \$96.80 \\ +\ 91.94 \\ \hline \end{array}$
16. $\begin{array}{r} \$33.59 \\ +\ 9.71 \\ \hline \end{array}$
17. $\begin{array}{r} \$.18 \\ .56 \\ +\ .07 \\ \hline \end{array}$
18. $\begin{array}{r} \$7.48 \\ 4.12 \\ +\ .93 \\ \hline \end{array}$
19. $\begin{array}{r} \$7.98 \\ 8.70 \\ +\ 6.82 \\ \hline \end{array}$
20. $\begin{array}{r} \$46.54 \\ 96.77 \\ +\ 68.69 \\ \hline \end{array}$
21. $\begin{array}{r} \$102.89 \\ 96.57 \\ +\ 78.48 \\ \hline \end{array}$

22. $55.89 + $70.81 = □
23. $45.76 + $2.75 = □
24. $84.67 + $85.94 = □
25. $55.18 + $32.71 = □
26. □ = $.99 + $18.76
27. □ = $96.59 + $75.63
28. □ = $65.49 + $14.57
29. □ = $76.49 + $23.98

Compare. Use > or < for ●.

30. $6.24 ● $.42
31. $8.97 ● $8.99
32. $4.31 ● $54.31
33. $75.38 ● $75.83

Think and Apply

DATA 34. **What if** you had $10.00? Find ads for 5 items you could buy. Which item costs the most? the least?

35. Eva wants to buy mittens for $8.75 and a set of books for $23.98. She has $40.00. Does she have enough? Explain.

36. Games were on sale for $8.00, $9.00, $12.00, and $15.00. Tasha spent $20.00 on 2 of these games. What were the prices of the 2 games she bought?

Common Error

37. Explain how to find the correct sum.

$\begin{array}{r} \$83.16 \\ +\ 67.24 \\ \hline \$897.184 \end{array}$ ← incorrect

SUMMING UP

Why is it important to line up the decimal points when you add money?

Using Problem Solving

Building the Greatest Sum

Marita is playing an addition game. The goal is to make the greatest sum. The digits 0–9 are on cards in a bag. The digits 6 and 3 have already been drawn. Marita wrote them on her chart.

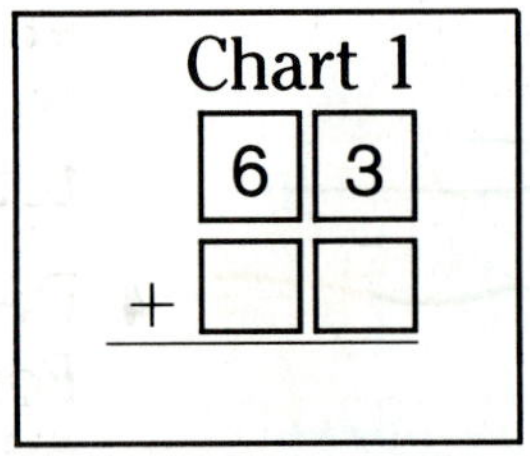

The next digit drawn was 2. Where should Marita write the 2? Think about this as you start to play Building the Greatest Sum.

Working together

Materials: a bag with cards for the digits 0–9, Workmat 5, a calculator

Follow these rules.

A. For each round, one player draws a digit card from the bag. All players write the digit in an open space on Chart 1 of their workmats. Once the digit is written, it cannot be erased.

B. Players take turns drawing digit cards. When the chart is complete, players add the numbers on their own charts. The player with the greatest sum wins the game.

C. Return the digit cards to the bag.

D. Repeat **A–C** for Charts 2, 3, and 4.

Sharing Your Ideas

1. How did you decide where to place each number?
2. How would the game be different if each number were returned to the bag after it was drawn?

THINK
EXPLORE
SOLVE
LOOK BACK

Extending Your Thinking

Create new ways to play!
Use a calculator if you wish.

CHOICES

3. Each player could find the total score for all games played. Find out if the player who won the most games also had the highest total score? Explain.
4. Decide on other rules for playing the game. For example, the player with the *least* sum might win. Or, the player with the sum closest to 1,000 might win.
5. Play at least five rounds of the game using your new rules.

Summing Up

6. Which of your new games did you enjoy the most? Why?
7. When you are trying to get the greatest sum, which digits should you try to write in the spaces on the far left?

Chapter Review

Add. pages 64–73

1. $\begin{array}{r} 42 \\ +13 \\ \hline \end{array}$
2. $\begin{array}{r} 16 \\ +25 \\ \hline \end{array}$
3. $\begin{array}{r} 47 \\ +36 \\ \hline \end{array}$
4. $\begin{array}{r} 91 \\ +72 \\ \hline \end{array}$
5. $\begin{array}{r} 59 \\ +68 \\ \hline \end{array}$
6. $\begin{array}{r} 8 \\ 7 \\ +6 \\ \hline \end{array}$
7. $\begin{array}{r} 23 \\ 42 \\ +14 \\ \hline \end{array}$
8. $\begin{array}{r} 68 \\ 45 \\ +\ 9 \\ \hline \end{array}$
9. $\begin{array}{r} 13 \\ 37 \\ +67 \\ \hline \end{array}$
10. $\begin{array}{r} 5 \\ 16 \\ +93 \\ \hline \end{array}$
11. 6 + 9 + 2 = □
12. □ = 3 + 17 + 46

Estimate each sum. pages 78–81

13. $\begin{array}{r} 72 \\ +27 \\ \hline \end{array}$
14. $\begin{array}{r} 38 \\ +35 \\ \hline \end{array}$
15. $\begin{array}{r} 305 \\ +569 \\ \hline \end{array}$
16. $\begin{array}{r} 156 \\ +612 \\ \hline \end{array}$
17. $\begin{array}{r} 249 \\ +125 \\ \hline \end{array}$

Find each sum. pages 82–89

18. $\begin{array}{r} 203 \\ +176 \\ \hline \end{array}$
19. $\begin{array}{r} 237 \\ +438 \\ \hline \end{array}$
20. $\begin{array}{r} 959 \\ +364 \\ \hline \end{array}$
21. $\begin{array}{r} 1,067 \\ +4,120 \\ \hline \end{array}$
22. $\begin{array}{r} 1,248 \\ +6,956 \\ \hline \end{array}$
23. $\begin{array}{r} \$1.60 \\ +\ \ .98 \\ \hline \end{array}$
24. $\begin{array}{r} \$4.87 \\ +\ 2.59 \\ \hline \end{array}$
25. $\begin{array}{r} \$17.20 \\ +\ \ 8.97 \\ \hline \end{array}$
26. $\begin{array}{r} \$18.46 \\ +\ \ 7.68 \\ \hline \end{array}$
27. $\begin{array}{r} \$95.88 \\ +\ 39.57 \\ \hline \end{array}$

Choose the correct words to complete the sentence.

28. One way to _____ sums is to add the values of the _____.

Words To Know
front digits estimate

Solve. pages 75–77

29. Look at the chart. Name two books that have a total of about 800 pages. a total of about 600 pages.

Books	Number of Pages
Science	302
Math	411
Music	237
Reading	423

30. Jason has 350 baseball cards, Steve has 86, and Alvin has 175. How many cards do Jason and Alvin have?

Chapter Test

Add.

1. 25 + 14
2. 84 + 65
3. 59 + 28
4. 63 + 98
5. 73 + 49
6. 9 + 8 + 4
7. 39 + 6 + 28
8. 35 + 67 + 46
9. 44 + 16 + 8
10. 15 + 53 + 77

Estimate each sum.

11. 65 + 13
12. 48 + 12
13. 408 + 379
14. 755 + 927
15. 82 + 65 = □
16. 26 + 31 = □
17. □ = 249 + 125

Find each sum.

18. 763 + 15
19. 568 + 121
20. 347 + 219
21. 877 + 649
22. 1,625 + 218
23. 4,192 + 1,306
24. 3,967 + 8,437
25. 3,009 + 5,993
26. \$6.27 + .15
27. \$1.53 + 2.68
28. \$19.65 + 8.35
29. \$56.51 + 24.69
30. 968 + 54 = □
31. □ = \$3.59 + \$.85

Solve.

32. Ann counted 1,010 math books and 1,250 science books. Then 389 new math books arrived. Now how many math books are there?
33. Lou's camping trip lasted 2 weeks. He spent \$3.80 for sunscreen lotion and \$8.50 for a hat. How much did he spend in all?

THINK Without adding, decide which sum is greater: 7,482 + 1,320 = □ or 7,482 + 2,320 = □. How do you know?

Computer Link

Adding Turtle Steps

What number is needed for the turtle to draw a path the same length?

```
HOME
FD 34
FD 15
HOME
FD __
```

Materials: Logo

A. Find the missing number to draw a path the same length. Start at HOME.

•	•	•
FD 39	FD 43	FD 55
FD 14	FD 28	FD 27
HOME	HOME	HOME
FD __	FD __	FD __

B. Find each missing number to draw two paths the same length. Compare the commands for each pair of paths. What do you notice?

• Path 1	Path 2	• Path 3	Path 4	• Path 5	Path 6
HOME	HOME	HOME	HOME	HOME	HOME
FD 51	FD 23	FD 32	FD __	FD 81	FD 22
FD 23	FD __	FD 47	FD 32	FD __	FD 81
HOME	HOME	HOME	HOME	HOME	HOME
FD __	FD __	FD __	FD __	FD __	FD __

C. Explain how the numbers for one path help you to find the numbers for the other path.

Sharing Your Results

1. How did you find the missing numbers?
2. Does your way work for any path? for any numbers?

Extending the Activity

You can use the Logo `PRINT` (`PR`) command to do math problems. Enter `PR 45 + 89`. Logo adds the numbers and prints `134`.

3. Predict what Logo will print. Then enter each command to check.

   a. `PR 117 + 98` b. `PR 83 + 129` c. `PR 65 + 235`

4. Work with a partner. What numbers can you use to make Path 2 the same length as Path 1? Find as many different number pairs as you can. Investigate with the turtle and the `PRINT` command. Record your work.

Path 1	Path 2
HOME	HOME
FD 34	FD __
FD 66	FD __

5. How many different pairs of numbers did you find?

6. Compare your results with those of other teams. Did everyone get the same number pairs?

Summing Up

7. **Look back** at **4–6**. Explain why different answers are possible.

EXTENSION

The Name Game

Jane says that her name is worth 30 cents. What could she mean?

A	= $.01	N	= $.14
B	= $.02	O	= $.15
C	= $.03	P	= $.16
D	= $.04	Q	= $.17
E	= $.05	R	= $.18
F	= $.06	S	= $.19
G	= $.07	T	= $.20
H	= $.08	U	= $.21
I	= $.09	V	= $.22
J	= $.10	W	= $.23
K	= $.11	X	= $.24
L	= $.12	Y	= $.25
M	= $.13	Z	= $.26

Use the chart in the same way Jane did.

1. Find the value of each letter in the name Veronica. What is the total value?
2. What is the total value of your first name? your last name? your full name?
3. Compare the total value of your first name with those of your classmates. Whose first name is worth the most?
4. Repeat step **3** for your last name and your full name.

Use the chart above to answer each.

5. Is it possible to find a 3-letter word that is worth $1.00? Explain.
6. What is the most a 3-letter word could be worth?
7. Without finding the values of each letter, decide if the word *addend* could be worth $1.00. Explain your thinking.

Build-A-Word

Work in a small group. Place a set of cards labeled A–Z in a bag. Each player draws 7 letters from the bag and uses as many of them as possible to form a word. The player with the word that has the greatest value scores a point. Repeat this 5 times.

Maintaining Skills

Choose the correct answer. Write A, B, C, or D.

1. $2 + 4 + 5 = \square$

- **A** 9
- **B** 11
- **C** 10
- **D** not given

2.
$$\begin{array}{r} 16 \\ -\ 9 \\ \hline \end{array}$$

- **A** 7
- **B** 8
- **C** 6
- **D** not given

3. What is the value of the digit 8 in 480?

- **A** 8
- **B** 800
- **C** 80
- **D** not given

4. Round 42 to the nearest ten.

- **A** 50
- **B** 40
- **C** 30
- **D** not given

5. Round 655 to the nearest hundred.

- **A** 700
- **B** 600
- **C** 400
- **D** not given

6. Choose the number for 5,000 + 200 + 3.

- **A** 523
- **B** 5,302
- **C** 5,023
- **D** not given

7. Compare. 23,068 ⬬ 2,368

- **A** $<$
- **B** $>$
- **C** $=$
- **D** not given

8. $15 + 23 = \square$

- **A** 28
- **B** 32
- **C** 38
- **D** not given

9.
$$\begin{array}{r} 712 \\ +245 \\ \hline \end{array}$$

- **A** 957
- **B** 956
- **C** 967
- **D** not given

10. $2{,}038 + 5{,}124 = \square$

- **A** 7,152
- **B** 7,262
- **C** 7,162
- **D** not given

Solve.

11. Terri takes 15 steps to cross the room. Marge needs to take 17 steps. Who has the longer step?

- **A** Both are the same.
- **B** Marge
- **C** Terri
- **D** not given

12. Raoul counted about 4 hand spans across his desk. About how many of his hand spans will go across his teacher's desk?

- **A** 10
- **B** 25
- **C** 3
- **D** 5

4 Subtracting

THEME Third Graders: Busy at School

Sharing What You Know

A third grader's life is very busy. There is school, after-school classes and activities, homework, and chores at home. What happens when you have too many things to do? Tell how taking time away from one activity would give you more time for another.

Using Language

At school it is sometimes fun to trade with friends. You might **exchange** your peanut butter sandwich for a friend's cheese sandwich. What other things do you exchange? When you use base-ten blocks to show subtraction, you sometimes need to **exchange** 1 ten for 10 ones. How is exchanging blocks in subtraction different from exchanging things with friends?

Words to Know: exchange, tens, subtract, hundreds, regroup, estimate, difference

Be a Problem Solver

Your class is going to the Media Center for half an hour. Six of you will work at the computers. What will the other students do? Make a plan so that everyone has an activity. **What if** you got to the library and 5 new computers had just been set up?

Describe how you would change your plan.

Activity

Exploring Subtraction

Two teams of disc jockeys took turns playing a total of 99 tapes for a school carnival. The teams drew number cards to find out how many tapes they could play each turn. How many tapes will be left for the last turn?

Working together

Materials: eight 3 x 5 cards, base-ten blocks

A. Make a set of 8 cards. Write one of these numbers on each: 3, 6, 8, 11, 12, 15, 17, 19. Place the cards face down in a pile.

B. Use base-ten blocks to model the 99 tapes.

C. Draw a number card for Team A. Take away that number of blocks from 99 blocks. Exchange if necessary.

D. Record the subtraction in a chart like the one at the right. How many tapes are left?

tens	ones
−	

E. Draw a card for Team B. Take away blocks from those that are left to show the number on the card. Record the subtraction.

F. Continue taking turns until all the cards have been drawn.

Sharing Your Results

1. How many tapes are left for the last turn?

2. How did you know when to exchange blocks?

Extending the Activity

Students are planning to have three teams of disc jockeys use 999 tapes in a month-long fund raiser. Help them plan how this will work.

Work in a small group.

3. Make a set of 10 number cards: 61, 90, 93, 97, 102, 104, 105, 108, 109, 116.
4. Take turns drawing cards for Teams A, B, and C. Draw a card for the first time. Then take away that number of blocks from 999 blocks. Exchange if necessary.
5. Record the subtraction in a chart like the one at the right. How many tapes are left?
6. Take turns for each of the teams until all the cards have been drawn.

hundreds	tens	ones
−		

Summing Up

7. How is subtracting from 999 like subtracting from 99? How is it different?
8. How is exchanging with 999 blocks the same as exchanging with 99 blocks? How is it different?
9. How did you decide when to exchange blocks?

Use base-ten blocks to subtract.

10.

tens	ones
9	2
−5	7

11.

hundreds	tens	ones
3	3	1
−	2	8

12.

hundreds	tens	ones
5	0	3
− 2	1	4

13. **Look back** at **10–12**. For each, tell when you exchanged. Then explain how you exchanged.

Activity

Understanding Subtraction

Here's a math challenge for you. What different subtraction examples can you make, using the digits 1, 2, 3, and 4? Try it!

Working together

Materials: two cubes, one for tens and one for ones, each numbered 1, 1, 2, 2, 3, and 4; base-ten blocks

A. Toss the cubes to create 20 different two-digit numbers. Record the numbers.

B. Choose pairs of numbers and write at least 8 different subtraction examples.

C. For each subtraction, show the greater number with blocks. Take away blocks for the other number. Exchange if necessary. Record your work.

tens	ones
3	2
−1	4

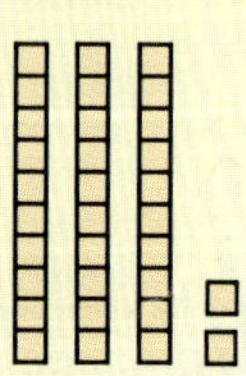

tens	ones
2	12
~~3~~	~~2~~
−1	4

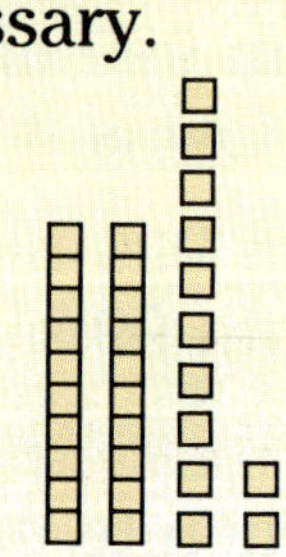

tens	ones
2	12
~~3~~	~~2~~
−1	4
1	8

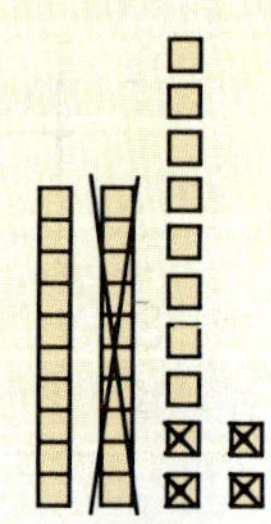

Sharing Your Results

1. How did you decide which two-digit numbers to use for the subtraction examples?
2. How did you know when to exchange blocks?
3. How does your record show that you exchanged 1 ten for 10 ones?

Extending the Activity

Now use three-digit numbers for your challenge.

Work with a partner.

Materials: three cubes, each numbered 0, 1, 2, 3, 4, 5; base-ten blocks

4. Toss the cubes to create 18 different three-digit numbers. Record the numbers.

5. Choose pairs of numbers and write at least 6 different subtraction examples.

6. Record the subtraction. Use base-ten blocks if you wish.

a.

h	t	o
3	0	4
−1	2	5

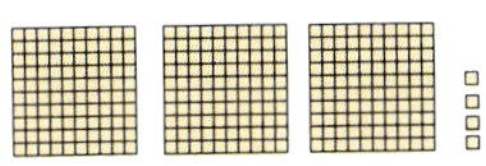

b.

h	t	o
2	10	
3	0	4
−1	2	5

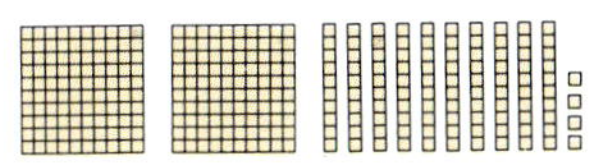

c.

h	t	o
	9	
2	10	14
3	0	4
−1	2	5

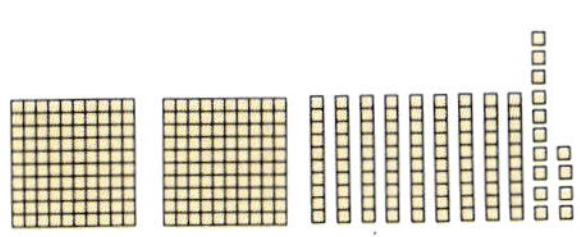

d.

h	t	o
	9	
2	10	14
3	0	4
−1	2	5
1	7	9

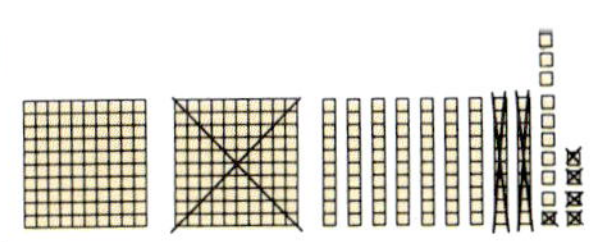

Summing Up

Use a subtraction example to help you explain each.

7. How can you tell when you do not need to regroup? When you do need to regroup?

8. How do you know when to regroup tens to ones? hundreds to tens?

GETTING STARTED

You have 7 ten-dollar bills and 3 one-dollar bills in play money. Would you trade for 6 tens and 13 ones? Why or why not?

Subtracting Two-Digit Numbers

Rock Hill School is planning a pet parade. The tally sheet shows 37 pets are signed up. If 19 of the pets are dogs, how many are not dogs?

37 − 19 = □

Step 1
Subtract ones.
Not enough ones.

tens	ones
3	7
−1	9

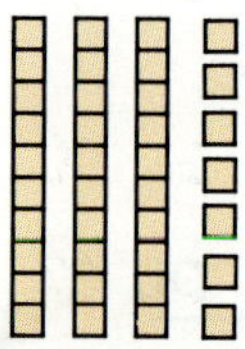

Step 2
Regroup 1 ten as 10 ones.

tens	ones
2	17
~~3~~	~~7~~
−1	9

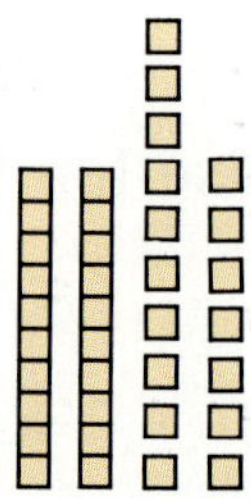

Step 3
Subtract ones.

tens	ones
2	17
~~3~~	~~7~~
−1	9
	8

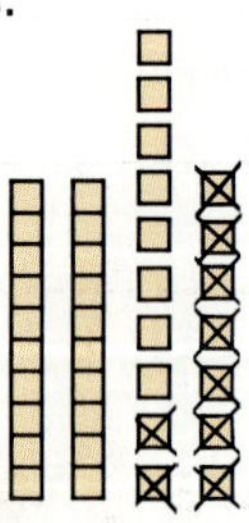

Step 4
Subtract tens.

tens	ones
2	17
~~3~~	~~7~~
−1	9
1	8

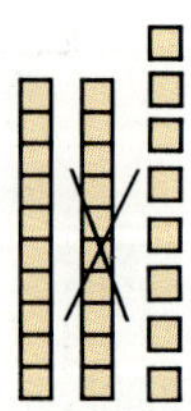

Check by adding.

$$\begin{array}{r} {}^{1} \\ 18 \\ +19 \\ \hline 37 \end{array}$$

There are 18 pets that are not dogs.

Check Your Understanding

Subtract. Check by adding.

1. $\begin{array}{r} 38 \\ -23 \\ \hline \end{array}$ **2.** $\begin{array}{r} 92 \\ -65 \\ \hline \end{array}$ **3.** $\begin{array}{r} 44 \\ -30 \\ \hline \end{array}$ **4.** $\begin{array}{r} 61 \\ -49 \\ \hline \end{array}$ **5.** $\begin{array}{r} 84 \\ -8 \\ \hline \end{array}$ **6.** $\begin{array}{r} 50 \\ -37 \\ \hline \end{array}$

Share Your Ideas How can you tell when to regroup in subtraction with two-digit numbers?

Practice

Subtract. Check by adding.

7.

tens	ones
4	5
−3	1

8.

tens	ones
7	5
−1	8

9.

tens	ones
5	7
−2	4

10.

tens	ones
1	8
−	9

11. $76 - 5$ **12.** $93 - 91$ **13.** $85 - 28$ **14.** $51 - 37$ **15.** $42 - 7$ **16.** $60 - 50$

17. $52 - 35$ **18.** $96 - 27$ **19.** $44 - 41$ **20.** $81 - 53$ **21.** $73 - 19$ **22.** $64 - 28$

23. $97 - 4$ **24.** $92 - 39$ **25.** $52 - 44$ **26.** $70 - 56$ **27.** $50 - 11$ **28.** $80 - 80$

29. 96 − 5 = ☐ **30.** 54 − 26 = ☐ **31.** 60 − 2 = ☐

32. 82 − 36 = ☐ **33.** 98 − 88 = ☐ **34.** 47 − 39 = ☐

35. ☐ = 97 − 66 **36.** ☐ = 72 − 72 **37.** ☐ = 17 − 9

Continue each pattern.

38. 97, 92, 87, ____, ____, ____ **39.** 33, 44, 55, ____, ____, ____

40. 85, 76, 67, ____, ____, ____ **41.** 62, 54, 46, ____, ____, ____

Think and Apply

42. There are 19 dogs, 12 cats, and 6 birds in the parade. How many more dogs than cats are in the parade? How many more cats than birds are there?

DATA **43.** List the pets your classmates have. Make a tally mark for each pet. What pet is named most often? least often?

Common Error

44. Mary made an error in example **b**. Explain how to subtract 27 − 19 correctly.

Write in your own words a rule for subtraction with two-digit numbers.

SUMMING UP

A small town reports that 15,000 people take part in fitness programs. Did someone actually count 15,000 people? Explain.

Estimating Differences

Clark wrote about the School Fitness Fair. He found that 87 students raced in the mini marathon. Last year only 42 students raced. About how many more students raced this year?

When you do not need an exact answer, estimate! Estimate 87–42.

Round to the nearest ten to make it easy to subtract.

$$\begin{array}{r} 87 \\ -42 \\ \hline \end{array} \xrightarrow{\text{rounds to}} \begin{array}{r} 90 \\ -40 \\ \hline 50 \end{array} \text{ estimated difference}$$

Clark reported that about 50 more students raced this year.

Another Example Estimate 461 − 350.

Round to the nearest hundred and subtract.

$$\begin{array}{r} 461 \\ -350 \\ \hline \end{array} \xrightarrow{\text{rounds to}} \begin{array}{r} 500 \\ -400 \\ \hline 100 \end{array} \text{ estimated difference}$$

Explain how you would estimate 820 − 560.

Check Your Understanding

Estimate each difference.

1. $\begin{array}{r} 68 \\ -21 \\ \hline \end{array}$ **2.** $\begin{array}{r} 53 \\ -44 \\ \hline \end{array}$ **3.** $\begin{array}{r} 73 \\ -25 \\ \hline \end{array}$ **4.** $\begin{array}{r} 85 \\ -19 \\ \hline \end{array}$ **5.** $\begin{array}{r} 835 \\ -350 \\ \hline \end{array}$ **6.** $\begin{array}{r} 940 \\ -630 \\ \hline \end{array}$

Share Your Ideas Explain why Clark could use estimates to report his findings.

Practice

Estimate each difference.

7. 72 − 31
8. 51 − 17
9. 84 − 50
10. 93 − 32
11. 76 − 10
12. 40 − 29
13. 309 − 126
14. 518 − 347
15. 276 − 103
16. 823 − 456
17. 905 − 200
18. 650 − 217
19. 73 − 19 = ☐
20. 67 − 18 = ☐
21. 808 − 334 = ☐
22. 432 − 175 = ☐
23. 361 − 120 = ☐
24. 156 − 99 = ☐

Compare. Write >, <, or = for ⬬. Choose estimation, paper and pencil, or a calculator. Explain your choices.

CHOICES

25. 38 − 10 ⬬ 42 − 21
26. 85 − 30 ⬬ 57 − 19
27. 67 − 15 ⬬ 87 − 31
28. 77 − 68 ⬬ 99 − 18
29. 913 − 225 ⬬ 613 − 503
30. 945 − 203 ⬬ 268 + 474

Think and Apply

31. Words like *approximately* and *about* are used when we estimate. What other words can you think of for estimation?

32. Explain how rounding can also be used to estimate a sum.

Visual Thinking

Estimate. Which shape has about the same amount of blue as the first shape?

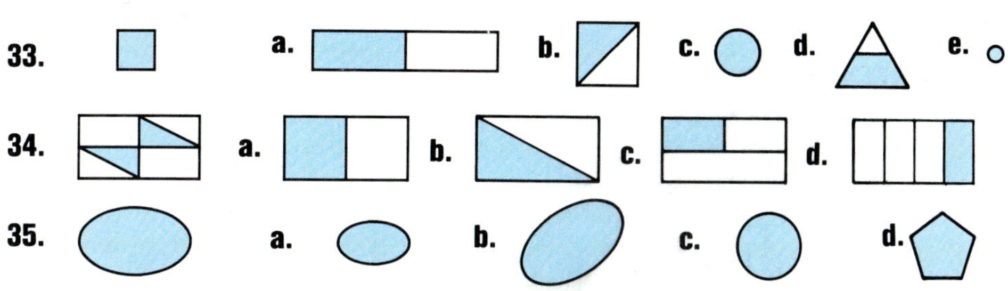

Explain how you can estimate differences in your head.

SUMMING UP

GETTING STARTED

Think about the number 485. What digit is in the hundreds place? the tens place? the ones place?

Subtracting Three-Digit Numbers

Sam's class wants to make Fluffy Puffies for 485 students. The class has made 218 so far. How many more are needed?

$485 - 218 = \square$

Estimate first. Then find the exact answer.

Estimate.

$$\begin{array}{r} 485 \\ -218 \\ \hline \end{array} \quad \text{rounds to} \quad \begin{array}{r} 500 \\ -200 \\ \hline 300 \end{array}$$

estimated difference

About 300 more are needed.

Step 1
Subtract ones.
Not enough ones.

$$\begin{array}{r} 485 \\ -218 \\ \hline \end{array}$$

Step 2
Regroup 1 ten as 10 ones.

$$\begin{array}{r} {\scriptstyle 7\,15} \\ 4\not{8}\not{5} \\ -218 \\ \hline \end{array}$$

8 tens 5 ones = 7 tens 15 ones

Step 3
Subtract ones.

$$\begin{array}{r} {\scriptstyle 7\,15} \\ 4\not{8}\not{5} \\ -218 \\ \hline 7 \end{array}$$

Step 4
Subtract tens.

$$\begin{array}{r} {\scriptstyle 7\,15} \\ 4\not{8}\not{5} \\ -218 \\ \hline 67 \end{array}$$

Step 5
Subtract hundreds.

$$\begin{array}{r} {\scriptstyle 7\,15} \\ 4\not{8}\not{5} \\ -218 \\ \hline 267 \end{array}$$

Exactly 267 more Fluffy Puffies are needed.

Check Your Understanding

Estimate first. Then subtract.

1. $\begin{array}{r} 993 \\ -451 \\ \hline \end{array}$ 2. $\begin{array}{r} 395 \\ -292 \\ \hline \end{array}$ 3. $\begin{array}{r} 306 \\ -100 \\ \hline \end{array}$ 4. $\begin{array}{r} 778 \\ -229 \\ \hline \end{array}$ 5. $\begin{array}{r} 653 \\ -305 \\ \hline \end{array}$ 6. $\begin{array}{r} 558 \\ -98 \\ \hline \end{array}$

Share Your Ideas Explain how to use estimation to check subtraction.

Practice

Estimate first. Then subtract.

7. 673 − 137	**8.** 947 − 436	**9.** 792 − 654	**10.** 320 − 310	**11.** 467 − 238	**12.** 957 − 631
13. 391 − 122	**14.** 944 − 315	**15.** 129 − 102	**16.** 751 − 134	**17.** 647 − 347	**18.** 98 − 79
19. 794 − 456	**20.** 647 − 136	**21.** 222 − 209	**22.** 836 − 209	**23.** 483 − 137	**24.** 592 − 348
25. 252 − 117	**26.** 931 − 513	**27.** 341 − 128	**28.** 692 − 346	**29.** 743 − 525	**30.** 184 − 122

31. 697 − 597 = □ **32.** 365 − 139 = □ **33.** 712 − 500 = □

34. 434 − 129 = □ **35.** 581 − 567 = □ **36.** 842 − 610 = □

Use a calculator to subtract.

37. five hundred eighty-two minus four hundred sixty-seven

38. one hundred thirty minus twenty-five

39. seven hundred sixteen minus nineteen

Think and Apply

40. By Thursday afternoon, Sam's class had made 325 Fluffy Puffies. They gave out 118 to students in other classes. How many Fluffy Puffies were left?

41. Kameo has a mixed bag of 350 red, blue, and green sequins. The label lists 120 blue and 115 green. How many sequins are red?

42. What fact is extra? A bag of pompoms costs $1.98. Sam used 195 pompoms. His friend, Sue, used 78 pompoms. Who used more pompoms? How many more?

43. Jason has 285 pipecleaners. He gave 7 to each classmate. Jason now has 89 pipecleaners left over. How many did he give out?

SUMMING UP

How is subtracting three-digit numbers like subtracting two-digit numbers? How is it different?

GETTING STARTED

Imagine you have 7 hundred-dollar bills in play money. Would you trade for 70 ten-dollar bills? for 700 one-dollar bills? Explain.

Regrouping Tens or Hundreds

Pine Valley School held a design-a-patch contest. Of 452 patches, 238 were shaped like circles. How many patches were not circles?

452 − 238 = □

Step 1 Subtract ones. Not enough ones. Regroup 1 ten as 10 ones.	**Step 2** Subtract ones.	**Step 3** Subtract tens.	**Step 4** Subtract hundreds.
4 12 452 −238 5 tens 2 ones = 4 tens 12 ones	4 12 452 −238 4	4 12 452 −238 14	4 12 452 −238 214

Of 452 patches, 214 were not circles.

Find 314 − 183.

Step 1 Subtract ones.	**Step 2** Subtract tens. Not enough tens. Regroup 1 hundred as 10 tens.	**Step 3** Subtract tens.	**Step 4** Subtract hundreds.
314 −183 1	2 11 314 −183 1 3 hundreds 1 ten = 2 hundreds 11 tens	2 11 314 −183 31	2 11 314 −183 131

Check Your Understanding

Subtract. Check by adding.

1. 863 − 672 = □
2. 419 − 266 = □
3. 771 − 213 = □

Share Your Ideas In subtraction, how can you tell where you need to regroup?

Practice

Subtract. Check by adding.

4. $\begin{array}{r} 298 \\ -172 \\ \hline \end{array}$
5. $\begin{array}{r} 516 \\ -420 \\ \hline \end{array}$
6. $\begin{array}{r} 915 \\ -345 \\ \hline \end{array}$
7. $\begin{array}{r} 832 \\ -614 \\ \hline \end{array}$
8. $\begin{array}{r} 487 \\ -391 \\ \hline \end{array}$
9. $\begin{array}{r} 367 \\ -57 \\ \hline \end{array}$
10. $\begin{array}{r} 129 \\ -82 \\ \hline \end{array}$
11. $\begin{array}{r} 754 \\ -138 \\ \hline \end{array}$
12. $\begin{array}{r} 390 \\ -171 \\ \hline \end{array}$
13. $\begin{array}{r} 537 \\ -209 \\ \hline \end{array}$
14. $\begin{array}{r} 417 \\ -32 \\ \hline \end{array}$
15. $\begin{array}{r} 995 \\ -494 \\ \hline \end{array}$

16. $697 - 189 = \square$
17. $620 - 507 = \square$
18. $700 - 116 = \square$
19. $221 - 212 = \square$
20. $\square = 444 - 333$
21. $\square = 731 - 408$

Complete. Find each output.

Rule: Subtract 354.

	Input	Output
22.	680	
23.	549	
24.	354	
25.	498	

Rule: Subtract 175.

	Input	Output
26.	875	
27.	655	
28.	984	
29.	390	

Think and Apply

30. There are 526 students at Pine Valley School. If 452 students entered the contest, how many did not enter?

31. Design a patch for your school. Estimate how many patches you will need for all the students in school.

Write in your own words a rule for subtraction with three-digit numbers. Be sure to include regrouping.

SUMMING UP

Mixed Review

1. $\begin{array}{r} 3 \\ +6 \\ \hline \end{array}$
2. $\begin{array}{r} 8 \\ +8 \\ \hline \end{array}$
3. $\begin{array}{r} 8 \\ -4 \\ \hline \end{array}$
4. $\begin{array}{r} 10 \\ -5 \\ \hline \end{array}$
5. $6 + 7 = \square$
6. $14 - 5 = \square$

Give the value of the underlined digit.

7. $4\underline{7}$
8. $\underline{8}3$
9. $\underline{7}39$
10. $2\underline{0}5$
11. $\underline{1},437$
12. $9,\underline{6}89$
13. $\underline{2}3,000$
14. $\underline{2}46,523$

Round to the nearest ten.

15. 61
16. 47
17. 78
18. 55

Round to the nearest hundred.

19. 619
20. 139
21. 480
22. 250
23. 751
24. 398
25. 949
26. 567
27. 135
28. 875

GETTING STARTED

Start with 645. Choose a number to subtract so you will need to regroup in the tens place. in the hundreds place.

Regrouping Twice

Students made 530 patches. They sold 148. How many were left?

530 − 148 = □

Step 1	Step 2	Step 3
Subtract ones. Not enough ones. Regroup. 2 10 5 3̸ 0̸ − 1 4 8 3 tens 0 ones = 2 tens 10 ones	Subtract ones. 2 10 5 3̸ 0̸ − 1 4 8 2	Subtract tens. Not enough tens. Regroup. 12 4 2̸ 10 5̸ 3̸ 0̸ − 1 4 8 2 5 hundreds 2 tens = 4 hundreds 12 tens
Step 4	**Step 5**	
Subtract tens. 12 4 2̸ 10 5̸ 3̸ 0̸ − 1 4 8 8 2	Subtract hundreds. 12 4 2̸ 10 5̸ 3̸ 0̸ − 1 4 8 3 8 2	

There were 382 patches left.

Check Your Understanding

Subtract. Check by adding.

1. 581 − 225 **2.** 712 − 597 **3.** 429 − 289 **4.** 125 − 66 **5.** 431 − 177 **6.** 644 − 352

7. 653 − 454 = □ **8.** 340 − 162 = □ **9.** 825 − 98 = □

Share Your Ideas Look back at **4**. Explain the number of digits in the answer.

Practice

Subtract. Check by adding.

10. $\begin{array}{r} 896 \\ -653 \\ \hline \end{array}$ 11. $\begin{array}{r} 445 \\ -120 \\ \hline \end{array}$ 12. $\begin{array}{r} 877 \\ -438 \\ \hline \end{array}$ 13. $\begin{array}{r} 526 \\ -235 \\ \hline \end{array}$ 14. $\begin{array}{r} 431 \\ -108 \\ \hline \end{array}$ 15. $\begin{array}{r} 685 \\ -85 \\ \hline \end{array}$

16. $\begin{array}{r} 735 \\ -495 \\ \hline \end{array}$ 17. $\begin{array}{r} 344 \\ -252 \\ \hline \end{array}$ 18. $\begin{array}{r} 843 \\ -86 \\ \hline \end{array}$ 19. $\begin{array}{r} 615 \\ -387 \\ \hline \end{array}$ 20. $\begin{array}{r} 920 \\ -368 \\ \hline \end{array}$ 21. $\begin{array}{r} 562 \\ -199 \\ \hline \end{array}$

22. 356 − 256 = □ 23. 825 − 617 = □ 24. 932 − 158 = □

25. □ = 328 − 136 26. □ = 730 − 298 27. □ = 850 − 263

Round to the nearest hundred and estimate each difference.

28. 719 − 468 = □ 29. 885 − 190 = □ 30. 624 − 282 = □

31. 859 − 471 = □ 32. 302 − 151 = □ 33. 553 − 391 = □

Subtract across and down.

34. (→ across, ↓ down)

999	750	
637	475	

35. (→ across, ↓ down)

875	349	
627	242	

36. (→ across, ↓ down)

631	278	
495	199	

Think and Apply

37. One year, Pine Valley School sold 475 patches in two days. If 280 were sold the first day, how many were sold the second day?

38. Use a calculator and the digits 1–9 only once.
- Enter any three-digit number.
- Reverse the digits. Subtract the lesser number from the greater.
- Reverse the digits and add the new number.
- Repeat this several times.
- Write about what happens.

Explain how you would regroup to subtract 327 − 189.

SUMMING UP

Midchapter Review

Subtract. pages 104–105

1. $\begin{array}{r} 57 \\ -35 \\ \hline \end{array}$
2. $\begin{array}{r} 73 \\ -41 \\ \hline \end{array}$
3. $\begin{array}{r} 68 \\ -40 \\ \hline \end{array}$
4. $\begin{array}{r} 82 \\ -9 \\ \hline \end{array}$
5. $\begin{array}{r} 66 \\ -37 \\ \hline \end{array}$
6. $\begin{array}{r} 86 \\ -18 \\ \hline \end{array}$
7. $\begin{array}{r} 42 \\ -27 \\ \hline \end{array}$
8. $\begin{array}{r} 60 \\ -12 \\ \hline \end{array}$
9. $\begin{array}{r} 90 \\ -90 \\ \hline \end{array}$
10. $\begin{array}{r} 45 \\ -39 \\ \hline \end{array}$

11. 52 − 11 = □
12. 81 − 19 = □
13. □ = 90 − 78

Estimate each difference. pages 106–107

14. $\begin{array}{r} 47 \\ -19 \\ \hline \end{array}$
15. $\begin{array}{r} 52 \\ -33 \\ \hline \end{array}$
16. $\begin{array}{r} 71 \\ -28 \\ \hline \end{array}$
17. $\begin{array}{r} 934 \\ -576 \\ \hline \end{array}$
18. $\begin{array}{r} 457 \\ -125 \\ \hline \end{array}$

19. 82 − 39 = □
20. 56 − 37 = □
21. □ = 679 − 218

Find each difference. pages 108–113

22. $\begin{array}{r} 824 \\ -622 \\ \hline \end{array}$
23. $\begin{array}{r} 997 \\ -289 \\ \hline \end{array}$
24. $\begin{array}{r} 418 \\ -246 \\ \hline \end{array}$
25. $\begin{array}{r} 837 \\ -787 \\ \hline \end{array}$
26. $\begin{array}{r} 626 \\ -538 \\ \hline \end{array}$
27. $\begin{array}{r} 413 \\ -152 \\ \hline \end{array}$
28. $\begin{array}{r} 512 \\ -438 \\ \hline \end{array}$
29. $\begin{array}{r} 958 \\ -170 \\ \hline \end{array}$
30. $\begin{array}{r} 314 \\ -92 \\ \hline \end{array}$
31. $\begin{array}{r} 451 \\ -274 \\ \hline \end{array}$

32. 633 − 619 = □
33. 716 − 452 = □
34. □ = 320 − 174

Use the words to complete the sentence.

35. In the subtraction problem 305 − 231 = □, you need to _____ in the _____ place in order to have enough _____ to subtract.

Words to Know
hundreds regroup tens

Solve.

36. Last year Shady Hill School had 785 students. This year it has 687 students. How many fewer students does the school have this year?

37. Of Shady Hill School's 687 students, 395 are girls. How many boys attend the school?

Exploring Problem Solving

THINK
EXPLORE
SOLVE
LOOK BACK

What's Missing?

Hector's and Susie's parents recorded their children's heights on each birthday.

Thinking Critically

Between which years did Hector and Susie grow the most? Who has grown more since birth?

Work in a small group to solve the problem.

HECTOR		SUSAN	
Age	Height in Inches	Age	Height in Inches
1	29	1	27
2	34	2	31
3	38	3	36
4	41	4	39
5	44	5	41
6	46	6	43
7	48	7	46
8	51	8	50

Analyzing and Making Decisions

1. Study the table. What does it tell you?
2. How can you tell how much they grew each year?
3. How much did they grow in their first year? Explain your answer.
4. How much did they grow each year? When did they grow the most?
5. Who has grown more since birth?

Look Back Double the height of someone who is 2 years old. You get a good estimate of the person's height as a adult. How tall do you think Hector and Susie will be as adults?

Problem Solving Strategies

What's Missing?

Ms. Reynolds has 13 boys and 15 girls on her class list. On Monday, 7 students were absent. How many boys were in class on Monday?

Sometimes you do not have all the information you need to answer a question.

Solving the Problem

Think What is the question? What information do you have?

Explore How could you find out how many boys were in the class on Monday? Do you know how many boys are on Ms. Reynolds' class list? Do you know how many boys were absent on Monday?

Solve Do you have enough information to solve the problem? Explain.

Look Back What can you tell about the number of boys in the class on Monday? Could only boys have been absent? Could no boys have been absent? What is the least and the greatest number of boys that could have been in class on Monday?

Share Your Ideas

1. What is a question that you could have answered with the information given?
2. You did not have enough information to solve this problem or the one on page 115. How could you have found the information to solve them?

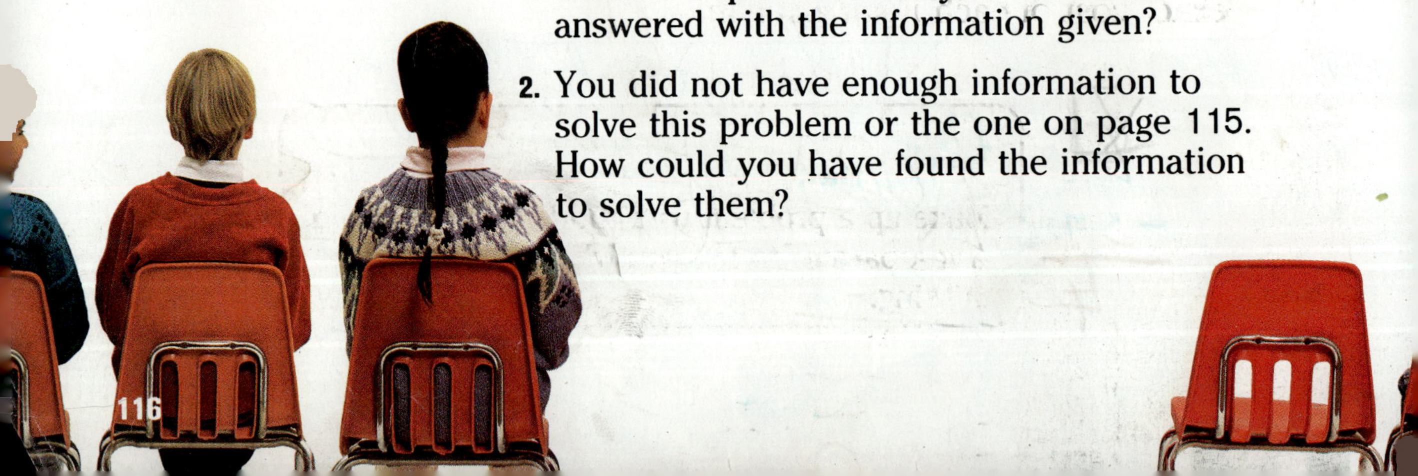

THINK
EXPLORE
SOLVE
LOOK BACK

Practice

Solve. Use a calculator where appropriate.

CHOICES

3. Three classes are collecting cans of food. Ms. Day's class collected 36 cans. Mrs. Hall's class collected 45. How many cans did all three classes collect?

4. Tom's class watched two T.V. specials at school. The first one was 30 minutes long. The second one was the same length. How many minutes did they watch T.V.?

5. Sarah has 70 baseball cards. Tricia has 48 cards. Who has more cards? How many more?

6. Whose cards cost more, Sarah's or Tricia's?

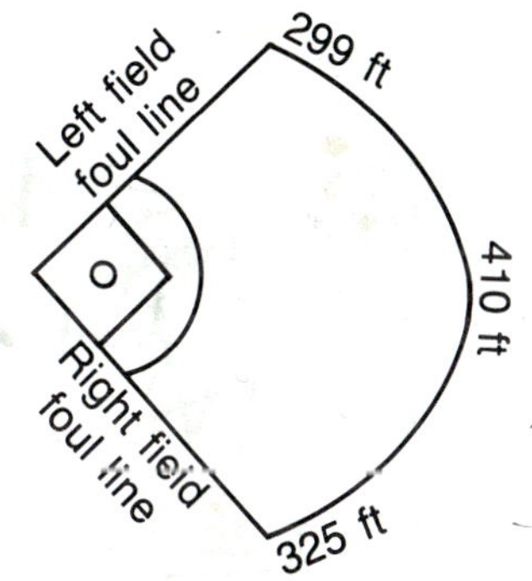

Mixed Strategy Review

7. How much farther is it down the right field foul line than down the left field foul line?

8. The Leopards' second baseman hit a home run over the left field fence. The Robins' third baseman hit the ball over the right field fence. Who hit the ball farther?

9. You have $90. You want to buy three items that cost $28, $41, and $14. Can you buy them? Show why or why not.

10. How can you solve **9** without adding the exact cost of each item? Explain.

Create Your Own

Make up a problem with too little information. Ask a partner to tell you what information is missing.

GETTING STARTED

What does the zero mean in each of these numbers?
360 306 3,061

Subtracting Across Zeros

Mill School is having a spelling marathon. The Aces need 205 points to win. They have 149 points. How many more points do they need?

205 − 149 = □

Subtract to find how many more are needed.

Step 1	Step 2	Step 3
Not enough ones. No tens. Regroup 1 hundred as 10 tens. 1 10 ~~2~~ ~~0~~ 5 − 1 4 9 2 hundreds 0 tens = 1 hundred 10 tens	Regroup 1 ten as 10 ones. 9 1 ~~10~~ 15 ~~2~~ ~~0~~ ~~5~~ − 1 4 9 10 tens 5 ones = 9 tens 15 ones	Subtract ones. Subtract tens. Subtract hundreds. 9 1 ~~10~~ 15 ~~2~~ ~~0~~ ~~5~~ − 1 4 9 5 6

The Aces need 56 more points to win.

Find 500 − 235.

Step 1	Step 2	Step 3
No ones. No tens. Regroup 1 hundred as 10 tens. 4 10 ~~5~~ ~~0~~ 0 − 2 3 5 5 hundreds 0 tens = 4 hundreds 10 ones	Regroup 1 ten as 10 ones. 9 4 10 10 ~~5~~ ~~0~~ ~~0~~ − 2 3 5 10 tens 0 ones = 9 tens 10 ones	Subtract ones. Subtract tens. Subtract hundreds. 9 4 ~~10~~ 10 ~~5~~ ~~0~~ ~~0~~ − 2 3 5 2 6 5

Check Your Understanding

Subtract. Estimate to check your work.

1. 205 − 147	**2.** 803 − 599	**3.** 400 − 151	**4.** 700 − 408	**5.** 307 − 98	**6.** 900 − 256

Share Your Ideas How did you subtract in **6**?

Practice

Subtract. Estimate to be sure your answers make sense.

7. $501 - 347$	**8.** $205 - 138$	**9.** $407 - 264$	**10.** $500 - 360$	**11.** $900 - 183$	**12.** $602 - 0$
13. $101 - 89$	**14.** $603 - 419$	**15.** $200 - 14$	**16.** $700 - 320$	**17.** $304 - 53$	**18.** $300 - 156$
19. $400 - 100$	**20.** $502 - 73$	**21.** $306 - 8$	**22.** $805 - 230$	**23.** $900 - 517$	**24.** $400 - 226$
25. $607 - 407$	**26.** $700 - 45$	**27.** $309 - 179$	**28.** $800 - 612$	**29.** $100 - 37$	**30.** $200 - 9$

31. 303 − 175 = ☐ **32.** 906 − 98 = ☐ **33.** 200 − 199 = ☐

34. 801 − 58 = ☐ **35.** 205 − 91 = ☐ **36.** 900 − 326 = ☐

37. 700 − 69 = ☐ **38.** 600 − 70 = ☐ **39.** 403 − 14 = ☐

40. 419 − 67 = ☐ **41.** 600 − ☐ = 215 **42.** 803 − ☐ = 246

Find the correct difference. Choose mental math, paper and pencil, or a calculator.

CHOICES

43. 701 − 100 = ☐
a. 610 **b.** 601
c. 801 **d.** 600

44. 500 − 250 = ☐
a. 250 **b.** 200
c. 150 **d.** 205

45. 375 = 600 − ☐
a. 125 **b.** 205
c. 300 **d.** 225

Think and Apply

46. A total of 510 words were spelled correctly in the spelling marathon. 106 words were "spelling traps." How many were not?

47. The Aces needed 205 points to win. Sam got 60 points. Bill got 45 points. How many points did Beth earn to reach the total needed by the team?

Make up a rule that tells how to regroup when the place to the left has a zero in it.

SUMMING UP

GETTING STARTED

How are these subtraction examples alike? How are they different?

```
  7,000     700
- 4,000   - 400
```

Subtracting Four-Digit Numbers

Last week the Reds had 1,213 points. After 3 more games they had 2,415 points. How many points did they bowl in those last games?

2,415 − 1,213 = □

Use what you know about subtracting three-digit numbers to subtract four-digit numbers.

Step 1 Subtract ones.	**Step 2** Subtract tens.	**Step 3** Subtract hundreds.	**Step 4** Subtract thousands.
2,415 − 1,213 2	2,415 − 1,213 02	2,415 − 1,213 202	2,415 − 1,213 1,202

The Reds bowled 1,202 points in those last games. Sometimes you need to regroup in different places.

Regrouping tens	Regrouping hundreds	Regrouping thousands
a. 5 12 over 6,3~~6~~~~2~~ − 2,157 4,205	**b.** 2 11 over 7,~~3~~~~1~~7 − 3,186 4,131	**c.** 8 14 over 9,~~4~~~~7~~5 − 850 8,625

Check Your Understanding

Subtract. Check by adding.

1. 7,512 − 2,400	**2.** 3,524 − 2,234	**3.** 5,861 − 5,453	**4.** 6,078 − 3,654	**5.** 8,999 − 798

Share Your Ideas How is subtracting four-digit numbers like subtracting three-digit numbers?

Practice

Subtract. Check by adding.

6. 8,918 − 6,542
7. 2,354 − 1,062
8. 1,740 − 800
9. 6,195 − 2,039
10. 6,743 − 3,182
11. 1,499 − 1,298
12. 2,644 − 1,635
13. 9,634 − 931
14. 5,760 − 3,960
15. 8,918 − 6,543
16. 6,776 − 2,368
17. 3,492 − 1,590
18. 7,903 − 172
19. 4,600 − 1,700
20. 2,340 − 260
21. 8,542 − 731 = □
22. 6,333 − 1,343 = □
23. 3,960 − 65 = □
24. 6,999 − 799 = □
25. 9,835 − 935 = □
26. 5,378 − 4,236 = □
27. 5,199 − 953 − 1,237 = □
28. 7,002 − 4,015 − 2,906 = □

Compare. Write >, <, or = for ●. Choose estimation, paper and pencil, or a calculator. Explain your choices.

CHOICES

29. 800 − 115 ● 698 − 230
30. 1,000 − 333 ● 5,000 − 3,000
31. 3,999 − 2,000 ● 8,250 − 2,099
32. 7,225 − 1,850 ● 9,000 − 8,020

Think and Apply

Use the chart to answer 33–35.

BOWLING TOURNAMENT TEAM SCORES	
Team	**Points**
Longhorns	4,025
Ten Pins	1,300
Reds	3,975

33. How many more points do the Reds have than the Ten Pins?
34. How many more points do the Longhorns have than the Reds?
35. Which team scored the most points? the fewest points?

SUMMING UP

Is this statement *always, sometimes,* or *never* true? Explain. When you subtract two four-digit numbers, the difference is a four-digit number.

GETTING STARTED

In which places will you need to regroup in this subtraction?

$$\begin{array}{r} 2{,}450 \\ -\ \ 590 \\ \hline \end{array}$$

Regrouping More Than Once

Use Enrico's map. How much farther away is Los Angeles from New York than from Denver?

2,788 − 1,189 = ☐

Step 1	Step 2	Step 3
Subtract ones. Not enough ones. Regroup.	Subtract ones.	Subtract tens. Not enough tens. Regroup.
7 18 2,7 8̸ 8̸ − 1,1 8 9	7 18 2,7 8̸ 8̸ − 1,1 8 9 9	17 6 7̸ 18 2,7̸ 8̸ 8̸ − 1,1 8 9 9

Step 4	Step 5
Subtract tens.	Subtract hundreds. Subtract thousands.
17 6 7̸ 18 2,7̸ 8̸ 8̸ − 1,1 8 9 9 9	17 6 7̸ 18 2,7̸ 8̸ 8̸ − 1,1 8 9 1,5 9 9

Los Angeles is 1,599 miles farther from New York than from Denver.

Check Your Understanding

Subtract. Check by adding.

1. 4,361 − 1,470
2. 5,682 − 85
3. 6,819 − 3,469
4. 7,549 − 5,488
5. 7,208 − 4,809

Share Your Ideas Can you change one digit in **2** so you would not have to regroup? Explain.

Practice

Subtract. Check by adding.

6. 6,828 − 1,503
7. 765 − 643
8. 2,499 − 2,298
9. 6,729 − 4,268
10. 1,349 − 299
11. 5,760 − 3,960
12. 6,963 − 129
13. 9,752 − 1,673
14. 8,614 − 3,632
15. 9,300 − 8,460
16. 8,146 − 2,091
17. 5,728 − 3,214
18. 4,901 − 1,573
19. 7,420 − 1,580
20. 6,395 − 4,769
21. 7,405 − 2,366
22. 3,081 − 1,254
23. 6,471 − 1,198
24. 9,270 − 5,495
25. 6,430 − 5,980

26. 7,000 − 5,249 = □
27. 5,322 − 404 = □
28. 8,500 − 6,521 = □
29. 5,914 − 999 = □
30. 3,727 − 1,848 = □
31. 6,285 − 6,193 = □
32. 9,930 − □ = 9,205
33. 9,000 − □ = 5,766

Think and Apply

34. Enrico estimated that Los Angeles is 3,000 miles away from Montreal. The actual mileage is 2,918. How far off was Enrico?

Use the table for 35 and 36.

	Philadelphia, PA	Houston, TX	San Diego, CA
Philadelphia, PA	—	1,648 miles	2,779 miles
Houston, TX	1,648 miles	—	1,578 miles
San Diego, CA	2,779 miles	1,578 miles	—

35. Find the roundtrip mileage between Houston and San Diego and between Houston and Philadelphia. Which is farther?

36. Create your own subtraction problems. Ask a classmate to solve them. Discuss how each problem was solved.

SUMMING UP

Write a subtraction example where you need to regroup once. Write an example with regrouping in more than one place.

GETTING STARTED

Which items could you buy with $2.00?
pencil case — $2.75 pencil — $ .20
book — $1.50 markers — $1.98

Subtracting Money

Mitch has $5.75. He spends $2.50 on a notebook. Can he also buy the pen for $3.00?

$5.75 − $2.50 = □

Subtract money the same way you subtract other numbers.

$5.75 **− 2.50**	Line up the decimal points.	**$5.75** **− 2.50** **$3.25**	Subtract. Write the dollar sign and decimal point in the answer.

Mitch will have $3.25 left. He can buy the pen.

Another Example $17.00 − $12.35 = □

Line up the decimal points.	Regroup. Subtract.	Check by adding.
$17.00 − 12.35	9 6 10 10 $1 7̸.1̸0̸ 0̸ − 12.35 $ 4.65	1 1 $ 4.65 + 12.35 $17.00

Explain how you would find $25.00 − $6.98.

Check Your Understanding

Subtract. Check by adding.

1. $4.50 − 1.25 **2.** $8.00 − 1.69 **3.** $15.95 − 3.70 **4.** $28.39 − 19.15 **5.** $50.00 − 23.50

6. $9.59 − $1.30 = □ **7.** $22.00 − $.59 = □

Share Your Ideas How is subtracting money like subtracting other numbers? How is it different?

Practice

Subtract. Check by adding.

8.	$3.56 − 2.40	**9.**	$32.06 − 9.75	**10.**	$40.68 − 3.68
11.	$24.70 − 12.95	**12.**	$27.47 − 18.29	**13.**	$17.20 − 11.99
14.	$74.92 − 46.87	**15.**	$25.49 − 6.72	**16.**	$78.00 − 39.50

17. $39.87 − $16.25 = ☐

18. $17.35 − $2.95 = ☐

19. $46.81 − $12.98 = ☐

20. $28.35 − $1.69 = ☐

Subtract. Use mental math.

21. $20.00 − $10.00 = ☐

22. $9.00 − $4.00 = ☐

23. $8.50 − $1.50 = ☐

24. $86.00 − $16.00 = ☐

Think and Apply

25. What's missing? Marge has $10.00. Can she buy two looseleaf binders?

26. Jim had $9.50. He bought a calculator for $6.25 and a pen for $3.00. How much does he have left?

27. Pat has a ten-dollar bill. She buys a pencil case for $5.95. How much change should she expect to get back?

Look back at **17**. Explain why it is important to line up the decimal points when you subtract money.

SUMMING UP

Mixed Review

1.	6 +8	**2.**	11 − 6
3.	9 +8	**4.**	15 − 6
5.	16 + 4	**6.**	18 − 9
7.	62 +31	**8.**	98 +25
9.	67 +78	**10.**	23 +46
11.	395 +142	**12.**	873 +749

Compare.
Use >, <, or =
for ⬬.

13. 78 ⬬ 87

14. 362 ⬬ 326

15. 0 ⬬ 100

16. 499 ⬬ 49

17. 26 ⬬ 36 − 12

18. 82 + 0 ⬬ 8

19. 59 ⬬ 26 + 35

20. 324 ⬬ 301 + 23

21. 460 − 30 ⬬ 490

GETTING STARTED

How is the subtraction 80 − 50 like the subtraction 8 − 5? How is it different? How can solving one help you solve the other?

Mental Math: Strategies for Subtraction

Grant School needs 1,800 programs for a talent show. They have 900. How many more programs are needed?

1,800 − 900 = □

▸ You can use basic facts.

Think
$$\begin{array}{r} 18 \text{ hundred} \\ -\ 9 \text{ hundred} \\ \hline 9 \text{ hundred} \end{array}$$

900 more programs are needed.

What if you want to find 61 − 27?

▸ You can add to the lesser number to reach the nearest ten. Next add the same amount to the other number. Then subtract mentally.

$$\begin{array}{r} 61 \\ -27 \\ \hline \end{array}$$

Think
The nearest ten is 30.
27 + 3 = 30
Add 3 to each number.

$$\begin{array}{lr} 61 + 3 \longrightarrow & 64 \\ 27 + 3 \longrightarrow & -30 \\ \hline & 34 \end{array}$$

▸ You also can choose to use paper and pencil or a calculator.

Check Your Understanding

Subtract. Choose mental math, paper and pencil or a calculator.

CHOICES

1. $\begin{array}{r} 1{,}400 \\ -\ 800 \\ \hline \end{array}$
2. $\begin{array}{r} 71 \\ -46 \\ \hline \end{array}$
3. $\begin{array}{r} 16{,}354 \\ -\ 4{,}968 \\ \hline \end{array}$
4. $\begin{array}{r} 1{,}000 \\ -\ 400 \\ \hline \end{array}$
5. $\begin{array}{r} 489 \\ -325 \\ \hline \end{array}$

Share Your Ideas Look back at **1–5**. Explain why you chose the method you used.

Practice

Subtract. Choose mental math, paper and pencil, or a calculator. Explain your choices.

CHOICES

6. $800 - 600$
7. $76 - 37$
8. $1,400 - 900$
9. $42 - 18$
10. $1,700 - 900$
11. $95 - 26$
12. $1,300 - 600$
13. $367 - 259$
14. $24 - 17$
15. $1,500 - 800$
16. $7,954 - 3,947$
17. $963 - 358$
18. $81 - 38$
19. $1,000 - 300$
20. $1,600 - 800$
21. $876 - 99 =$ □
22. $85 - 67 =$ □
23. $1,500 - 400 =$ □
24. $685 - 492 =$ □
25. $4,623 - 2,755 =$ □
26. $1,200 - 300 =$ □

Compare. Write >, <, or = for ●. Choose estimation, paper and pencil, or a calculator. Explain your choices.

CHOICES

27. $800 - 305$ ● $600 - 213$
28. $85 - 46$ ● $92 - 87$
29. $485 - 109$ ● $999 - 230$
30. $732 - 532$ ● $795 - 199$

Think and Apply

31. Jack bought 7 tickets to the talent show. The cost of all the tickets was $10.50. He gave the ticket seller $20. How much change did he get?

32. The school's goal is to raise $1,600 from the ticket sales. Already, $1,346.50 has been raised. How much more does the school need to raise?

Test Taker

Improve your test scores. First do the easy problems. Then do the difficult problems.

33. Which problems are easier than the others?

a. $7,324 - 4,536$

b. $4,087 - 1,257$

c. $2,519 - 1,536$

d. $3,864 - 2,513$

SUMMING UP

Explain how you knew when to choose mental math in **6–10**.

Using Problem Solving

Interview: Calculators in the School Office

Mary Lou McDonough is the bookkeeper at the Garrison Mill School in Marietta, Georgia. She keeps a record of the money collected for field trips, book fairs, and other school activities. "I would be lost without my calculator," claims Ms. McDonough.

Here is a record of the amounts of money collected by Ms. McDonough during a book fair.

	$20 Bills	$10 Bills	$5 Bills	$1 Bills	Coins
Wednesday	$180	$170	$155	$18	$3.15
Thursday	$260	$190	$135	$ 7	$ .98
Friday	$220	$210	$175	$12	$1.50

CHOICES

Use a calculator if you wish.

A. How much money was collected each day? Was more money collected on Wednesday or Thursday? How much more?

B. Explain how estimation can help you decide if the answer is reasonable.

Sharing Your Ideas

1. What is the total amount of money collected on the three days?
2. On which day was the most money collected? How much more than the least amount was collected on this day?
3. How would you check your work if you were Ms. McDonough?

THINK
EXPLORE
SOLVE
LOOK BACK

Practice

Ms. McDonough is working at the book fair. Pretend that you are her helper.

Record of Books Sold						
	Mystery	Adventure	Sports	Biography	Other Nonfiction	Total
Wednesday	44	55	30	8	34	
Thursday	22	41	57	17	51	
Friday	37	90	29	12	32	

4. How many books were sold on each day?

5. Which type of book sold the most copies? How many were sold?

6. Which two types of books sold about the same amount?

7. After the expenses were paid, the school had a $697.60 profit. It will be spent on new books and computer software for the Media Center. If $435 is spent on books, how much is left for software?

Summing Up

8. Create two problems using the information in this chart. Give the problems to a classmate to solve.

Book Fair Attendance	
Wednesday	139
Thursday	256
Friday	360

9. Why do you think Ms. McDonough uses a calculator in her work?

Chapter Review

Subtract. pages 101–105, 108–113

1.	2.	3.	4.	5.
75 − 32	96 − 37	65 − 41	83 − 56	22 − 2

6.	7.	8.	9.	10.
364 − 123	485 − 295	477 − 68	521 − 197	481 − 263

11. 88 − 31 = ☐ 12. 268 − 149 = ☐ 13. 617 − 220 = ☐

Estimate each difference. pages 106–107

14.	15.	16.	17.	18.
23 − 18	95 − 36	71 − 48	80 − 51	67 − 24

19.	20.	21.	22.	23.
321 − 101	679 − 140	932 − 799	808 − 586	412 − 212

Find each difference. pages 118–125

24.	25.	26.	27.	28.
600 − 205	809 − 114	6,800 − 3,158	2,048 − 408	4,705 − 2,716

29.	30.	31.	32.	33.
\$9.50 − 2.25	\$6.17 − 3.17	\$43.36 − 2.89	\$30.00 − 22.75	\$98.56 − 73.99

Use the words to complete the sentence.

34. We _____ to find the _____ of two numbers.

Words to Know
difference subtract

Solve. pages 115–117

35. Students had 1,000 concert tickets to sell. They sold 842 tickets. How many are left to sell?

36. The teacher has \$50.00 for classroom poetry books. This is his order so far. How much more can he spend?

Chapter Test

Subtract.

1. $\begin{array}{r} 86 \\ -75 \\ \hline \end{array}$ 2. $\begin{array}{r} 96 \\ -64 \\ \hline \end{array}$ 3. $\begin{array}{r} 67 \\ -18 \\ \hline \end{array}$ 4. $\begin{array}{r} 93 \\ -47 \\ \hline \end{array}$ 5. $\begin{array}{r} 83 \\ -3 \\ \hline \end{array}$

6. $\begin{array}{r} 693 \\ -420 \\ \hline \end{array}$ 7. $\begin{array}{r} 780 \\ -656 \\ \hline \end{array}$ 8. $\begin{array}{r} 974 \\ -43 \\ \hline \end{array}$ 9. $\begin{array}{r} 534 \\ -199 \\ \hline \end{array}$ 10. $\begin{array}{r} 613 \\ -248 \\ \hline \end{array}$

11. $564 - 261 = \square$ 12. $871 - 193 = \square$ 13. $469 - 88 = \square$

Estimate each difference.

14. $\begin{array}{r} 92 \\ -38 \\ \hline \end{array}$ 15. $\begin{array}{r} 80 \\ -53 \\ \hline \end{array}$ 16. $\begin{array}{r} 38 \\ -27 \\ \hline \end{array}$ 17. $\begin{array}{r} 71 \\ -54 \\ \hline \end{array}$ 18. $\begin{array}{r} 73 \\ -25 \\ \hline \end{array}$

19. $\begin{array}{r} 568 \\ -105 \\ \hline \end{array}$ 20. $\begin{array}{r} 732 \\ -189 \\ \hline \end{array}$ 21. $\begin{array}{r} 179 \\ -25 \\ \hline \end{array}$ 22. $\begin{array}{r} 800 \\ -341 \\ \hline \end{array}$ 23. $\begin{array}{r} 699 \\ -236 \\ \hline \end{array}$

Find each difference.

24. $\begin{array}{r} 500 \\ -425 \\ \hline \end{array}$ 25. $\begin{array}{r} 607 \\ -324 \\ \hline \end{array}$ 26. $\begin{array}{r} 3{,}947 \\ -1{,}439 \\ \hline \end{array}$ 27. $\begin{array}{r} 7{,}621 \\ -3{,}999 \\ \hline \end{array}$

28. $\begin{array}{r} \$8.67 \\ -\ 1.35 \\ \hline \end{array}$ 29. $\begin{array}{r} \$7.14 \\ -\ 2.09 \\ \hline \end{array}$ 30. $\begin{array}{r} \$25.62 \\ -\ 19.89 \\ \hline \end{array}$ 31. $\begin{array}{r} \$87.35 \\ -\ 77.92 \end{array}$

Solve.

32. The Bluestars need 500 points to win the track meet. So far they have earned 160 points and 325 points. How many more points do they need to win?

33. Angelo had $20.00. He bought pens for $5.50, a fancy eraser for $2.00, and a book of poetry. How much did he have left?

Think Use the information to create your own subtraction problem. Then solve it.

216 seats
179 students
7 teachers

Computer Link

Subtracting Turtle Steps

The turtle is making a round trip from HOME to a point on the screen. What number can you use to get the turtle back HOME?

```
HOME
FD 76
RT 180
FD 21
FD ___
```

RT 180 turns the turtle around.

Materials: Logo

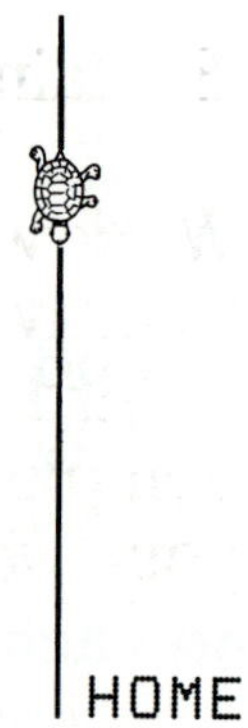

A. Find the number of steps for the turtle to get back HOME.

- HOME
FD 42
RT 180
FD 20
FD ___
- HOME
FD 57
RT 180
FD 34
FD ___
- HOME
FD 83
RT 180
FD 19
FD ___
- HOME
FD 88
RT 180
FD 49
FD ___

B. How many steps are in each round trip?

C. **What if** the turtle takes 20 steps away from HOME and turns around. Name some ways you can get the turtle back HOME. Use two FD moves to complete each trip.

Sharing Your Results

1. Explain how you found the number of steps in **A** and **B**
2. **Look back** at **C**. How many ways did you find for the turtle to get HOME?

Extending the Activity

Janet is drawing a rectangle that is 37 turtle steps long and 22 steps wide. What number does she need to complete the rectangle?

```
FD 37
RT 90
FD 22
RT 90
FD 37
RT 90
FD 10
FD ___
```

3. Explain how you found the missing number.

4. Work with a partner. Enter Logo commands to draw two sides of a rectangle and part of a third side. Then ask your partner to complete the rectangle. Take turns. Record your work.

5. Find the missing number in the `FD` command that completes the figure.

```
HOME
FD 20  RT 90  FD 30  LT 90  FD 10
RT 90  FD 50  RT 90  FD 40  RT 90
FD 30  RT 90  FD 10  LT 90  FD ___
```

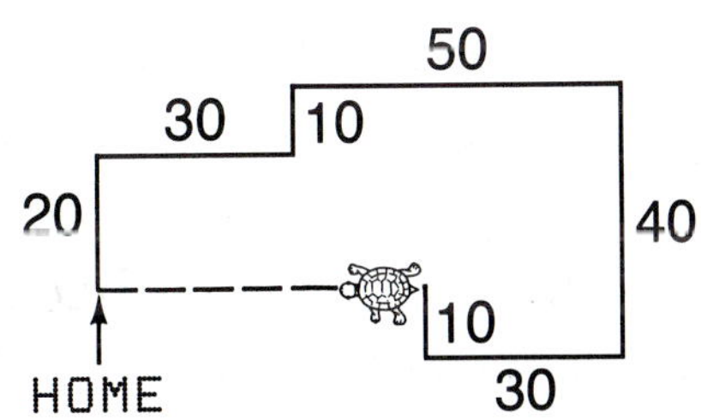

Summing Up

6. **Look back** at **5**. Explain how you found the number of steps.

7. Find the missing commands that complete the figure.

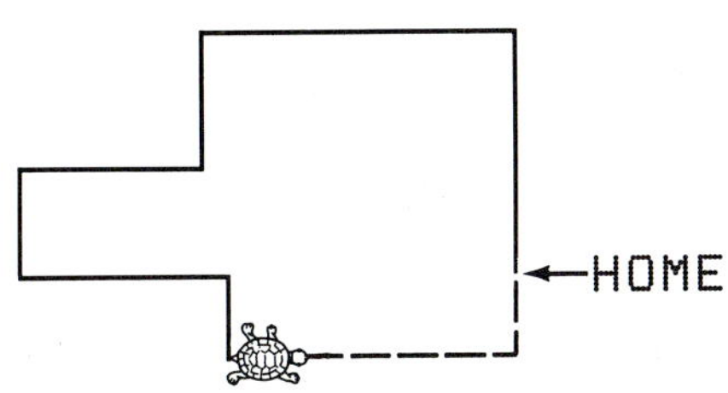


```
HOME FD 45 LT 90 FD 60 LT 90 FD 25 RT 90 FD 35 LT 90 FD 20
LT 90 FD 40 RT 90 FD 15 LT 90 _____ _____ _____
```

EXTENSION

Mental Math: Subtracting Across Zeros

What if you have this subtraction?

How many times do you need to regroup to find the difference?

```
   9
 3 10 10
 4  0  0
-2  7  9
 1  2  1
```

Here is another way to subtract across zeros.

Subtract 1 from each number.

Original Subtraction		New Subtraction
400	Subtract 1	399
− 279	Subtract 1	− 278
		121

Is the answer the same as the original? Can the new subtraction be solved using mental math? Try it again.

Original Subtraction		New Subtraction
5,000	Subtract 1	4,999
− 1,874	Subtract 1	− 1,873
		3,126

Take turns with a partner.

Subtract. Use the new method.

1. $700 - 195$
2. $900 - 367$
3. $1{,}000 - 698$
4. $8{,}000 - 2{,}896$

Create Your Own subtraction which can be solved using the new method. Ask your partner to solve it. Then discuss different ways you can check your answers.

Family Math

In Chapters **1–4** your child has been adding and subtracting whole numbers. This project of food planning and shopping will help put these skills to work in a meaningful way.

Enjoy doing the project together.

Food for Thought

1. Together plan 3 meals with your family. Let each person choose at least one dish.
2. Make a shopping list.
3. Cut out coupons from newspapers and supermarket circulars and keep a list of them. Be careful that you clip coupons only for the food you need.
4. Shop for the food together. Take the coupons with you.
5. How much did the food cost? What would the total cost have been without the coupons?
6. How much did you save with the coupons?
7. Enjoy your meals.

Daily Minimum for 1 Person

6 ounces of protein (meat, dried beans, fish, . . .)

4 vegetables and fruit

3 breads

2 glasses of milk or milk products

1 tablespoon of fat (margarine, salad dressing, . . .)

Cumulative Review

Choose the correct answer. Write A, B, C, or D.

1. 8 + 4 = □

   A 12 **C** 11

   B 13 **D** not given

2. Find the missing number.
   5 + 8 = □ + 5

   A 4 **C** 5

   B 8 **D** not given

3. $\begin{array}{r} 14 \\ -\ 10 \\ \hline \end{array}$

   A 14 **C** 4

   B 24 **D** not given

4. What is another fact in this fact family? 15 − 9 = 6

   A 15 − 6 = 9 **C** 15 − 8 = 7

   B 9 + 7 = 16 **D** not given

5. What is the value of the digit 6 in 746?

   A 6 **C** 600

   B 60 **D** not given

6. What ten comes after 170?

   A 180 **C** 160

   B 171 **D** not given

7. Write the number
   3,000 + 80 + 6.

   A 3,846 **C** 3,406

   B 3,086 **D** not given

8. Compare. 4,518 ● 4,519

   A = **C** $<$

   B $>$ **D** not given

9. Compare. 62,084 ● 60,284

   A $<$ **C** =

   B $>$ **D** not given

10. What is the value of 2 five-dollar bills and 3 dimes?

    A $10.30 **C** $1.30

    B $2.30 **D** not given

11. $\begin{array}{r} 85 \\ +\ 49 \\ \hline \end{array}$

    A 124 **C** 136

    B 135 **D** not given

12. 33 + 18 + 25 = □

    A 73 **C** 76

    B 66 **D** not given

13. $\begin{array}{r} 692 \\ +\ 107 \\ \hline \end{array}$

    A 799 **C** 899

    B 699 **D** not given

Choose the correct answer. Write A, B, C, or D.

14. 8,521 + 3,169 = ☐

A 11,680 C 11,700
B 11,690 D not given

15. $92.18 + 45.36

A $137.44 C $137.54
B $127.54 D not given

16. $.59 + $2.98 = ☐

A $3.57 C $2.57
B $3.60 D not given

17. Estimate. 57 − 26 = ☐

A 60 C 40
B 20 D 30

18. 226 − 43

A 223 C 123
B 183 D not given

19. 400 − 235 = ☐

A 165 C 175
B 235 D not given

20. 8,000 − 2,987

A 5,123 C 5,013
B 6,013 D not given

Solve.

21. Scott has a puppy that stands 10 paper clips high. About how many small clips placed end to end would reach to the top of a doorway?

A 25 C 18
B 70 D not given

22. Drew lined her paper to divide it into 12 equal parts. What was the fewest number of lines she needed?

A 3 C 15
B 12 D not given

Solve.

23. Ten classmates walk to school and 17 ride their bikes. Nine classmates eat lunch at home. How many more classmates ride to school than walk?

A 12 C 10
B 18 D not given

24. Jay walks to school in 15 minutes. Ali walks to school in 12 minutes. Troy lives 5 blocks away. How many more minutes does it take Jay to walk to school than it takes Ali?

A 3 minutes C 12 minutes
B 7 minutes D not given

5 Time • Graphing

THEME Wonderful Things Around Us

Sharing What You Know

What time does the zoo open? When will the show start? What time should I meet you? How much longer will you be? We talk about time all the time. When is it important for you to keep track of time? How do you do it?

Using Language

What if you were 15 minutes late for the show at the zoo? Talk about how 15 minutes can change your whole day. A **quarter hour** is the same as 15 minutes. A **quarter** is a coin worth 25 cents. A **quarter** of a year is 3 months. How are all these things alike?

Words to Know: hour, half hour, quarter hour, minute, elapsed time, bar graph, pictograph, ordered pair, point

Be a Problem Solver

Someone accidentally made the hands the same size on this clock. Can you still tell what time it is? How?

What if there were no clocks in the world? Write about what it would be like.

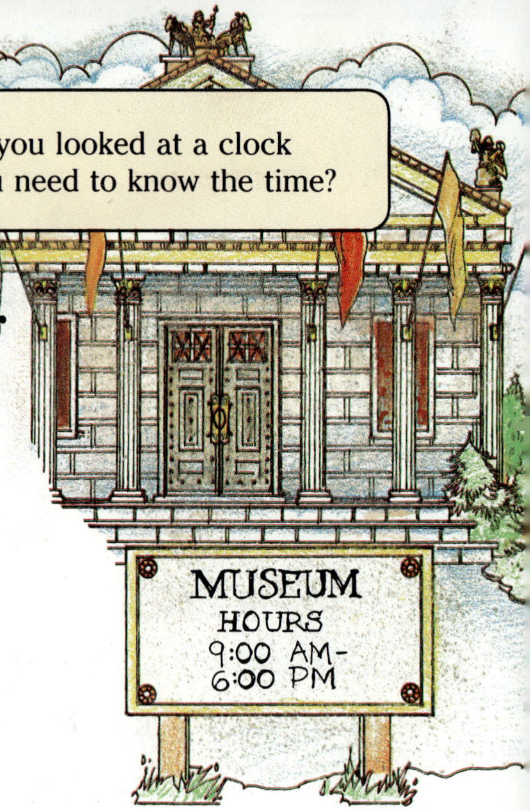

GETTING STARTED Think about when you looked at a clock today. Why did you need to know the time?

Hour, Half Hour, Quarter Hour

The museum has an exhibit of "Wonderful Things Around Us." Cara knows the museum opens at 9 o'clock.

The short hand shows hours. The long hand shows minutes.

"It is only eight thirty," she said.

30 minutes after eight, or half past eight

By nine fifteen, Cara was looking at a display of beautiful crystals.

15 minutes after nine, or quarter after nine

60 minutes = 1 hour
30 minutes = 1 half hour
15 minutes = 1 quarter hour
The hours between 12 midnight and 12 noon are A.M.
The hours between 12 noon and 12 midnight are P.M.

Check Your Understanding

Write each time in two ways.

1. 

2. 

3. 

4. 

Share Your Ideas How is using a clock face to tell time different from using a digital clock? Which is easier? Why?

Practice

Write each time in two ways.

5. 

6. 

7. 

8. 

9. 

10. 

11. 

12. 

Match the times.

13. three thirty in the afternoon
14. eight o'clock in the evening
15. 2:15 in the morning
16. quarter after four in the morning
17. half past three in the morning

a. 8:00 P.M.
b. 4:15 A.M.
c. 3:30 A.M.
d. 3:30 P.M.
e. 2:15 A.M.

Choose the most sensible time for each.

18. eating breakfast
    a. 8:30 P.M. b. 8:00 A.M.
    c. 4:15 P.M. d. 1:00 A.M.

19. going to bed
    a. 10:15 A.M. b. 9:30 A.M.
    c. 9:00 P.M. d. 3:00 P.M.

Think and Apply

Copy and complete Cara's schedule.

20. Cara visited displays from nine o'clock to twelve o'clock.

21. She had lunch from quarter after twelve to one o'clock.

22. Starting at half past one, she visited displays until half past two.

23. Then she left for home. She got there at quarter after three.

	Starting Time	Ending Time
Visit displays		
Lunch		
Visit displays		
Trip home		

Why is it important for you to be able to tell time?

SUMMING UP

GETTING STARTED

Count by fives to 60. Then count backwards by fives. How can counting this way help you tell time?

Minutes

It's amazing! Scientists have found that a bumblebee flaps its tiny wings 18,000 times a minute when it flies.

How do we tell time by the minute? These clocks show five minutes after three.

The minute hand moves from one mark to the next mark in 1 minute.
It moves from one number to the next in 5 minutes.
It moves around the clock in 60 minutes, or 1 hour.

What time do these clocks show?

twelve minutes after three, or three twelve

twenty minutes to four, or three forty

Where would the hands be at 3:27?

Check Your Understanding

Write each time in two ways.

1. 

2. 

3. 

4. 

Share Your Ideas Look at a clock right now. What time is it? Write the time in two ways.

Practice

Write each time in two ways.

5. 

6. 

7. 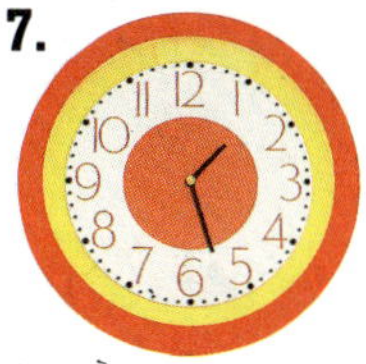

8. 

9. 

10. 

11. 

12. 

13. 

14. 

15. 

16. 

Complete the table.

Time	7:30	7:35	7:45	7:50
17. minutes after 7	30			
18. minutes to 8	30	25		

Think and Apply

19. Luis went to the bee exhibit at eleven forty. He left at twelve fifteen. He went to the rock exhibit at one thirty-five. Write the times, using A.M. or P.M.

20. How many minutes are there in one hour? in two hours? How many five-minute periods are there in one hour? in two hours?

Logical Thinking

21. Look at this table. How do the sizes of the animals compare with their heartbeats?

Animal	Number of Heartbeats in One Minute
Hummingbird	250
Cat	120
Adult person	72
Elephant	40
Whale	20

Draw clocks to show the time you usually get up in the morning, the time you go to school, and the time you get home from school.

SUMMING UP

GETTING STARTED

What were you doing one hour ago? What will you be doing one hour from now?

Elapsed Time

At 5:00 P.M., after the rain, a beautiful rainbow appeared in the sky. It lasted only 1 hour. At what time was the rainbow gone?

5:00

The minute hand moves once around the clock in 1 hour. The hour hand moves from one number to the next.

6:00

The rainbow was gone at 6:00 P.M.

Ana read about rainbows. She began at 7:10 P.M. and finished at 7:30 P.M. How many minutes did she read?

7:10

Count the minutes.

7:30

Ana read for 20 minutes.

Check Your Understanding

Tell what time it will be.

1. in 2 hours

2. in 6 minutes

3. in 7 hours

Tell how much time has passed.

4.

5.

Share Your Ideas What time does math class start? What time does it end? How long is it?

Practice

Tell what time it will be.

6. in 3 hours

7. in 10 minutes

8. in 30 minutes

9. in 20 minutes

Tell how much time has passed.

10.

11.

12. start 7:30 end 10:30
13. start 8:45 end 1:45

Complete. Follow the rule.

Rule: Add 40 minutes.

	Input	Output
14.	6:00	
15.	9:30	
16.	3:15	

Rule: Subtract 30 minutes.

	Input	Output
17.	10:00	
18.	4:10	
19.	8:55	

Think and Apply

20. How long did each of Ana's activities take?

Ana's Schedule	
Walk	3:40-4:10 P.M.
Play	4:10-5:00 P.M.
Eat	5:00-5:30 P.M.
Read	5:30-8:30 P.M.

Give some reasons why it may be important to know how long something takes.

SUMMING UP

Mixed Review

1. $2 + 3 + 6 = \square$
2. $7 + 0 + 8 = \square$
3. $17 - 8 = \square$
4. $13 - 9 = \square$
5. $12 - 0 = \square$

6. $\begin{array}{r} 35 \\ 19 \\ +68 \\ \hline \end{array}$
7. $\begin{array}{r} 107 \\ 24 \\ +953 \\ \hline \end{array}$

8. $\begin{array}{r} 84 \\ -69 \\ \hline \end{array}$
9. $\begin{array}{r} 700 \\ -516 \\ \hline \end{array}$

10. $\begin{array}{r} 3{,}071 \\ -\ \ 298 \\ \hline \end{array}$
11. $\begin{array}{r} 5{,}623 \\ +6{,}088 \\ \hline \end{array}$

Give the value of the underlined digit.

12. $\underline{3}06$
13. $1{,}29\underline{5}$
14. $72{,}4\underline{6}1$
15. $8\underline{9}6{,}005$
16. $\underline{6}50{,}009$
17. $4\underline{2}{,}687$

Round to the nearest hundred.

18. 580
19. 739
20. 4,023
21. 1,261
22. 9,512
23. 75
24. 351
25. 6,998

Activity

Exploring Estimating Time

Estimate how long it will take you to write the numbers 1 to 100. Will it take a minute, less than a minute, or more than a minute? Record your estimate.

Working together

Materials: a timer, or a watch with a second hand

A. Use the timer. Tell your partner when to start writing the numbers 1 to 100.

B. Record the start time and the finish time.

C. Together, figure out the time it took. Did it take a minute, less than a minute, or more than a minute?

D. Estimate again. Change roles. Repeat the experiment.

Sharing Your Results

1. Discuss your results with other classmates. Were your estimates close to the actual numbers? Why or why not?

2. Explain how doing the experiment the first time gave you a better idea of how long a minute is.

3. Explain how you made your estimate the second time. Could this method help you to estimate the time it takes to do other things? Explain.

Extending the Activity

It takes about one minute to wash your hands.
A gymnastics class takes about one hour.

A **minute** is the unit used to measure a short amount of time.

An **hour** is the unit used to measure longer amounts of time.

Which unit of time would you use to measure each activity?

4. tie your shoes
5. eat breakfast
6. play a baseball game
7. sleep at night
8. visit a zoo
9. walk to school

Work in a small group.
Conduct these experiments.

10. Estimate how long it takes to do 30 jumping jacks. Compare your estimate with those of others. How do you explain any great differences?
Try it. Time 30 jumping jacks.

11. Estimate how many times you can write your name in one minute.
Explain how you estimated.
Try the experiment.
How many times did you write your name?
How does this number compare with your estimate?

Summing Up

12. **Look back** at your estimates and records. Were you better at estimating in some activities? Explain.

GETTING STARTED

How would you know the date without using a calendar?

Calendar

A calendar shows the 12 months of the year in order. Which month comes first in the year? last? In which month is your birthday? Which month comes after your birthday month? Which month comes before May? Which month is the seventh month?

December

Sunday	Monday	Tuesday	Wednesday	Thursday	Friday	Saturday
1	2	3	4	5	6	7
8	9	10	11	12	13	14
15	16	17	18	19	20	21
22	23	24	25	26	27	28
29	30	31				

This calendar of December shows the days of the week. Which day comes after Monday, December ninth?

The second Wednesday is December 11. The fourth Sunday is December 22. What date is the third Friday?

Check Your Understanding

Name the day of the week for each date.

1. December 21 2. December 2 3. December 27

Name the date for each day.

4. the second Tuesday in December
5. the fourth Thursday in December
6. three weeks after Sunday, December 8
7. one week after Thursday, December 12

Share Your Ideas Name the months of the year in order. If March 1 is on a Friday, explain how you can find out what day April 1 is.

Practice

Use the calendar at the right for 9–18.

8. What month of the year is June?
9. How many days are in June?
10. On which day does the month begin? end?

June

Sunday	Monday	Tuesday	Wednesday	Thursday	Friday	Saturday
						1
2	3	4	5	6	7	8
9	10	11	12	13	14	15
16	17	18	19	20	21	22
23/30	24	25	26	27	28	29

Name the day of the week for each date.

11. June 13
12. June 21
13. June 9
14. June 25

Name each date.

15. the fourth Tuesday
16. the first Wednesday
17. the last Saturday
18. the third Monday

Make a calendar for the current month.

19. How many days does this month have?
20. Mark today's date with an **X**.
21. Draw a ring around the date one week from today.
22. Draw a square around the second Thursday.
23. Make up three questions that can be answered using the calendar.

Think and Apply

DATA
24. For a class display about birthdays, use a large sheet of paper. List the months. Then have students write their names next to their birthday month. Now answer **25** and **26**.
25. Which month has the most birthdays? the fewest birthdays?
26. Would the results be the same in another class? Explain.

Explain how you could find out how many school days there are left in this school year.

SUMMING UP

Midchapter Review

1. 
2. 
3. 
4. 

Tell what time it will be. pages 144–145

5. in 1 hour 
6. in 15 minutes 
7. in 4 hours 

8. in 5 minutes 

Tell how much time has passed. pages 144–145

9. start 7:30 P.M. end 8:30 P.M.
10. start 8:15 A.M. end 11:15 A.M.

Match the times. pages 140–141

11. 1:00 A.M. a. half past nine in the evening
12. 9:30 P.M. b. quarter to eight in the morning
13. 7:45 A.M. c. one o'clock in the morning

Complete. pages 148–149

14. May 8 is Monday. May 10 is _____.
15. May 19 is Friday. May 22 is _____.

Choose the correct word to complete each sentence.

16. 60 minutes is a(n) ______________.
17. 30 minutes is a(n) ______________.
18. 15 minutes is a(n) ______________.

Words to Know
half hour hour quarter hour

Solve.

19. Bob went to the bee exhibit at 10:20. He stayed there for 15 minutes. What time did he leave?

20. Tuesday is June 11. On June 17, Jane starts her summer vacation. What day of the week is that?

Exploring Problem Solving

THINK
EXPLORE
SOLVE
LOOK BACK

How Much Does Howie Weigh?

"Howie, our gerbil, weighs more than 1 ounce and less than 13 ounces. I have these weights: 1, 1, 2, 2, 3, and 4 ounces. How can you use this set of weights to find out what Howie weighs?" asked Mrs. Carter.

"I can make all the weights from 1 ounce to 13 ounces by using the weights," said Bob.

"Can anyone check Bob's work by using this set of weights: 1, 2, 3, and 7 ounces?" asked Mrs. Carter.

"I'll do it," said Ellen.

Thinking Critically

Were Bob and Ellen able to weigh Howie? How could they make all the weights? Work in a small group as you solve the problem.

Analyzing and Making Decisions

1. What must Bob and Ellen do?
2. Examine the weights given to Bob. How could you combine the weights to total 5 ounces? to total 6 ounces?
3. How can you show 5 and 6 ounces with the weights Ellen has?
4. Show how you can get all the sums from 1 to 13 by using each set of weights. Would making a list help?

Look Back Which set of weights would you rather use? Explain.

Problem Solving Strategies

Making a List

Tim's class is visiting the museum. The students are put into four small groups called A, B, C, and D. They are to have a guided tour of these four rooms: the Animal Room, the Gem Room, the Dinosaur Room, and the Insect Zoo. Only one group can be in a room at a time. Each group must visit all four rooms. In what order can each group tour the rooms?

Sometimes a problem has many facts. Making a list will help you to keep them in order so you can solve the problem.

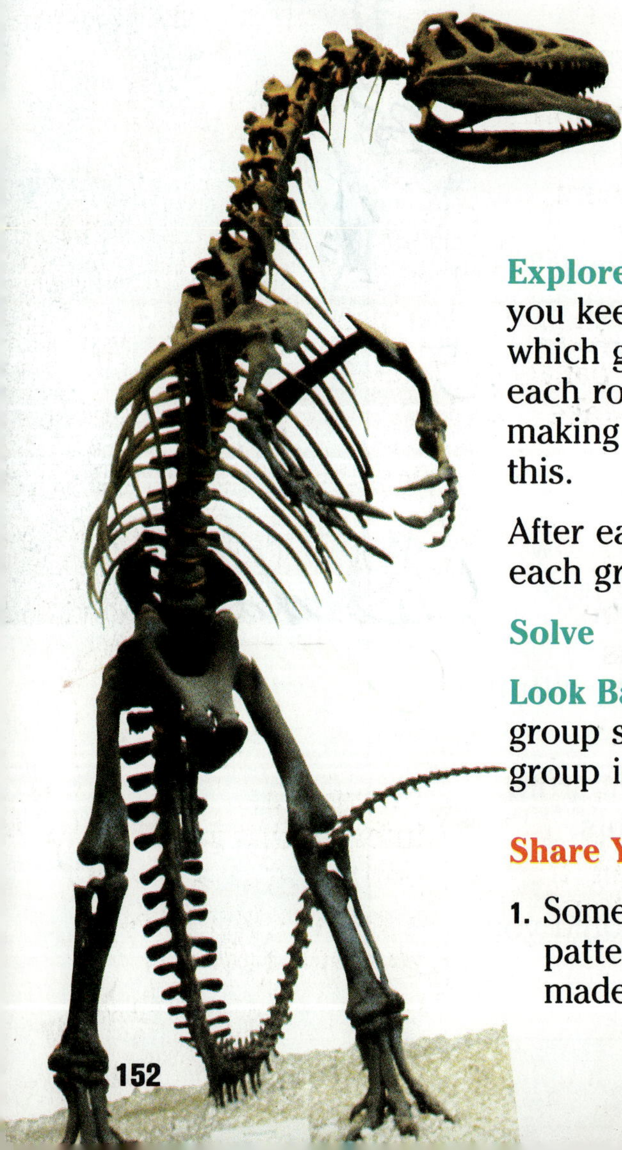

Solving the Problem

Think What is the problem? Can there be more than one group in a room at a time? Does every group need to be in a room?

Explore How can you keep track of which group is in each room? Try making a list like this.

Group A	Group B	Group C	Group D
Animal	Dinosaur	Gem	

After each group has been in a room, have each group move to another room.

Solve Show your completed list.

Look Back Does your list show that every group sees every room? Is there only one group in each room at a time?

Share Your Ideas

1. Sometimes making a list can help you see a pattern. Did you see a pattern when you made your list? Explain.

THINK
EXPLORE
SOLVE
LOOK BACK

Practice

Solve.

2. Tom needs to put numbers on 11 T-shirts for his ball team. He bought stencils for the numbers 1, 2, and 3. He wants to make only 1- or 2-digit numbers. Can he make a different number for each of the shirts? Explain.

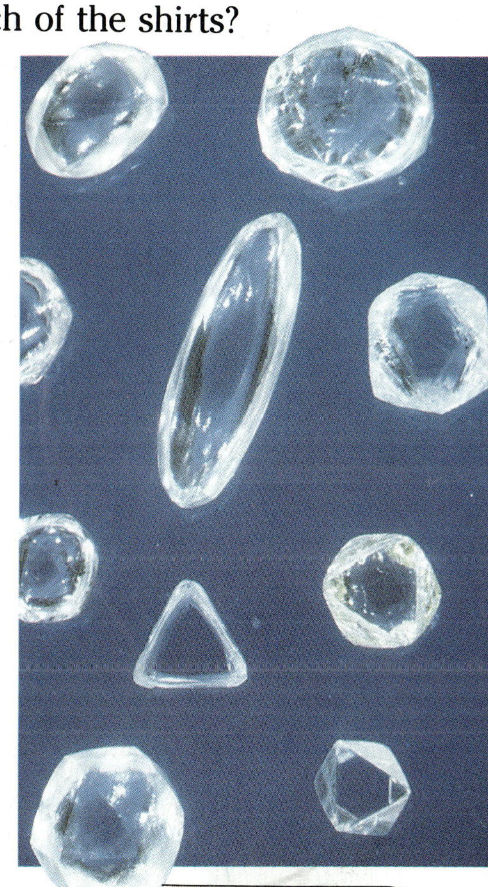

3. The Taco Burger Restaurant advertised 7 different kinds of beef tacos. How can this be if they have only onions, lettuce, and tomatoes? Explain.
4. Each of the 12 runners wore a number from 1 through 12. Runners 5, 8, 3, 10, and 4 crossed the finish line. How many runners have not crossed the finish line? What numbers are they wearing?
5. Mrs. Scott asked her children in what order they wanted to take the 3 rides at the park. How many different ways are there to take the 3 rides?

Mixed Strategy Review

6. On Thursday, the 27th of the month, the class trip committee met to prepare its report. "We will give our report next Thursday at the same time," said the chairperson. On what date will they give the report?
7. George's piano lessons are for one hour. If he finished a lesson at four o'clock, what time did he start?

Create Your Own

Change the information in problem **5** or **6**. Write a new problem using your information.

GETTING STARTED

Would you use a table or a graph to show how students get to school every day? Explain.

Understanding Graphs

As part of a science project, Helen wanted to take a nature walk each day. "But where will I find the time?" Helen wondered.
When can Helen take her nature walk?

This **bar graph** shows how Helen spends her time.

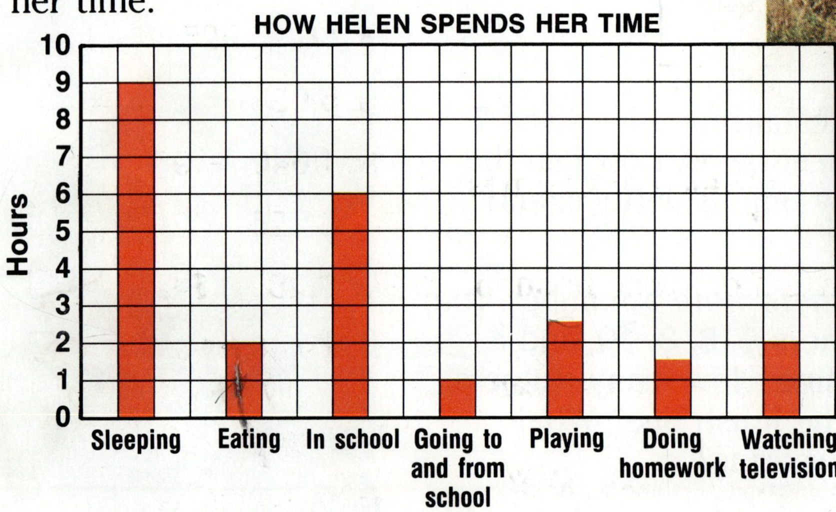

How do you use the scale on the side?
How many hours are spent sleeping?
How many hours are spent playing? in school?
Did Helen spend more time doing homework or watching television?

When do you think Helen can take the time to walk?

Check Your Understanding

1. On which activity did Helen spend the most time? the least time?
2. How is one half hour shown on the graph?
3. How many hours does Helen play and watch television? sleep and eat?

Share Your Ideas Write two questions that can be answered by using the graph.

Practice

On the weekend, Helen helps her mother at home. This bar graph shows how much time Helen's mother takes for each task.

4. Which task takes Helen's mother the most time? the least time?
5. Does her mother take more time to shop or do laundry?
6. How many hours does she take to clean and cook?
7. How many hours do all her tasks take?

Think and Apply

Think about how you spend your time.

8. List your activities on a school day.
9. Estimate how many hours you spend at each.
10. Make a bar graph showing the information.
11. Compare your graph to Helen's graph.
    - How is it the same?
    - How is it different?

> What is the advantage of showing information on a graph instead of just making a list?

SUMMING UP

Mixed Review

1. $\begin{array}{r} 5{,}601 \\ -\ \ 958 \\ \hline \end{array}$
2. $\begin{array}{r} 7{,}326 \\ +1{,}854 \\ \hline \end{array}$
3. $\begin{array}{r} 6{,}301 \\ +7{,}599 \\ \hline \end{array}$
4. $\begin{array}{r} 9{,}080 \\ -6{,}507 \\ \hline \end{array}$
5. $\begin{array}{r} \$7.50 \\ +\ \ 3.89 \\ \hline \end{array}$
6. $\begin{array}{r} \$20.00 \\ -\ \ \ 5.95 \\ \hline \end{array}$
7. 6,327 − 489 = ☐
8. 86 + 925 = ☐
9. 5,023 − 753 = ☐
10. 1,640 − 95 = ☐
11. \$1.25 + \$.16 = ☐
12. \$.28 + \$4.79 = ☐
13. \$9.00 − \$.33 = ☐

Estimate.

14. 95 + 36 = ☐
15. 71 − 49 = ☐
16. 320 + 44 = ☐
17. 165 − 54 = ☐

Compare. Use >, <, or = for ⬬.

18. 58 ⬬ 68
19. 2,490 ⬬ 2,489
20. 9,023 ⬬ 9,023
21. 305 ⬬ 350
22. 17 + 8 ⬬ 25

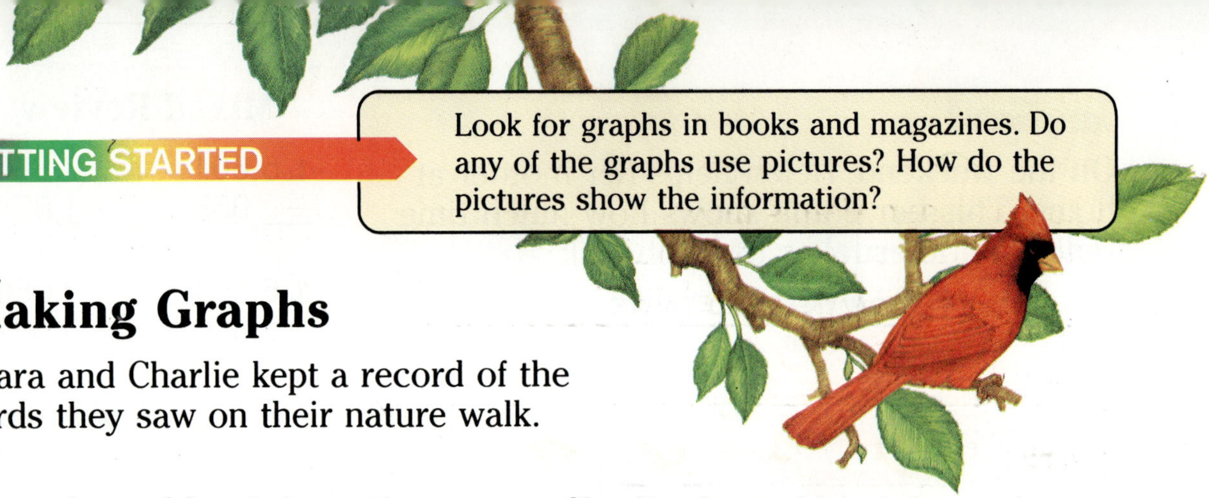

GETTING STARTED

Look for graphs in books and magazines. Do any of the graphs use pictures? How do the pictures show the information?

Making Graphs

Clara and Charlie kept a record of the birds they saw on their nature walk.

Clara showed her information on a bar graph.

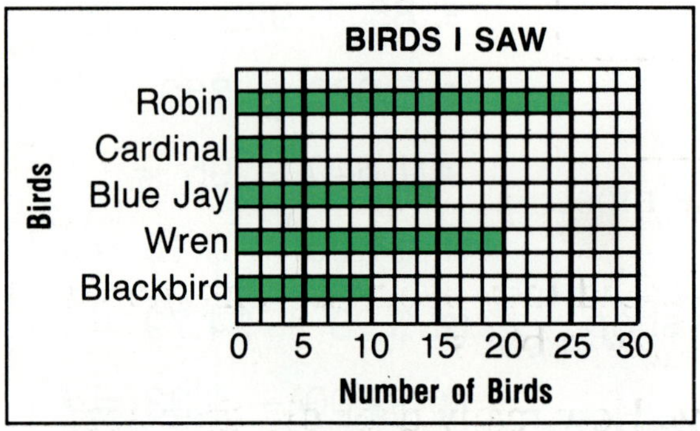

Charlie showed his information on a **pictograph**.

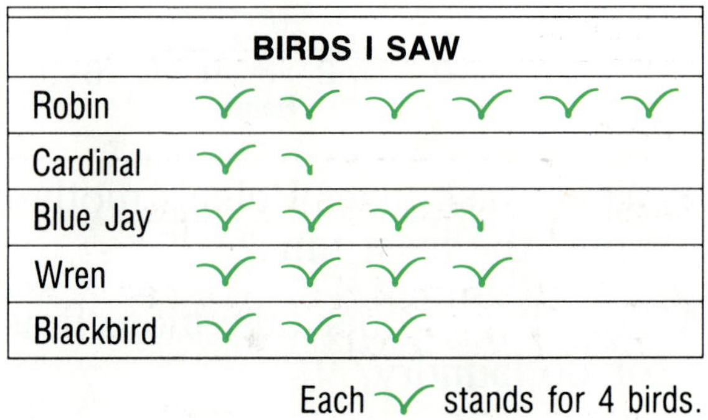

Look at the bar graph.

- How do you use the numbers along the bottom?
- What do the bars show?

Look at the pictograph.

- How do you use the symbol?
- What does mean?

Check Your Understanding

1. How are the two graphs alike?

2. How are the two graphs different?

3. How can you tell how many robins Clara saw?

4. What is the order of the birds Clara saw, from least number to greatest number?

5. Did Charlie see more wrens or robins?

6. Did Charlie ever see one half of a bird? Explain.

Share Your Ideas What are the advantages of each kind of graph? Why do you think graphs like these are useful?

Practice

Clara and her friends made a graph of the animals they saw on their walk.

ANIMALS WE SAW

Each (symbol) **stands for 10 animals.**

7. Which animal did they see the most? the least?
8. Did they see more chipmunks or rabbits?
9. What does (symbol) stand for?
10. How many deer did they see?
11. How many more squirrels than rabbits did they see?
12. How many animals did they see in all?

Think and Apply

Take a walk outdoors or around the school.

13. Make a tally of interesting things you see.
14. Use graphing software to record and display your information.
    - What title will you give the graph?
    - What will you write along the side and bottom of the graph?
    - If you choose a bar graph, which numbers will you use?
    - If you choose a pictograph, how many will each symbol stand for?
    - What kind of graph will you make?

Talk about the graphs in your own words. How were they made? What do they show? Make a class display of the graphs.

SUMMING UP

Activity

Exploring Collecting, Organizing, and Displaying Information

Which letter of the alphabet do you think is used most frequently?

Record your guess. Now explore to see if your prediction is correct.

Working together

Record your work.

A. Make a list of the letters of the alphabet.

B. Choose a page from a book in your classroom. Use any three sentences on that page.

C. Make a tally mark for each letter in the sentences.

D. Count each set of tally marks to find the number of times each letter was used.

Team members:

Book ______ **Page** _____

Letter	Tallies	Total
A	~~||||~~ ||||	9
B	|||	
C	||||	

Sharing Your Results

1. Which letter did you find was used most often? Did your results match your prediction? Explain.
2. Compare your results with other teams' results. Did other teams have the same results?
3. Why might the results of the teams be different?

Extending the Activity

4. Record all team results on a class table. Add to find the class total for each letter.

Letter	Team Results	Class Total
A	9, 6, 4, 6, 8, 5, 7	45
B		

5. Discuss what your table shows. Which letter was used most?

6. List the letters in order from those used most to those used least.

7. Which results do you think are closer to what you would find in *all* books: one team's results or the total class results? Explain.

8. Do the results match what you think is true just by looking through some books? Explain.

Summing Up

Now predict whether the results below would be about the same, or different from the first experiment. Explain.

9. using different pages in the same book

10. using pages from a different book

11. using pages from a book written in another language

GETTING STARTED

Look at some maps in your social studies book, or some road maps. Tell how to find the location of different places.

Ordered Pairs

Some ships were caught in a dense fog. Each ship radioed its position. The positions were located by pairs of numbers on a grid.

Where is the red ship?
Start at 0.
Go 2 spaces to the right.
Go 5 spaces up.
The red ship is at (2, 5).

▸An **ordered pair** of numbers (2, 5) names the point.

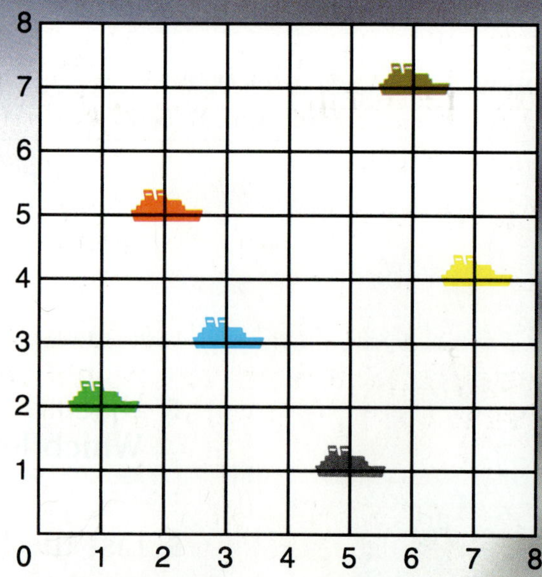

To find the blue ship:
- Start at 0.
- Go 3 spaces to the right.
- Go 3 spaces up.

The blue ship is at (3, 3).

To find the green ship:
- Start at 0.
- Go 1 space to the right.
- Go 2 spaces up.

The green ship is at (1, 2).

Use an ordered pair to name the location of the grey ship.

Check Your Understanding

Use the grid. Use an ordered pair to locate each.

1. the brown ship
2. the gold ship
3. the ship closest to the green ship

Share Your Ideas Do you think that the ordered pair (2, 3) names the same point as the ordered pair (3, 2)? Explain.

Practice

Use ordered pairs to name the locations of each ship on this grid.

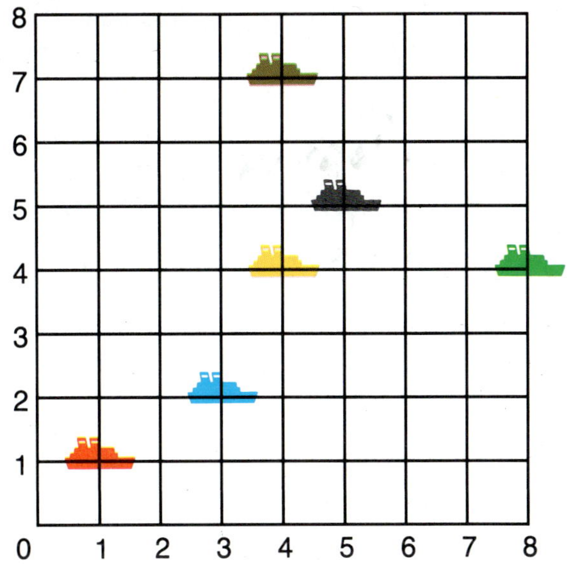

4. the red ship
5. the blue ship
6. the green ship
7. the grey ship
8. the brown ship
9. the gold ship
10. Use grid paper. Draw and label your own map. Draw ships at any points you choose. Name each point with an ordered pair.

Think and Apply

Each letter of the alphabet is at a point on this grid.

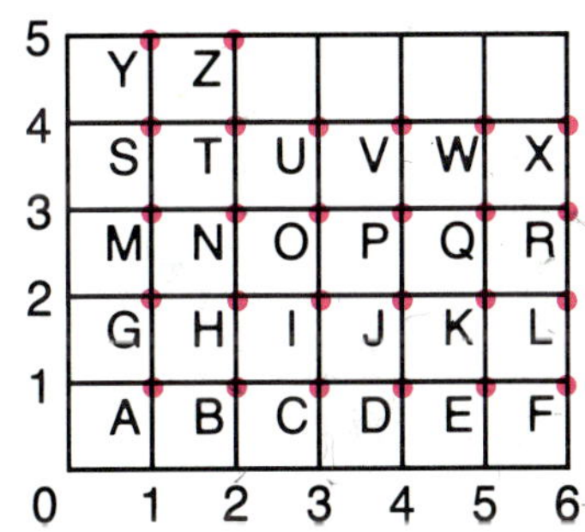

11. Find the letter at each of these points.
(6, 3) (2, 1) (2, 4) (6, 3) (3, 3) (5, 1)
12. Unscramble the letters in **11** to spell a boy's name.
13. Use ordered pairs to name the letters in your first name.
    - Scramble the ordered pairs. Then write them on a card.
    - Place the cards from all the students in a bag.
    - Each student draws a card and figures out the name.
14. What name is this?

Look back at **13**. Explain how you found the ordered pairs of your first name.

SUMMING UP

Using Problem Solving

Making a Map or Model

How can you show someone else what your classroom looks like? You can make a model or a map.

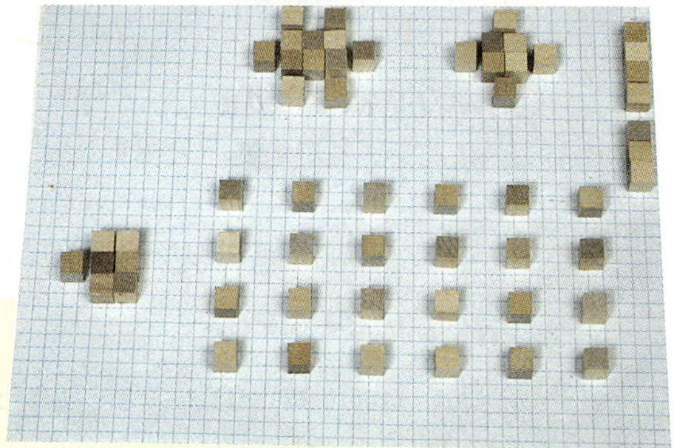

Working together

Materials: base-ten blocks, Workmat 6 or centimeter grid paper, pencils

A. Make a model of your classroom. Look at your workmat. Is it about the shape of your room? If not, cut it so that it is. This will be the floor of your model.

B. Use a block to represent one student's desk. How many student's desks fit across your room? Can that many model desks fit across your model? What should you do if they cannot?

C. How much bigger is a teacher's desk than a student's desk? How many blocks should you use to show a teacher's desk? How tall are the bookcases? How tall did you make them in your model?

D. Have you represented all the furniture in your room? Look at your room. Look at your model. Is everything in the correct place?

Sharing Your Ideas

1. What was the easiest step in making your model? What was the hardest step?

2. What would you need to do if you had to use bigger blocks?

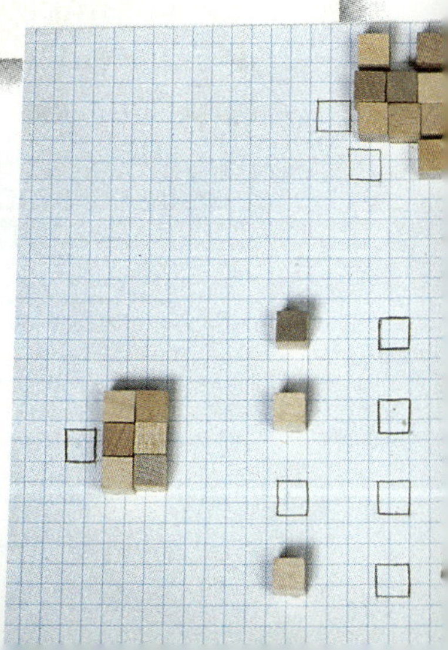

Extending Your Thinking

Now you can make a map to show what your classroom looks like. Pretend you could look down from the ceiling. That is what a map of your classroom could look like.

Work in your group and help make a map of your classroom.

3. One person in the group makes the map by drawing the outline of each block used in the model on the workmat.

4. After the outline for each piece of furniture is drawn, you will have a map of your classroom. Check your map. Does it show where the desks are? What else does it show?

5. Where is the door in your classroom? Make a mark on your map to show where it is.

   Now try this with your class.

6. Select two map readers. Have them close their eyes. Ask one person to hide an object. Have another person mark the map to show where the object is hidden. The readers use the map to find the object.

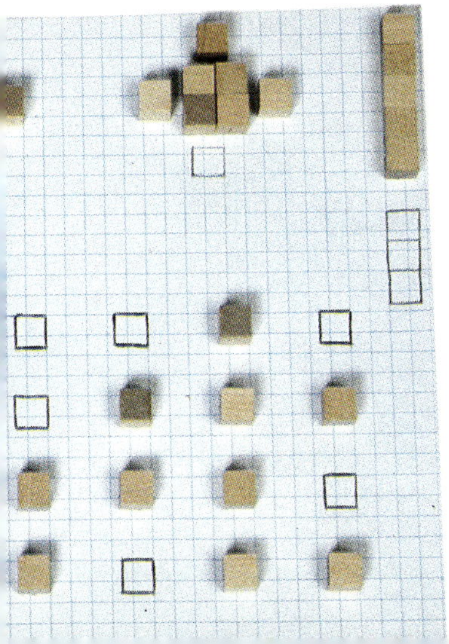

Summing Up

7. How are your map and model alike? How are they different?

8. Do you always have to make a map from a model? Explain.

9. Someone said, "The better the map is, the easier it is to find the hidden object." Explain.

Chapter Review

Write the time it will be. pages 140–145

1. in 4 hours

2. in 20 minutes

3. in 35 minutes

Tell how much time has passed. pages 144–145

4. 2:30–3:30 P.M.
5. 4:00–4:45 A.M.
6. 11:50 A.M.–12:15 P.M.

Choose minutes or hours to measure the time spent. pages 146–147

7. washing your hands
8. reading a 300-page book
9. drinking a glass of juice

Complete. pages 148–149

10. June 4 is Thursday. June 7 is _____.
11. June 24 is Monday. June 30 is _____.

Choose the correct word to complete each sentence.

Words to Know
point pictograph bar graph

12. A _____ uses symbols.
13. A _____ has a scale on the side or the bottom.
14. An ordered pair of numbers names a _____ on a grid.

Use an ordered pair of numbers to name the location of each. pages 160–161

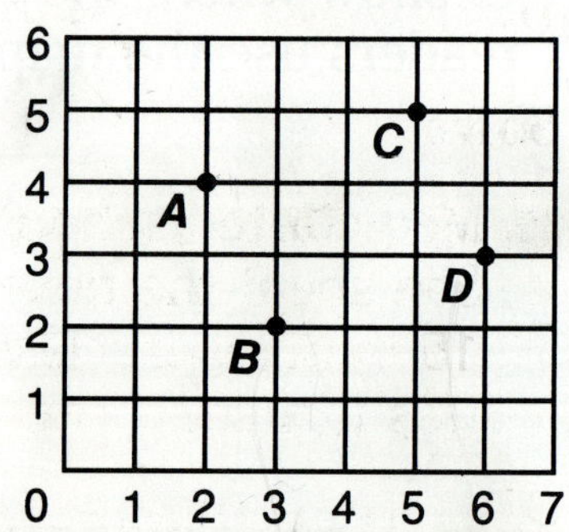

15. A
16. D
17. B
18. C

Solve. pages 151–153

19. Dave came home from school at 3:30 P.M. Two hours later he had supper. What time was it then?
20. Tao, Ben, Al, and Pia work in pairs as crossing guards. Make a list of possible pairs. How many are there?

Chapter Test

Write the time it will be.

1. in 5 minutes

2. in 3 hours

3. in 45 minutes

Tell how much time has passed.

4. 8:00–11:00 A.M.
5. 5:10–5:40 P.M.
6. 11:30 P.M.–12:05 A.M.

Choose minutes or hours to measure the time spent.

7. brushing your teeth
8. writing a book
9. taking a shower

Complete.

10. May 8 is Tuesday. May 10 is ______.
11. May 5 is Sunday. May 12 is ______.

Use a graph.

12. How many spruce are there?
13. Are there more maple or pine?
14. Which tree is found most often? How many are there?

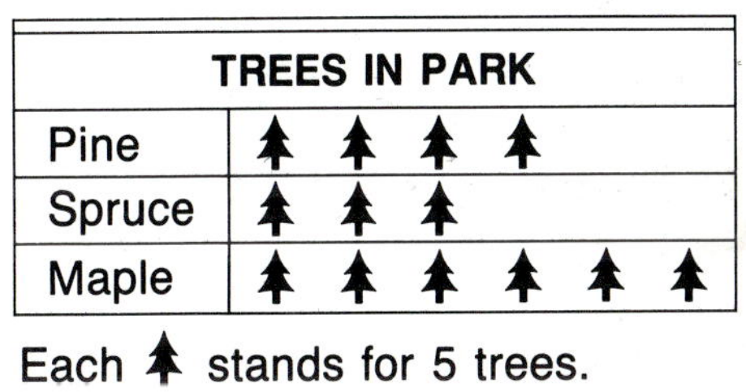

Use an ordered pair of numbers to name the location of each.

15. E
16. G
17. F
18. H

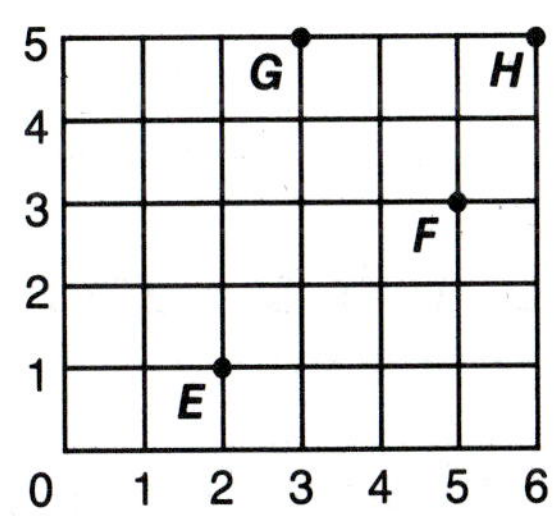

Solve.

19. Meg started school at 9:00 A.M. She spent the next 3 hours and 15 minutes in class. What time was it then?
20. In a lab, Ann, Meg, and Pete work in pairs. How many pairs are possible? Make a list to show all the pairs.

THINK How would you find out which sport was the favorite of all the students in your class? Make a list of the steps you would take.

Computer Link

Computer Counting

How long does it take you to count to 100? Do you think you can count faster than a computer can print the numbers on the screen?

Materials: Logo, clock

A. Work with a partner. Use a clock to time how long it takes your partner to count aloud from 1 to 100.

B. Define the procedure `TIME` shown below. Then enter `TIME` to have the computer show the numbers 1 to 100 on the screen. Use a clock to time how long it takes.

```
TO TIME
MAKE "COUNT 1
REPEAT 100 [PRINT :COUNT
MAKE "COUNT :COUNT + 1]
END
```

Type as one line. ↗ (REPEAT 100 [PRINT :COUNT MAKE "COUNT :COUNT + 1])

Starts at 1.
Prints each number.
Adds 1 each time.

C. Estimate how long it would take a person to count to 1,000.

D. Estimate how long it would take the computer to count to 1,000. Edit the `TIME` procedure. Change the number after `REPEAT` to `1000`. Then time the computer. Compare your estimate to the time you recorded.

Sharing Your Results

1. Was your estimate for the computer to count to 1,000 close to the time you recorded?
2. Do you think your estimate for a person to count to 1,000 is accurate? Why or why not?

Extending the Activity

How long do you think it would take the computer to count to 1,000 by twos? by fives? by tens?

3. Estimate how long it would take the computer to count by twos from 2 to 1,000.

4. Change the commands in TIME to count by twos. Enter TIME and use a clock to time the computer. Was your estimate close to the time you recorded?

```
TO TIME
MAKE "COUNT 2
REPEAT 500 [PRINT :COUNT
MAKE "COUNT :COUNT + 2]
END
```

Type as one line. (REPEAT 500 [PRINT :COUNT MAKE "COUNT :COUNT + 2])

Starts at 2.
Prints each number.
Adds 2 each time.

5. Change the commands in the procedure TIME to count from 5 to 1,000 by fives and then from 10 to 1,000 by tens. Estimate first. Then compare your estimate to the time you recorded.

Summing Up

6. Were your estimates for how long it would take the computer to count by fives and by tens close? Why or why not?

7. Explain how to get a good estimate for counting by tens if you know how long it takes to count by fives.

EXTENSION

Using Time Lines

Make a time line of your life!

Go back in time to when you were born. What important things have happened to you? When did they happen?

You can show the important events in your life on your own personal time line.

1. Use a strip of paper about 3 feet long.
2. Mark one end "born." Mark today's date near the other end. Plan to extend and mark more events on the time line.
3. Where would you put last summer?
4. Draw pictures of some important events in your life. Put them where they belong on the time line.
5. Discuss your time line with others. How did you decide on the important events and where to place them?

Class Project: A School-Year Time Line

Make a time line of your school year!

- What great times have you had as a class?
- Use pictures to show the important events.
- Add to the time line as events happen.
- Display your time line for all to see.

Maintaining Skills

Choose the correct answer. Write A, B, C, or D.

1. $\begin{array}{r} 25 \\ 35 \\ +47 \\ \hline \end{array}$

A 117 **C** 107

B 97 **D** not given

2. 362 + 149 = ☐

A 401 **C** 501

B 511 **D** not given

3. $\begin{array}{r} 872 \\ -641 \\ \hline \end{array}$

A 231 **C** 241

B 221 **D** not given

4. 546 − 357 = ☐

A 189 **C** 299

B 211 **D** not given

5. $\begin{array}{r} 4{,}398 \\ -2{,}199 \\ \hline \end{array}$

A 2,201 **C** 2,299

B 2,209 **D** not given

6. What time is it?

A 11:15 **C** 12:45

B 11:45 **D** not given

7. How many elm leaves did students collect?

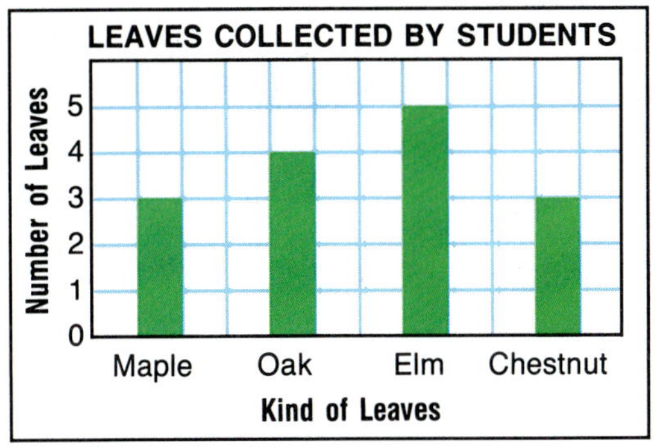

A 5 **C** 4

B 3 **D** not given

Solve.

8. Mrs. Garcia's class went on a picnic on Tuesday. Four children were absent. How many children went on the picnic?

A 4 **C** 5

B 15 **D** not enough information

9. Johnny handed out rulers to 23 children in his art class. He had 10 rulers left. How many did he start with?

A 13 **C** 43

B 33 **D** not enough information

6 Multiplication Facts 0–5

THEME Fascinating Creatures

Sharing What You Know

Another world lies below the surface of the sea. It is a world filled with strange and fascinating creatures. Describe the creatures you see in the pictures.

Using Language

Look again at the pictures. What colors do you see? Which colors are **repeated**? Repeat means to do again and again. In mathematics, adding 2 plus 2 plus 2 is **repeated addition**. When you multiply, you would say "three groups of two." How is repeated addition like multiplication?

Words to Know: repeated addition, multiplication, multiply, factor, product, order property

Be a Problem Solver

A pet store sells angel fish for $3 each. Sunfish cost $5 for the first one and $2 for each additional one. What can Susie buy for $20?

Design an aquarium. Decide what kinds of fish you would like. List the things you need for your aquarium.

Activity

Exploring Multiplication

Kittens make fascinating pets. What do you notice about the number of kittens in each basket?

Working together

Item	Number of Groups	Number in Each Group	Number in All

Materials: 50 counters, Workmat 7

Record your work.

A. Choose an animal.
Use counters to show how the animals are grouped.
- How many groups are there?
- How many animals are in each group?
- How many animals are there in all?

B. Repeat the activity for each kind of animal.

Sharing Your Results

1. How did you find the number in all for each kind of animal?
2. Look at your workmat. Compare these groups of animals. How is each the same? different?

   a. the puppies and the kittens
   b. the ducks and the ladybugs
   c. the kittens and the ducks

Extending the Activity

Look around your classroom.
What items come in groups with
the same number in each group?

Work in a small group.
Record your work on Workmat 7.

Item	Number of Groups	Number in Each Group	Number in All
desks	6	4	

3. List items that come in groups with the same number in each group.

4. For each item on your list, record
   - the number of groups.
   - the number of items in each group.

5. Look at the numbers on your workmat. Discuss different ways to find how many there are in all for each item. Record the number in all for each item.

Summing Up

6. Choose one of the items on your workmat.
   - Draw a picture to show the equal groups.
   - Write a few sentences to explain how you found the number in all.

7. Think of situations outside of school when you have equal groups. Why might you need to know how many in all?

Activity

Understanding Multiplication

What if you are making pebble pets for the craft show? How many pebbles will you need?

Working together

Materials: pebbles, glue, paint, Workmat 8

Take turns. Record your work.

A. Decide whether to make 2, 3, 4, or 5 pebble pets.

B. Decide whether to use 2, 3, 4, or 5 pebbles for each pet.

C. Make a group of pebbles for each pet.

D. Add to find the number of pebbles in all.

Number of Pebble Pets	Pebbles for Each Pet	Number of Pebbles in All
3	4	$4 + 4 + 4 = 12$

Sharing Your Results

Look back at the numbers you recorded.

1. What do you notice about the addends in each addition sentence?

2. What do you notice about the number of addends in each sentence?

3. How are the sentences different from each other?

Extending the Activity

When you have equal groups, you can add or you can multiply to find how many in all.

$5 + 5 + 5 = \square$

Think 3 groups of 5

Write $3 \times 5 = 15$

factor ↑ factor ↑ product ↑

$$\begin{array}{r} 5 \\ \times 3 \\ \hline 15 \end{array} \quad \begin{array}{l} \leftarrow \text{factor} \\ \leftarrow \text{factor} \\ \leftarrow \text{product} \end{array}$$

Read Three times five equals fifteen

4. Look at your workmat. Write a multiplication sentence for each addition sentence.

Write a multiplication sentence for each.

5. 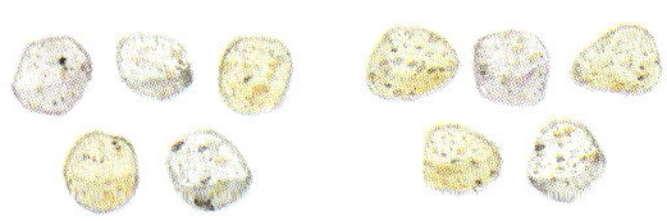

2 groups of 5

6. 

3 rows of 4

7. 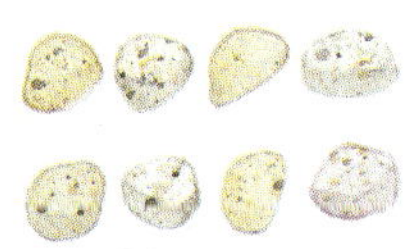

2 rows of 4

Think about what each multiplication sentence means. Draw a picture for each sentence. Then find each product.

8. $6 \times 2 = \square$

9. $3 \times 4 = \square$

10. $2 \times 7 = \square$

11. $4 \times 6 = \square$

Summing Up

12. You can use addition to find how many in all. When you have equal groups, you can also use multiplication to find how many in all. What are the advantages of using multiplication?

GETTING STARTED

Which foot do you put your sock on first? What would happen if you changed this order?

Order in Multiplication

Which bowl has the greater number of fish?

- Use grid paper. Trace around 3 rows of 5 squares. Then trace around 5 rows of 3 squares.
- Cut out both shapes and compare. Place one shape on the other. Do the shapes match?
- How many squares are in each?

  $3 \times 5 = \square$ $\qquad$ $5 \times 3 = \square$

  Does $3 \times 5 = 5 \times 3$?
- Try this with 2 rows of 6 and 6 rows of 2. Does $2 \times 6 = 6 \times 2$?

The Order Property for Multiplication

The order in which numbers are multiplied does not change the product.

Check Your Understanding

Write a multiplication fact for each.

1. 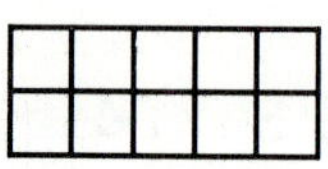

2 rows of 5

2. 

2 groups of 3

3. 

3 groups of 2

Complete.

4. If $2 \times 4 = 8$, then $4 \times 2 = \square$.

5. If $4 \times 6 = 24$, then $6 \times 4 = \square$.

6. If $5 \times 2 = 10$, then $2 \times 5 = \square$.

Share Your Ideas Explain what this means. When you know one multiplication fact, you really know two facts.

Practice

Write a multiplication fact for each.

7. 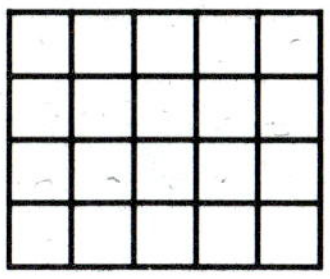
4 rows of 5

8. 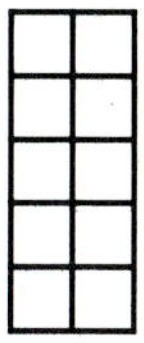
5 rows of 2

9. 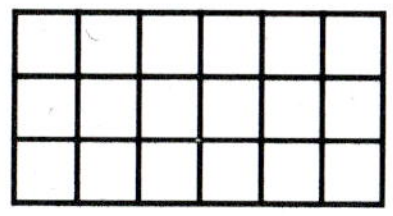
3 rows of 6

10. 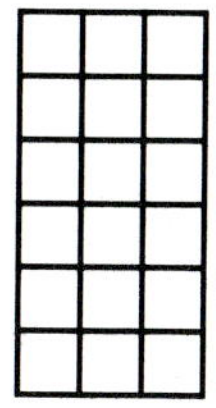
6 rows of 3

11. 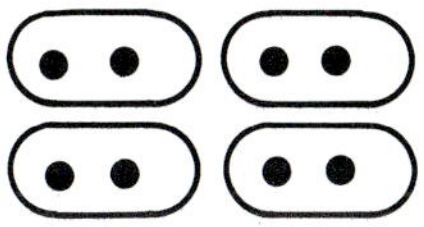
4 groups of 2

12. 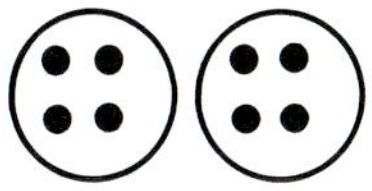
2 groups of 4

13. 
3 groups of 7

14. 
7 groups of 3

Complete.

15. If $2 \times 9 = 18$, then $9 \times 2 = \square$.

16. If $3 \times 4 = 12$, then $4 \times 3 = \square$.

17. If $7 \times 3 = 21$, then $3 \times 7 = \square$.

18. If $1 \times 5 = 5$, then $5 \times 1 = \square$.

19. If $6 \times 7 = 42$, then $7 \times 6 = \square$.

20. If $4 \times 8 = 32$, then $8 \times 4 = \square$.

21. $9 \times 6 = 6 \times \square$

22. $8 \times 7 = 7 \times \square$

Think and Apply

23. Draw a picture to show each.
$4 \times 3 = 3 \times 4$ $\qquad$ $2 \times 7 = 7 \times 2$

24. You have 4 tanks with 5 fish in each. Your friend has 5 tanks with 4 fish in each. Who has the greater number of fish? Explain.

25. Ben has 6 fish bowls. He wants to put two goldfish in each bowl. If he buys 10 goldfish, will he have enough? Explain.

Explain how you can use the order property to write a new fact for each.
$4 \times 8 = 24$ $\quad$ $4 \times 9 = 36$ $\quad$ $5 \times 6 = 30$

SUMMING UP

GETTING STARTED

Which of the following tell about the dots? 

4 twos	2 + 2 + 2 + 2	2 + 4
4 × 2	8	four	2 × 4

2 as a Factor

How many crows are there in all?

2 × 7 = ☐

There are 2 groups of 7 crows.

2 × 7 = 14 number in all

↑ number of groups ↖ number in each group

$$\begin{array}{r} 7 \\ \times 2 \\ \hline 14 \end{array}$$

7 ← number in each group
×2 ← number of groups
14 ← number in all

There are 14 crows in all.

You can also use doubles.
There are 2 groups of 7.

Think 7 + 7 = 14
So 2 × 7 = 14

When you multiply with 2, you can think of a related addition double.

What if you want to find 5 × 2?
You can use the Order Property.

Think 5 × 2 = 2 × 5
2 × 5 = 10
So 5 × 2 = 10

5	← factor →	2
×2	← factor →	×5
10	← product →	10

Check Your Understanding

Find each product.

1. 4 × 2
2. 3 × 2
3. 6 × 2
4. 2 × 7
5. 2 × 8
6. 2 × 2
7. 2 × 9

8. 5 × 2 = ☐
9. 3 × 2 = ☐
10. 2 × 9 = ☐
11. 2 × 8 = ☐

Share Your Ideas Explain how related addition doubles can help you to know multiplication facts with 2.

Practice

Find each product.

12. $\begin{array}{r} 5 \\ \times 2 \\ \hline \end{array}$	**13.** $\begin{array}{r} 3 \\ \times 2 \\ \hline \end{array}$	**14.** $\begin{array}{r} 2 \\ \times 2 \\ \hline \end{array}$	**15.** $\begin{array}{r} 2 \\ \times 4 \\ \hline \end{array}$	**16.** $\begin{array}{r} 2 \\ \times 6 \\ \hline \end{array}$	**17.** $\begin{array}{r} 7 \\ \times 2 \\ \hline \end{array}$	**18.** $\begin{array}{r} 1 \\ \times 2 \\ \hline \end{array}$
19. $\begin{array}{r} 8 \\ \times 2 \\ \hline \end{array}$	**20.** $\begin{array}{r} 4 \\ \times 2 \\ \hline \end{array}$	**21.** $\begin{array}{r} 2 \\ \times 7 \\ \hline \end{array}$	**22.** $\begin{array}{r} 2 \\ \times 5 \\ \hline \end{array}$	**23.** $\begin{array}{r} 2 \\ \times 9 \\ \hline \end{array}$	**24.** $\begin{array}{r} 2 \\ \times 8 \\ \hline \end{array}$	**25.** $\begin{array}{r} 6 \\ \times 2 \\ \hline \end{array}$

26. $3 \times 2 = \square$ **27.** $2 \times 2 = \square$ **28.** $2 \times 5 = \square$ **29.** $6 \times 2 = \square$

30. $\square = 2 \times 1$ **31.** $\square = 2 \times 9$ **32.** $\square = 4 \times 2$ **33.** $\square = 7 \times 2$

Choose the correct numbers to complete each pattern.

34. 2, 4, 6, ____, ____, ____.
- **a.** 7, 8, 9 **b.** 10, 11, 12
- **c.** 8, 10, 12 **d.** 7, 9, 11

35. 16, 14, 12, ____, ____, ____.
- **a.** 15, 18, 21 **b.** 10, 8, 6
- **c.** 11, 10, 9 **d.** 10, 9, 8

Compare. Use $>$, $<$, or $=$ for ●.

36. 2×9 ● 2×8

37. 2×7 ● 7×2

38. 2×3 ● 3×2

39. 2×6 ● 5×2

40. $2 + 2 + 2 + 2$ ● 4×2

41. 2×4 ● $2 + 2 + 2$

Think and Apply

42. Write a multiplication sentence for the number of crows in 2 groups of 8 crows. in 6 groups of 2 crows.

43. Explain what each means.
- There are **twice** as many crows now.
- **Double** the number of crows.

Visual Thinking

44. How can you build this square using 2 triangles? 4 triangles? 8 triangles?

SUMMING UP

Explain how you can use a related addition double to find the product of 2×9.

GETTING STARTED

What things can you think of that come in groups of three?

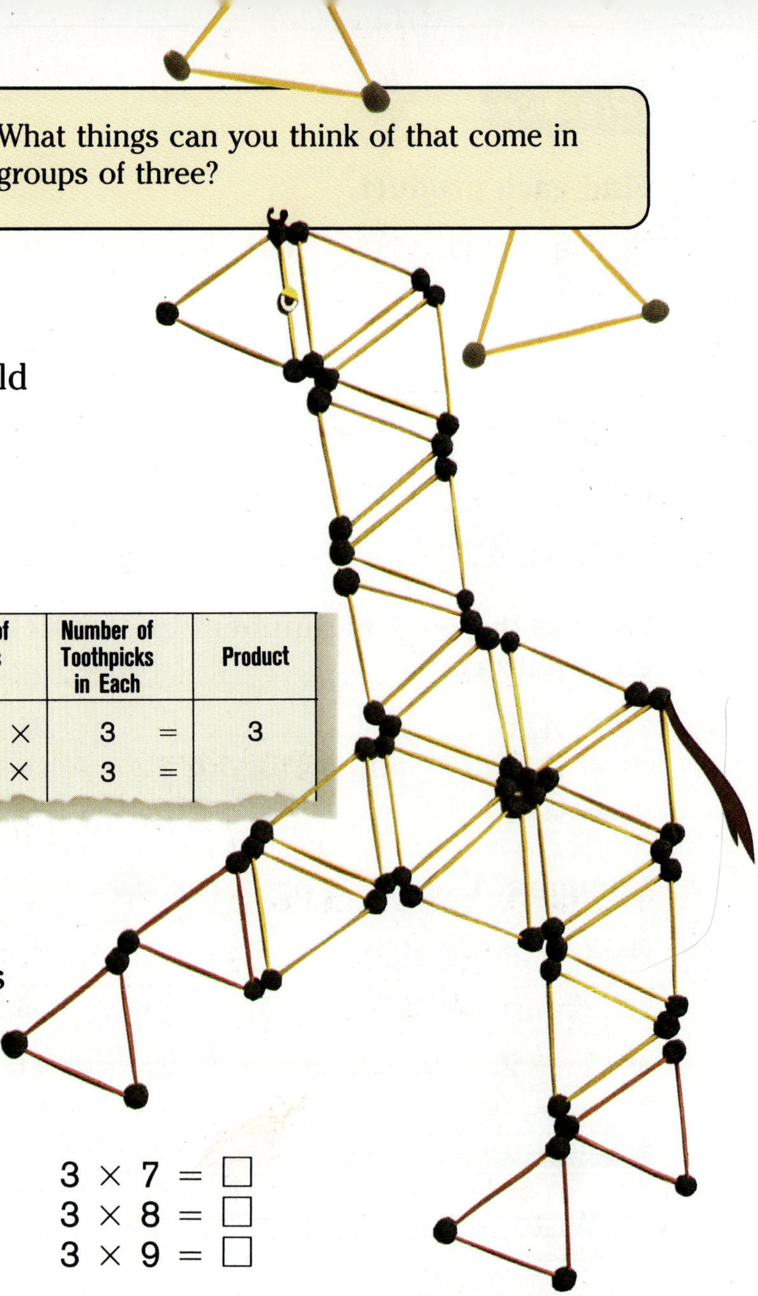

3 as a Factor

Are 20 toothpicks enough to build a creature made of 6 triangles?

- Arrange toothpicks or draw pictures to show 6 triangles.
- As you show each triangle, record your work in a chart like this.
- How many toothpicks do you need to build 6 triangles?

Number of Triangles	Number of Toothpicks in Each	Product
1 ×	3 =	3
2 ×	3 =	

Since 6 × 3 = 18, 20 toothpicks are enough.

How could you find the products for 7 × 3, 8 × 3, and 9 × 3?

How can you use your chart to find these products?

3 × 1 = ☐ 3 × 4 = ☐ 3 × 7 = ☐
3 × 2 = ☐ 3 × 5 = ☐ 3 × 8 = ☐
3 × 3 = ☐ 3 × 6 = ☐ 3 × 9 = ☐

Check Your Understanding

Multiply.

1. 3 × 4 **2.** 3 × 5 **3.** 3 × 3 **4.** 3 × 8 **5.** 3 × 7 **6.** 1 × 3 **7.** 9 × 3

8. 2 × 3 = ☐ **9.** 6 × 3 = ☐ **10.** 9 × 3 = ☐ **11.** 3 × 5 = ☐

Share Your Ideas Look back at your chart. How does counting by threes help you to know multiplication facts with 3?

Practice

Complete.

12. $\begin{array}{r} 3 \\ \times 1 \\ \hline \end{array}$

13. $\begin{array}{r} 3 \\ \times 3 \\ \hline \end{array}$

14. $\begin{array}{r} 2 \\ \times 4 \\ \hline \end{array}$

15. $\begin{array}{r} 6 \\ \times 3 \\ \hline \end{array}$

16. $\begin{array}{r} 8 \\ \times 3 \\ \hline \end{array}$

17. $\begin{array}{r} 5 \\ \times 3 \\ \hline \end{array}$

18. $\begin{array}{r} 3 \\ \times 7 \\ \hline \end{array}$

19. $\begin{array}{r} 3 \\ \times 9 \\ \hline \end{array}$

20. $3 \times 4 = \square$ 21. $5 \times 3 = \square$

22. $\square = 6 \times 3$ 23. $\square = 2 \times 8$ 24. $\square = 3 \times 7$

25. $3 + 3 + 3 = \square \times \square$

26. $\square \times \square = 3 + 3 + 3 + 3 + 3$

Find the next three numbers in each pattern.

27. 3, 6, 9, . . .

28. 18, 15, 12, . . .

Complete. Follow the rule.

Rule: Multiply by 3.

	Input	Output
29.	2	
30.	6	
31.	4	
32.	7	

Rule: Multiply by 3.

	Input	Output
33.		27
34.		24
35.		15
36.		9

Think and Apply

37. Use a calculator. Start at 0. Add 3 five times. Record each result.
    - What do you notice about the numbers?
    - How can you use addition to find the answer to a multiplication fact with 3?

38. You have 10 toothpicks. Some are red and some are blue. List all possible color combinations for the 10 toothpicks.

SUMMING UP

Explain how you can find the product of 9×3 if you cannot recall what it is.

Mixed Review

1. $\begin{array}{r} 8 \\ +9 \\ \hline \end{array}$

2. $\begin{array}{r} 14 \\ -\ 6 \\ \hline \end{array}$

3. $\begin{array}{r} 7 \\ +5 \\ \hline \end{array}$

4. $\begin{array}{r} 18 \\ -\ 9 \\ \hline \end{array}$

5. $\begin{array}{r} 26 \\ -17 \\ \hline \end{array}$

6. $\begin{array}{r} 85 \\ +67 \\ \hline \end{array}$

7. $\begin{array}{r} 49 \\ -27 \\ \hline \end{array}$

8. $\begin{array}{r} 65 \\ +48 \\ \hline \end{array}$

9. $\begin{array}{r} 399 \\ +145 \\ \hline \end{array}$

10. $\begin{array}{r} 682 \\ -327 \\ \hline \end{array}$

11. $\begin{array}{r} 213 \\ -125 \\ \hline \end{array}$

12. $\begin{array}{r} 563 \\ -\ 75 \\ \hline \end{array}$

13. $\begin{array}{r} \$1.56 \\ +\ 2.97 \\ \hline \end{array}$

14. $\begin{array}{r} \$7.80 \\ -\ 3.85 \\ \hline \end{array}$

15. $\begin{array}{r} 62 \\ 8 \\ +10 \\ \hline \end{array}$

16. $\begin{array}{r} 73 \\ 49 \\ +25 \\ \hline \end{array}$

Draw a clock to show each.

17. half past two
18. three thirty
19. quarter to one
20. nine fifty
21. six forty-five
22. eleven fifteen

Activity

Exploring 1 and 0

There are 3 trees and 1 koala in each tree. How many koalas are there? $3 \times 1 = \square$

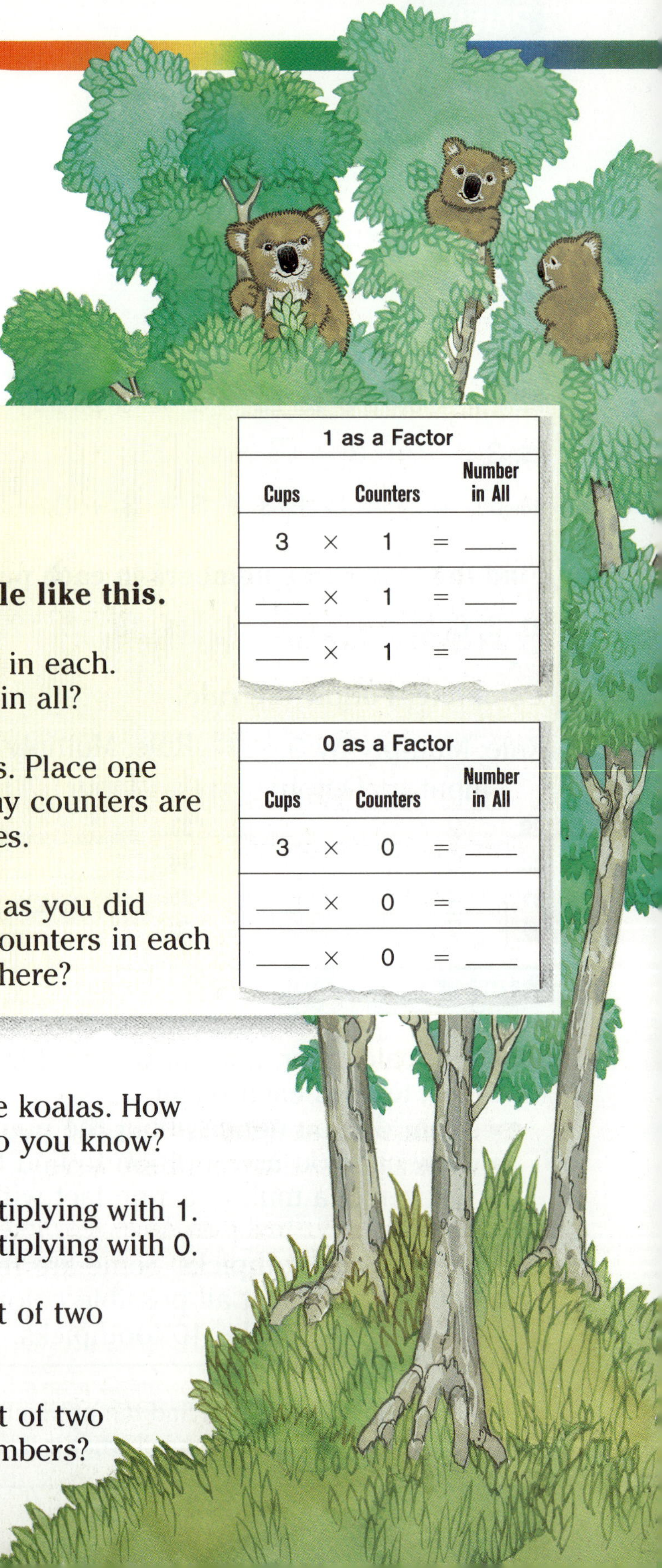

Working together

Materials: 9 counters 9 cups

Record your work. Make a table like this.

1 as a Factor				
Cups		Counters		Number in All
3	×	1	=	___
___	×	1	=	___
___	×	1	=	___

0 as a Factor				
Cups		Counters		Number in All
3	×	0	=	___
___	×	0	=	___
___	×	0	=	___

A. Use 3 cups. Place one counter in each. How many counters are there in all?

B. Use a different number of cups. Place one counter in each cup. How many counters are there? Repeat this several times.

C. Use the same number of cups as you did in **A** and **B**. This time place 0 counters in each cup. How many counters are there?

Sharing Your Results

1. **Look back** at the trees and the koalas. How many koalas are there? How do you know?
2. Give a rule that tells about multiplying with 1. Give a rule that tells about multiplying with 0.
3. How can you tell if the product of two numbers will be 0?
4. How can you tell if the product of two numbers will be one of the numbers?

Extending the Activity

- The product of any number and 1 is that number.
- The product of any number and 0 is 0.

Work on your own.
Use the rules above to complete each sentence.
Tell which rule you used for each.

5. $8 \times 1 = \square$ **6.** $7 \times 0 = \square$ **7.** $1 \times 300 = \square$

8. $0 \times 0 = \square$ **9.** $1 \times 0 = \square$ **10.** $1 \times 1 = \square$

11. $0 \times 39 = \square$ **12.** $47 \times 1 = \square$ **13.** $845 \times 0 = \square$

Compare. Use >, <, or = for each ●.
Choose mental math, paper and pencil, or a calculator. Explain your choices.

CHOICES

14. 6×0 ● 9×1 **15.** 46×0 ● 0×89

16. 1×70 ● 1×55 **17.** 351×1 ● 1×651

18. 1×175 ● 0×932 **19.** $7 \times 1 \times 2$ ● $3 \times 9 \times 0$

Find each missing number.

20. $2 \times 4 \times \square = 8$ **21.** $6 \times 3 \times \square = 0$

22. $1 \times 60 \times \square = 60$ **23.** $\square \times 4 \times 8 = 0$

24. $0 = 385 \times 7 \times \square$ **25.** $986 = 986 \times 1 \times \square$

Summing Up

26. **Look back at 20–25.** How did you know what the missing numbers were?

27. How do the rules you have learned help you to multiply any numbers when 0 or 1 are factors?

Midchapter Review

Find each product. pages 176–181

1. $\begin{array}{r}2\\ \times 3\\ \hline\end{array}$ $\begin{array}{r}3\\ \times 2\\ \hline\end{array}$

2. $\begin{array}{r}4\\ \times 2\\ \hline\end{array}$ $\begin{array}{r}2\\ \times 4\\ \hline\end{array}$

3. $\begin{array}{r}4\\ \times 3\\ \hline\end{array}$ $\begin{array}{r}3\\ \times 4\\ \hline\end{array}$

4. $\begin{array}{r}6\\ \times 3\\ \hline\end{array}$
5. $\begin{array}{r}5\\ \times 2\\ \hline\end{array}$
6. $\begin{array}{r}7\\ \times 2\\ \hline\end{array}$
7. $\begin{array}{r}5\\ \times 3\\ \hline\end{array}$
8. $\begin{array}{r}2\\ \times 8\\ \hline\end{array}$
9. $\begin{array}{r}3\\ \times 5\\ \hline\end{array}$
10. $\begin{array}{r}8\\ \times 3\\ \hline\end{array}$

11. $\begin{array}{r}3\\ \times 3\\ \hline\end{array}$
12. $\begin{array}{r}2\\ \times 6\\ \hline\end{array}$
13. $\begin{array}{r}9\\ \times 2\\ \hline\end{array}$
14. $\begin{array}{r}2\\ \times 7\\ \hline\end{array}$
15. $\begin{array}{r}3\\ \times 6\\ \hline\end{array}$
16. $\begin{array}{r}3\\ \times 7\\ \hline\end{array}$
17. $\begin{array}{r}3\\ \times 9\\ \hline\end{array}$

18. $2 \times 8 = \square$
19. $3 \times 7 = \square$
20. $9 \times 2 = \square$
21. $8 \times 3 = \square$

22. $3 \times 9 = \square$
23. $2 \times 6 = \square$
24. $5 \times 2 = \square$
25. $3 \times 4 = \square$

Multiply. pages 182–183

26. $\begin{array}{r}0\\ \times 1\\ \hline\end{array}$
27. $\begin{array}{r}8\\ \times 0\\ \hline\end{array}$
28. $\begin{array}{r}9\\ \times 1\\ \hline\end{array}$
29. $\begin{array}{r}6\\ \times 0\\ \hline\end{array}$
30. $\begin{array}{r}1\\ \times 1\\ \hline\end{array}$
31. $\begin{array}{r}0\\ \times 0\\ \hline\end{array}$
32. $\begin{array}{r}0\\ \times 7\\ \hline\end{array}$

33. $4 \times 0 = \square$
34. $3 \times 1 = \square$
35. $0 \times 5 = \square$
36. $1 \times 6 = \square$

37. $\square = 2 \times 0$
38. $\square = 1 \times 8$
39. $\square = 9 \times 0$
40. $\square = 1 \times 5$

Choose the correct words to complete the sentence.

41. In the fact $3 \times 2 = 6$, 3 and 2 are ______ and 6 is the ______.

Words to Know
product factors

Solve.

42. Jane uses 3 toothpicks to make each triangle. How many toothpicks does she use to make 4 triangles?

43. There were 2 groups of 5 crows and 3 groups of 4 crows. How many crows were there in all?

Exploring Problem Solving

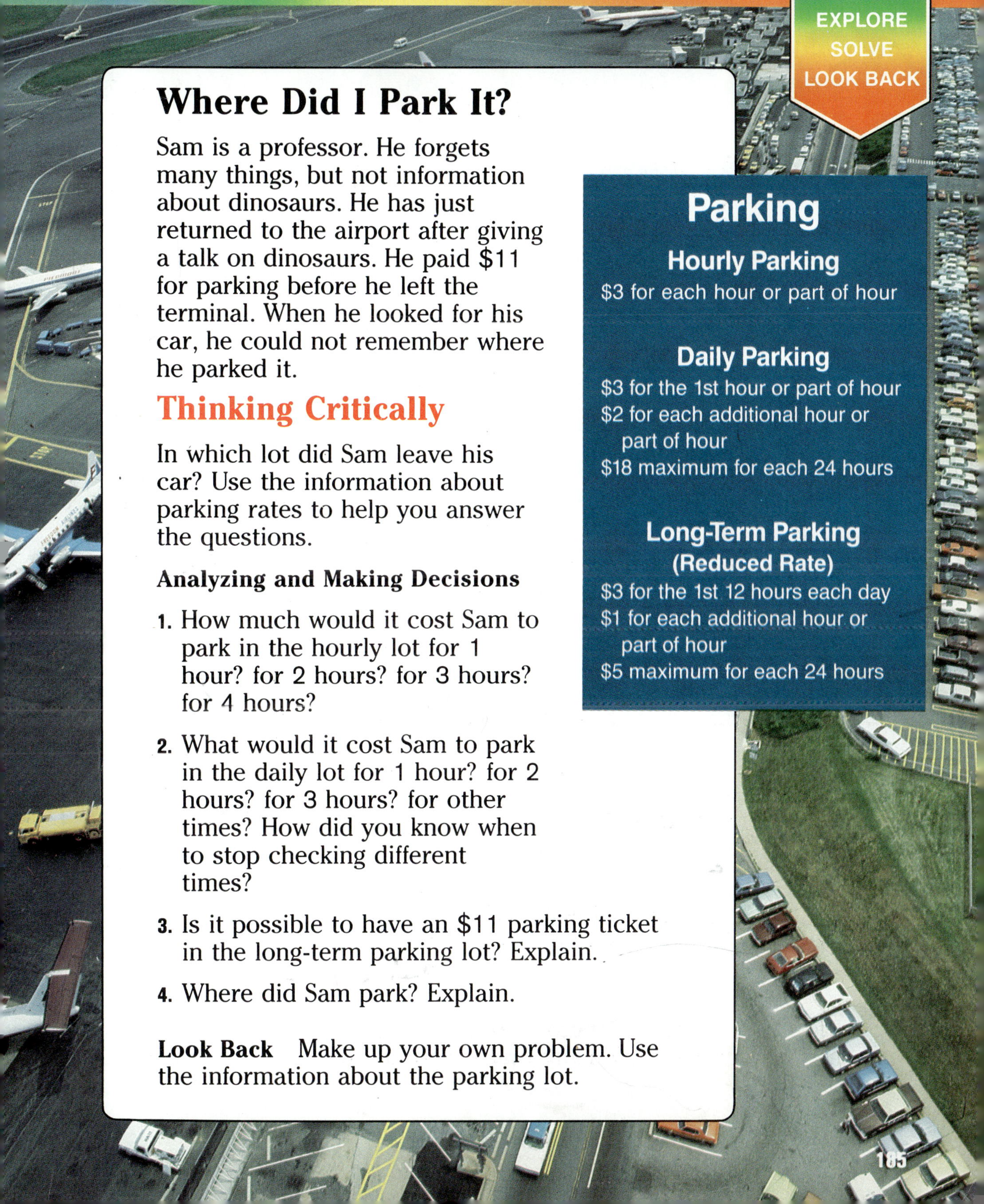

THINK
EXPLORE
SOLVE
LOOK BACK

Where Did I Park It?

Sam is a professor. He forgets many things, but not information about dinosaurs. He has just returned to the airport after giving a talk on dinosaurs. He paid $11 for parking before he left the terminal. When he looked for his car, he could not remember where he parked it.

Parking

Hourly Parking
$3 for each hour or part of hour

Daily Parking
$3 for the 1st hour or part of hour
$2 for each additional hour or part of hour
$18 maximum for each 24 hours

Long-Term Parking (Reduced Rate)
$3 for the 1st 12 hours each day
$1 for each additional hour or part of hour
$5 maximum for each 24 hours

Thinking Critically

In which lot did Sam leave his car? Use the information about parking rates to help you answer the questions.

Analyzing and Making Decisions

1. How much would it cost Sam to park in the hourly lot for 1 hour? for 2 hours? for 3 hours? for 4 hours?
2. What would it cost Sam to park in the daily lot for 1 hour? for 2 hours? for 3 hours? for other times? How did you know when to stop checking different times?
3. Is it possible to have an $11 parking ticket in the long-term parking lot? Explain.
4. Where did Sam park? Explain.

Look Back Make up your own problem. Use the information about the parking lot.

Problem Solving Strategies

Finding Patterns

Mr. Hill has two new creatures for his exhibit. He is not sure where to put each one. Find a pattern for the order of the animals on display. Where should Mr. Hill put the fox and the elephant?

Patterns occur in many problems. When they do, finding the pattern can help you solve the problem.

Solving the Problem

Think What is the question?

Explore How can you decide what animal to put next? How are the animals placed in line? How are they alike and different? Which ones have fur? Which do not? What patterns do you see?

Solve Where would you put the fox and the elephant? Explain.

Look Back Did you follow the pattern?

Share Your Ideas

1. **What if** we knew that the snake was a cobra? Look at the first letter in each animal's name. Where would the elephant and fox go if Mr. Hill used this pattern?

THINK
EXPLORE
SOLVE
LOOK BACK

Practice

Solve. Use a calculator where appropriate.

CHOICES

Complete each pattern for 2–3.

2. 10, 12, 16, 22, ____, ____

3. A, Z, B, Y, C, ____, ____

Use the table to solve 4 and 5.

4. You arrive at the airport at 10:15 P.M. At what time does the next shuttle to Washington leave?

5. You arrive at the airport at 2:00 P.M. At what time does the next shuttle to Boston leave?

Shuttle planes to Washington:
Planes leave every hour starting at 7:00 A.M.
Last plane leaves at 9:00 P.M.

Shuttle planes to Boston:
Planes leave every two hours starting at 7:15 A.M.
Last plane leaves at 9:15 P.M.

Mixed Strategy Review

6. Storage room A has 2 cartons of soap powder. Storage room B has 3 cartons of soap powder. Each carton has 4 boxes of soap powder. How many cartons are there in both storage rooms?

7. Eric had 8 animal posters. He bought 3 more posters. Now he has more than Steve. How many posters does Steve have?

Create Your Own
Make a plane schedule. Show a pattern in it. Write a problem, using your schedule.

Use the information below to solve 8 and 9.

In the World Series, Clark hit one home run, two doubles, and three singles. Johnson hit three home runs and two doubles. Evans hit one triple and five singles.

8. A single is a one-base hit, a double is a two-base hit, a triple is a three-base hit, and a home run is a four-base hit. Which player had the most total bases?

9. Which player got the most hits?

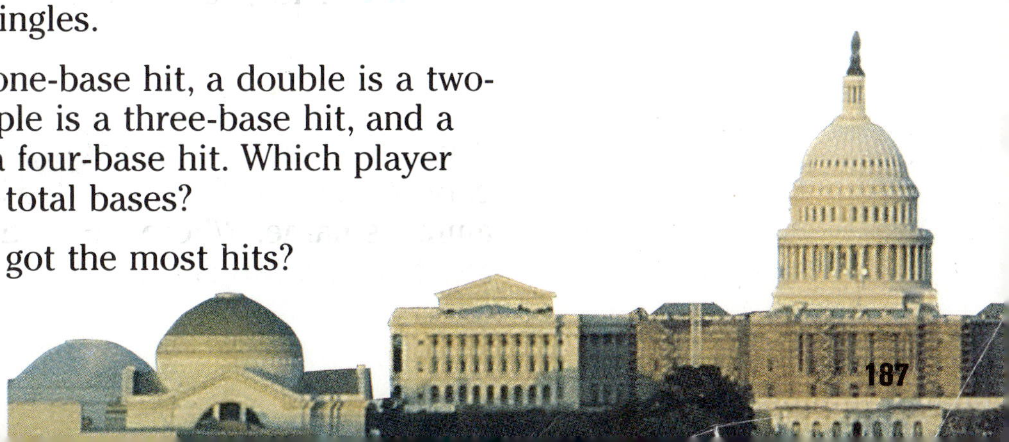

GETTING STARTED

Choose some numbers. Name their doubles.

4 as a Factor

How many owls are there?

$4 \times 3 = \square$

There are 4 groups of 3 owls.

$4 \times 3 = 12$ $\quad \begin{array}{r} 3 \\ \times 4 \\ \hline 12 \end{array}$

There are 12 owls.

You can also use doubles. 4 is the double of 2.

Think $2 \times 3 = 6$
4×3 is double 2×3
So $4 \times 3 = 6 + 6$
$4 \times 3 = 12$

When you multiply with 4, you can think of a multiplication fact with 2 and double it.

What if you want to find 6×4?
You can use the Order Property.

Think $6 \times 4 = 4 \times 6$
Now 4×6 is double 2×6
$2 \times 6 = 12$
So $4 \times 6 = 12 + 12$
$6 \times 4 = 24$

$\begin{array}{r} 6 \\ \times 4 \\ \hline 24 \end{array} \quad \begin{array}{r} 4 \\ \times 6 \\ \hline 24 \end{array}$

Check Your Understanding

Find each product.

1. $\begin{array}{r} 7 \\ \times 4 \\ \hline \end{array}$
2. $\begin{array}{r} 1 \\ \times 4 \\ \hline \end{array}$
3. $\begin{array}{r} 2 \\ \times 4 \\ \hline \end{array}$
4. $\begin{array}{r} 4 \\ \times 3 \\ \hline \end{array}$
5. $\begin{array}{r} 4 \\ \times 5 \\ \hline \end{array}$
6. $\begin{array}{r} 9 \\ \times 4 \\ \hline \end{array}$
7. $\begin{array}{r} 8 \\ \times 4 \\ \hline \end{array}$
8. $4 \times 6 = \square$
9. $7 \times 4 = \square$
10. $9 \times 4 = \square$
11. $4 \times 4 = \square$

Share Your Ideas Explain how multiplication facts with 2 as a factor help you to know multiplication facts with 4 as a factor.

Practice

Multiply.

12. $\begin{array}{r} 2 \\ \times 4 \\ \hline \end{array}$ 13. $\begin{array}{r} 5 \\ \times 4 \\ \hline \end{array}$ 14. $\begin{array}{r} 4 \\ \times 4 \\ \hline \end{array}$ 15. $\begin{array}{r} 4 \\ \times 3 \\ \hline \end{array}$ 16. $\begin{array}{r} 4 \\ \times 1 \\ \hline \end{array}$ 17. $\begin{array}{r} 0 \\ \times 4 \\ \hline \end{array}$ 18. $\begin{array}{r} 6 \\ \times 4 \\ \hline \end{array}$

19. $\begin{array}{r} 4 \\ \times 7 \\ \hline \end{array}$ 20. $\begin{array}{r} 4 \\ \times 0 \\ \hline \end{array}$ 21. $\begin{array}{r} 3 \\ \times 3 \\ \hline \end{array}$ 22. $\begin{array}{r} 9 \\ \times 4 \\ \hline \end{array}$ 23. $\begin{array}{r} 2 \\ \times 2 \\ \hline \end{array}$ 24. $\begin{array}{r} 8 \\ \times 4 \\ \hline \end{array}$ 25. $\begin{array}{r} 7 \\ \times 4 \\ \hline \end{array}$

26. $\begin{array}{r} 6 \\ \times 2 \\ \hline \end{array}$ 27. $\begin{array}{r} 4 \\ \times 5 \\ \hline \end{array}$ 28. $\begin{array}{r} 4 \\ \times 2 \\ \hline \end{array}$ 29. $\begin{array}{r} 3 \\ \times 4 \\ \hline \end{array}$ 30. $\begin{array}{r} 5 \\ \times 3 \\ \hline \end{array}$ 31. $\begin{array}{r} 1 \\ \times 4 \\ \hline \end{array}$ 32. $\begin{array}{r} 4 \\ \times 8 \\ \hline \end{array}$

33. $4 \times 6 = \square$ 34. $4 \times 9 = \square$ 35. $8 \times 4 = \square$ 36. $3 \times 9 = \square$

37. $\square = 4 \times 1$ 38. $\square = 7 \times 4$ 39. $\square = 4 \times 4$ 40. $\square = 9 \times 4$

Choose the correct answer.

41. $(4 \times 5) + (4 \times 5) = \square$
    a. 20 b. 40
    c. 9 d. 18

42. $(2 \times 6) + (2 \times 6) = \square$
    a. 12 b. 8×6
    c. 4×6 d. 16

Think and Apply

43. Make up a funny math problem about owls in the forest. Use the fact $4 \times 7 = 28$.

44. There are 4 groups of 6 owls and 5 groups of 4 hawks. Are there more owls or hawks? Explain.

45. **Look back** at **33–40**. Where 4 is a factor, are the products odd or even? Explain.

Logical Thinking

Are there more sides in:

46. 5 squares or 7 triangles?
47. 7 squares or 8 triangles?
48. 3 squares or 4 triangles?

Explain two different ways to find the product for $4 \times 8 = \square$.

SUMMING UP

GETTING STARTED

Trace your hand four times. Find how many fingers there are in all. Tell how you found out.

5 as a Factor

Joey drew 4 imaginary animals. Each animal had 5 eyes. How many eyes did he draw?

Count by fives:

5	10	15	20
1 group of 5	2 groups of 5	3 groups of 5	4 groups of 5

$4 \times 5 = 20$

Joey drew 20 eyes in all.

Multiplying by fives is like counting by fives. What patterns do you see in the products?

Now predict the products for other groups of 5. Use 5, 6, 7, 8, and 9 groups of 5.

Explain how you can find these products.

$5 \times 1 = \square$	$5 \times 4 = \square$	$5 \times 7 = \square$
$5 \times 2 = \square$	$5 \times 5 = \square$	$5 \times 8 = \square$
$5 \times 3 = \square$	$5 \times 6 = \square$	$5 \times 9 = \square$

Check Your Understanding

Multiply.

1. 5×2 **2.** 5×6 **3.** 4×5 **4.** 8×5 **5.** 5×0 **6.** 1×5 **7.** 5×5

Share Your Ideas Explain how counting by fives helps you to know multiplication facts with 5.

Practice

Multiply.

8. $\begin{array}{r} 5 \\ \times 3 \\ \hline \end{array}$ 9. $\begin{array}{r} 5 \\ \times 2 \\ \hline \end{array}$ 10. $\begin{array}{r} 4 \\ \times 5 \\ \hline \end{array}$ 11. $\begin{array}{r} 5 \\ \times 5 \\ \hline \end{array}$ 12. $\begin{array}{r} 6 \\ \times 5 \\ \hline \end{array}$ 13. $\begin{array}{r} 5 \\ \times 1 \\ \hline \end{array}$ 14. $\begin{array}{r} 4 \\ \times 4 \\ \hline \end{array}$

15. $\begin{array}{r} 7 \\ \times 5 \\ \hline \end{array}$ 16. $\begin{array}{r} 8 \\ \times 3 \\ \hline \end{array}$ 17. $\begin{array}{r} 5 \\ \times 8 \\ \hline \end{array}$ 18. $\begin{array}{r} 0 \\ \times 5 \\ \hline \end{array}$ 19. $\begin{array}{r} 3 \\ \times 5 \\ \hline \end{array}$ 20. $\begin{array}{r} 1 \\ \times 5 \\ \hline \end{array}$ 21. $\begin{array}{r} 5 \\ \times 4 \\ \hline \end{array}$

22. $\begin{array}{r} 5 \\ \times 9 \\ \hline \end{array}$ 23. $\begin{array}{r} 2 \\ \times 5 \\ \hline \end{array}$ 24. $\begin{array}{r} 0 \\ \times 1 \\ \hline \end{array}$ 25. $\begin{array}{r} 5 \\ \times 6 \\ \hline \end{array}$ 26. $\begin{array}{r} 5 \\ \times 5 \\ \hline \end{array}$ 27. $\begin{array}{r} 9 \\ \times 5 \\ \hline \end{array}$ 28. $\begin{array}{r} 5 \\ \times 0 \\ \hline \end{array}$

29. $6 \times 5 = \square$ 30. $9 \times 5 = \square$ 31. $7 \times 5 = \square$ 32. $5 \times 5 = \square$

33. $\square = 5 \times 8$ 34. $\square = 2 \times 6$ 35. $5 \times \square = 20$ 36. $\square \times 5 = 45$

Complete the table.

	Number of Pentagons ⬠	0	1	2	3	4	5	6	7	8	9
37.	Number of Sides in All										

Think and Apply

38. Draw a picture to show that 8 rows of 5 animals and 5 rows of 8 animals are equal.

39. **What if** you drew 8 animals with 5 eyes each? How many eyes would you have drawn?

40. Emma is drawing imaginary animals. The first animal has 1 eye. The second has 3 eyes. The third has 5 eyes. If the pattern continues, how many eyes will the next animal have?

Test Taker

Sometimes you can save time by first finding the answers that are not possible. Which answers are not possible? Now choose the correct answer.

$9 \times 5 = \square$

a. 5 **b.** 54 **c.** 500 **d.** 45

a and **c** are not possible. **a** is too small. **c** is too large. **b** and **d** are possible. **d** is correct.

41. Write an example in which 2 answers are not possible.

SUMMING UP

Look back at **8–11**. What is the digit in the ones place of each product? Write a rule that tells about the product when you multiply with 5.

GETTING STARTED

Name some everyday things that come in equal groups. Think about foods, clothing, toys, things for the house.

0 Through 5 as Factors

The Nature Club explores the park for birds' nests. They find 5 blue jay nests with 4 eggs in each. Then they see 3 cardinal nests with 5 eggs in each and 4 woodcock nests with 3 eggs in each. If all the eggs hatch, how many baby birds will there be?

Multiply to find how many of each.

5 × 4 = 20 blue jays

3 × 5 = 15 cardinals

4 × 3 = 12 woodcocks

Now add to find how many baby birds there will be.

20 + 15 + 12 = 47

If all the eggs hatch, there will be 47 baby birds.

Check Your Understanding

Multiply.

1. $\begin{array}{r} 7 \\ \times 1 \\ \hline \end{array}$ **2.** $\begin{array}{r} 8 \\ \times 2 \\ \hline \end{array}$ **3.** $\begin{array}{r} 9 \\ \times 3 \\ \hline \end{array}$ **4.** $\begin{array}{r} 5 \\ \times 6 \\ \hline \end{array}$ **5.** $\begin{array}{r} 9 \\ \times 2 \\ \hline \end{array}$ **6.** $\begin{array}{r} 0 \\ \times 6 \\ \hline \end{array}$ **7.** $\begin{array}{r} 1 \\ \times 9 \\ \hline \end{array}$

8. 0 × 8 = □ **9.** 5 × 9 = □ **10.** 7 × 0 = □ **11.** 0 × 0 = □

Share Your Ideas Look back at **1–11**. When did you use the rules for multiplying with 1 and 0? When did you use doubles?

Practice

Multiply.

12. $\begin{array}{r} 4 \\ \times 1 \\ \hline \end{array}$

13. $\begin{array}{r} 2 \\ \times 2 \\ \hline \end{array}$

14. $\begin{array}{r} 3 \\ \times 8 \\ \hline \end{array}$

15. $\begin{array}{r} 0 \\ \times 5 \\ \hline \end{array}$

16. $\begin{array}{r} 6 \\ \times 3 \\ \hline \end{array}$

17. $\begin{array}{r} 5 \\ \times 7 \\ \hline \end{array}$

18. $\begin{array}{r} 4 \\ \times 5 \\ \hline \end{array}$

19. $\begin{array}{r} 2 \\ \times 4 \\ \hline \end{array}$

20. $7 \times 2 = \square$

21. $1 \times 9 = \square$

22. $\square = 4 \times 4$

23. $\square = 8 \times 5$

24. $\square = 3 \times 5$

25. $\square = 2 \times 3$

Compare. Use >, <, or = for each ⬬. Explain your thinking.

26. 2×9 ⬬ 4×9

27. 5×7 ⬬ 3×7

28. 6×5 ⬬ 7×5

Think and Apply

29. Ted has 3 pages of bird pictures. Each page has 2 rows. Each row has 4 pictures. How many pictures are there in all?

30. Lori had 36 seeds to display. She made 2 rows of 18 seeds. How else could she arrange the seeds with the same number in each row?

Look back at **15**, **20**, and **21**. Make a drawing to show each.

SUMMING UP

Mixed Review

1. $\begin{array}{r} 5 \\ +6 \\ \hline \end{array}$

2. $\begin{array}{r} 16 \\ -\ 7 \\ \hline \end{array}$

3. $\begin{array}{r} 18 \\ -\ 9 \\ \hline \end{array}$

4. $\begin{array}{r} 9 \\ +6 \\ \hline \end{array}$

5. $\begin{array}{r} 79 \\ +67 \\ \hline \end{array}$

6. $\begin{array}{r} 37 \\ -19 \\ \hline \end{array}$

7. $\begin{array}{r} 64 \\ -25 \\ \hline \end{array}$

8. $\begin{array}{r} 46 \\ +58 \\ \hline \end{array}$

9. $\begin{array}{r} 398 \\ -109 \\ \hline \end{array}$

10. $\begin{array}{r} 467 \\ +358 \\ \hline \end{array}$

11. $\begin{array}{r} 5,270 \\ +6,360 \\ \hline \end{array}$

12. $\begin{array}{r} 4,612 \\ -2,166 \\ \hline \end{array}$

13. $\begin{array}{r} \$62.79 \\ +\ \ 4.98 \\ \hline \end{array}$

14. $\begin{array}{r} \$50.08 \\ -\ \ 6.18 \\ \hline \end{array}$

15. $\begin{array}{r} 38 \\ 19 \\ +27 \\ \hline \end{array}$

16. $\begin{array}{r} 86 \\ 19 \\ +\ 7 \\ \hline \end{array}$

17. $\begin{array}{r} 31 \\ 10 \\ +99 \\ \hline \end{array}$

18. $\begin{array}{r} 72 \\ 15 \\ +\ 9 \\ \hline \end{array}$

Find the value of the digit 6.

19. 46
20. 1,609
21. 89,063
22. 26,528
23. 61,540
24. 679,134

Using Problem Solving

How Many Animals?

The gray fox is the only fox that climbs trees. Baby gray foxes are called kits. A mother gray fox usually has from 3 to 5 kits at a time.

Solve these problems about gray foxes.

A. A ranger knows that 5 gray foxes have had kits. Two of the mothers had 3 kits each. Three of the mothers had 4 kits each. How many new kits are there?

B. There are 6 mother foxes in another area. The ranger knows that each mother had at least 3 kits. Also, no mother fox had more than 5 kits. What is the least number of new kits that could be in the area? What is the greatest number of new kits that could be in the area?

C. Three mother foxes had kits. There were 14 new kits. No mother had more than 5 new kits. How many mothers could have had 3 kits? How many mothers could have had 4 kits? How many mothers could have had 5 kits?

Sharing Your Ideas

1. Look at problem B about 6 mother gray foxes. How were you able to find the number of the fewest kits? How were you able to find the number of the most kits?

2. Were you able to find the answers to these questions by only multiplying? Explain.

THINK
EXPLORE
SOLVE
LOOK BACK

Practice

The red fox has between 1 and 10 kits in a litter.

3. There are four mother red foxes in an area. What are the most kits they might have? What are the fewest kits they might have?

4. One scientist thought that the four red foxes would have 8 kits. Another thought they would have 24 kits. Who do you think made the better prediction? Explain.

5. In another area there were 5 mother red foxes. Altogether they had 26 kits. Each had more than 4 kits. Four mothers had the same number of kits. The other had one more kit than the rest. How many kits did each mother have?

6. Four mother red foxes have 6 kits each. This is the same number of kits as six mother red foxes would have if each had how many kits?

7. Think of two other puzzles about animals and share them with a partner.

Summing Up

8. **Look back** at **4.** What prediction might you make?

9. How did you use multiplication to solve these problems?

Chapter Review

Find each product. pages 172–183, 188–193

1. $\begin{array}{r} 5 \\ \times 4 \\ \hline \end{array}$ 2. $\begin{array}{r} 4 \\ \times 4 \\ \hline \end{array}$ 3. $\begin{array}{r} 5 \\ \times 1 \\ \hline \end{array}$ 4. $\begin{array}{r} 2 \\ \times 2 \\ \hline \end{array}$ 5. $\begin{array}{r} 2 \\ \times 0 \\ \hline \end{array}$ 6. $\begin{array}{r} 1 \\ \times 3 \\ \hline \end{array}$ 7. $\begin{array}{r} 2 \\ \times 1 \\ \hline \end{array}$

8. $\begin{array}{r} 0 \\ \times 4 \\ \hline \end{array}$ 9. $\begin{array}{r} 4 \\ \times 3 \\ \hline \end{array}$ 10. $\begin{array}{r} 5 \\ \times 2 \\ \hline \end{array}$ 11. $\begin{array}{r} 3 \\ \times 5 \\ \hline \end{array}$ 12. $\begin{array}{r} 5 \\ \times 5 \\ \hline \end{array}$ 13. $\begin{array}{r} 4 \\ \times 2 \\ \hline \end{array}$ 14. $\begin{array}{r} 3 \\ \times 3 \\ \hline \end{array}$

15. $\begin{array}{r} 6 \\ \times 5 \\ \hline \end{array}$ 16. $\begin{array}{r} 8 \\ \times 3 \\ \hline \end{array}$ 17. $\begin{array}{r} 3 \\ \times 9 \\ \hline \end{array}$ 18. $\begin{array}{r} 7 \\ \times 3 \\ \hline \end{array}$ 19. $\begin{array}{r} 4 \\ \times 6 \\ \hline \end{array}$ 20. $\begin{array}{r} 0 \\ \times 0 \\ \hline \end{array}$ 21. $\begin{array}{r} 7 \\ \times 5 \\ \hline \end{array}$

Multiply.

22. $5 \times 9 = \square$ 23. $4 \times 8 = \square$ 24. $3 \times 7 = \square$ 25. $2 \times 5 = \square$

26. $2 \times 6 = \square$ 27. $5 \times 8 = \square$ 28. $4 \times 7 = \square$ 29. $3 \times 6 = \square$

30. $\square = 7 \times 5$ 31. $\square = 6 \times 4$ 32. $\square = 9 \times 3$ 33. $\square = 8 \times 2$

34. $2 \times 4 = \square$ / $4 \times 2 = \square$ 35. $4 \times 7 = \square$ / $7 \times 4 = \square$ 36. $6 \times 1 = \square$ / $1 \times 6 = \square$ 37. $8 \times 5 = \square$ / $5 \times 8 = \square$

38. $\square = 3 \times 4$ / $\square = 4 \times 3$ 39. $\square = 8 \times 3$ / $\square = 3 \times 8$ 40. $\square = 2 \times 9$ / $\square = 9 \times 2$ 41. $\square = 0 \times 7$ / $\square = 7 \times 0$

Choose the correct words to complete the sentence.

42. In multiplication, you _____ two _____ to get a _____.

Words to Know
multiply factors product

Solve. pages 185–187

43. Elephants have 4 legs. How many legs do 5 elephants have?

44. Lee is drawing creatures. The first creature has 2 legs. The second has 4 legs. The third has 6 legs. If Lee follows his pattern, how many legs will the fourth creature have?

Chapter Test

Multiply.

1. $\begin{array}{r} 7 \\ \times 2 \\ \hline \end{array}$ 2. $\begin{array}{r} 3 \\ \times 0 \\ \hline \end{array}$ 3. $\begin{array}{r} 2 \\ \times 8 \\ \hline \end{array}$ 4. $\begin{array}{r} 3 \\ \times 6 \\ \hline \end{array}$ 5. $\begin{array}{r} 3 \\ \times 3 \\ \hline \end{array}$ 6. $\begin{array}{r} 2 \\ \times 2 \\ \hline \end{array}$

7. $\begin{array}{r} 3 \\ \times 5 \\ \hline \end{array}$ 8. $\begin{array}{r} 5 \\ \times 4 \\ \hline \end{array}$ 9. $\begin{array}{r} 4 \\ \times 1 \\ \hline \end{array}$ 10. $\begin{array}{r} 8 \\ \times 3 \\ \hline \end{array}$ 11. $\begin{array}{r} 9 \\ \times 4 \\ \hline \end{array}$ 12. $\begin{array}{r} 5 \\ \times 9 \\ \hline \end{array}$

13. $\begin{array}{r} 5 \\ \times 6 \\ \hline \end{array}$ 14. $\begin{array}{r} 9 \\ \times 3 \\ \hline \end{array}$ 15. $\begin{array}{r} 4 \\ \times 7 \\ \hline \end{array}$ 16. $\begin{array}{r} 5 \\ \times 5 \\ \hline \end{array}$ 17. $\begin{array}{r} 8 \\ \times 1 \\ \hline \end{array}$ 18. $\begin{array}{r} 0 \\ \times 9 \\ \hline \end{array}$

19. $0 \times 6 = \square$ 20. $4 \times 3 = \square$ 21. $2 \times 2 = \square$ 22. $8 \times 5 = \square$

Find each product.

23. $5 \times 7 = \square$ 24. $2 \times 8 = \square$ 25. $0 \times 2 = \square$ 26. $3 \times 1 = \square$

$7 \times 5 = \square$ $8 \times 2 = \square$ $2 \times 0 = \square$ $1 \times 3 = \square$

27. $\square = 4 \times 6$ 28. $\square = 3 \times 7$ 29. $\square = 8 \times 4$ 30. $\square = 6 \times 3$

$\square = 6 \times 4$ $\square = 7 \times 3$ $\square = 4 \times 8$ $\square = 3 \times 6$

Solve.

31. A squirrel hid 3 acorns in each of 4 places. Then it hid 2 acorns in each of 5 places. How many acorns did it hide altogether?

32. Jeff is drawing spotted cats. The first cat has 2 spots. The second has 4 spots. The third has 8 spots. If Jeff follows his pattern, how many spots will the fourth cat have?

THINK How many dinosaurs will fit in 5 boxes?

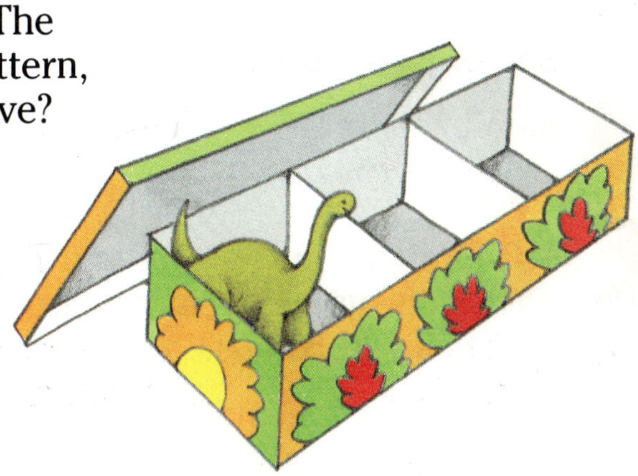

EXTENSION

Making Tables and Diagrams

What if you have 2 pairs of slacks and 3 T-shirts? How many different outfits can you make?

A table can help you to organize your thinking.

1. Copy and complete the Table of Outfits.
2. Draw each possible outfit in the proper space.
3. How many outfits did you make?
4. What do you notice about the number of slacks, shirts, and outfits?
5. Is it true that

TABLE OF OUTFITS

Numbers of Slacks	×	Number of Shirts	=	Number of Outfits	?

A tree diagram can also show the combinations.

The red slacks with each T-shirt:

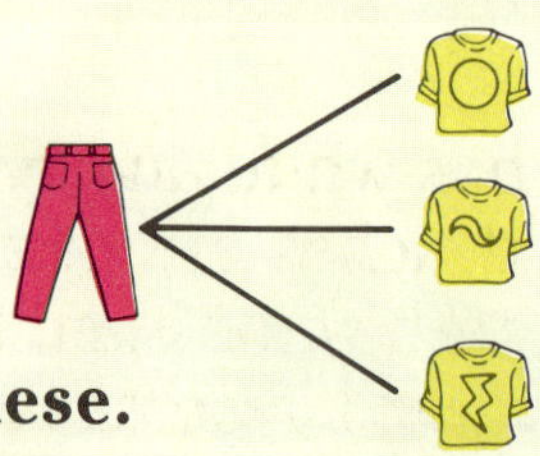

6 outfits in all

The blue slacks with each T-shirt:

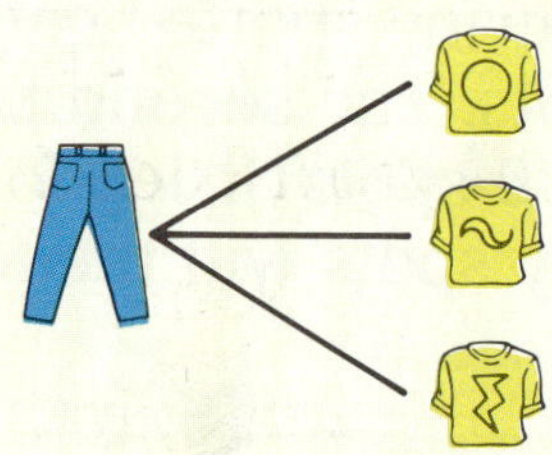

Try these.

6. How many outfits can you make with 3 skirts and 4 sweaters? 2 pairs of slacks and 5 tops?
7. How can you find the number of outfits without making tables or diagrams?

Maintaining Skills

Choose the correct answer. Write A, B, C, or D.

1. $\begin{array}{r} 97 \\ -36 \\ \hline \end{array}$

   A 133   **C** 61
   B 51   **D** not given

2. $62.42 − $18.65 = □

   A $56.23   **C** $54.87
   B $43.77   **D** not given

3. What time will it be in 12 minutes?

   A 4:28   **C** 4:46
   B 4:18   **D** not given

Use the pictograph to solve 4–5.

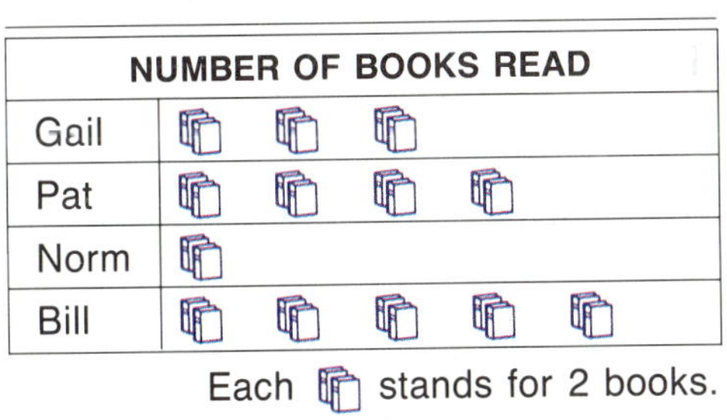

4. Who read the most books?

   A Gail   **C** Norm
   B Pat   **D** not given

5. How many books did Norm read?

   A 2   **C** 4
   B 3   **D** not given

6. $7 \times 3 = \square$

   A 20   **C** 27
   B 21   **D** not given

7. $0 \times 7 = \square$

   A 1   **C** 0
   B 7   **D** not given

8. $\begin{array}{r} 4 \\ \times 8 \\ \hline \end{array}$

   A 32   **C** 24
   B 36   **D** not given

Solve.

9. Bill stencils racing cars with the digits 3, 4, and 5. Each car can have one or two digits. How many different cars can he label?

   A 12   **C** 9
   B 3   **D** not given

10. The store has white and rye bread. Sandwiches are made with tuna, turkey, or ham. Each sandwich uses 1 type of bread and 1 type of filling. How many different kinds of sandwiches can be made?

   A 3   **C** 6
   B 5   **D** not given

7 Multiplication Facts 6–9

THEME Mountain Adventure

Sharing What You Know

Brrr! It's cold outside! You know the Arctic is cold. You know the Antarctic is freezing, too. But did you know that when you climb up a mountain, the temperature goes down? Mountain climbers have to know about temperature and distance as they plan for a climb. What else would they have to know? How could mathematics help them plan?

Using Language

Mountain climbers often see tracks in the snow. They can tell from the **pattern** of tracks if they are following people, animals, or machines. In mathematics, a **pattern** is an arrangement of numbers or shapes according to a plan. A pattern can help you skip count or multiply. What follows in the pattern 3, 6, 9, . . . ?

Words to Know: factors, product, square numbers, order property, patterns, grouping property

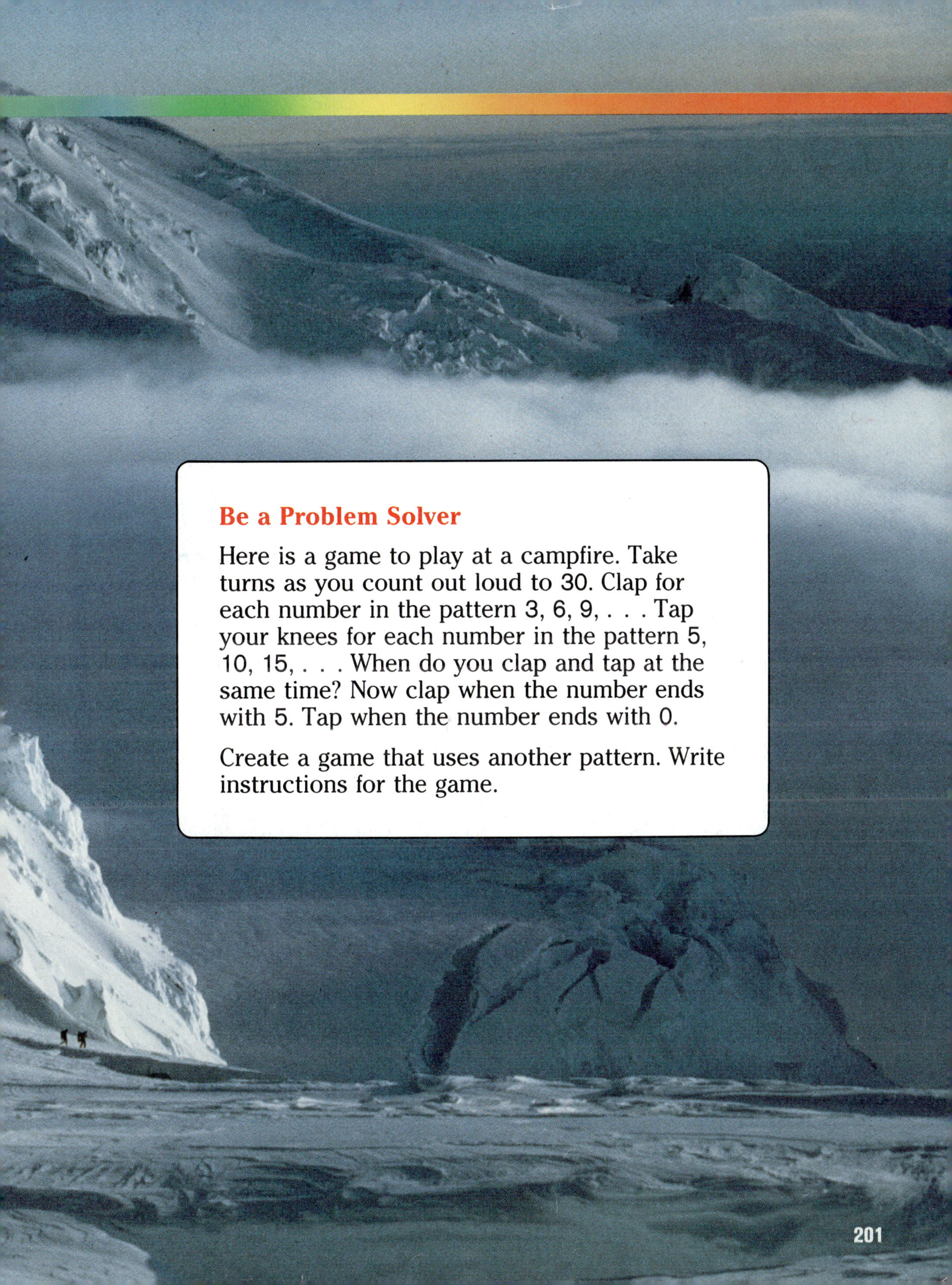

Be a Problem Solver

Here is a game to play at a campfire. Take turns as you count out loud to 30. Clap for each number in the pattern 3, 6, 9, . . . Tap your knees for each number in the pattern 5, 10, 15, . . . When do you clap and tap at the same time? Now clap when the number ends with 5. Tap when the number ends with 0.

Create a game that uses another pattern. Write instructions for the game.

Activity

Exploring Square Numbers

Some numbers are called square numbers. Explore why they have this name.

Working together

Record your work.
Materials: 30 square tiles, Workmat 9

A. Make several rows of 2 tiles each. How many rows of 2 tiles form a square? How many tiles did you need to form the square?

Number of Rows	Number in Each Row	Total Number	Multiplication Fact

B. Explore. Make squares with rows of 3, 4, and 5 tiles.

C. How many tiles would you need to form a square with rows of one tile?

Sharing Your Results

Look back at your record.

1. What do you notice about the number of rows and the number of tiles in each row? Now write a multiplication fact to show the total number for each square.

2. How can you continue this pattern to make other squares?

Extending the Activity

When you recorded the number of tiles in each square, you wrote a square number.

16 is a square number.
25 is a square number.

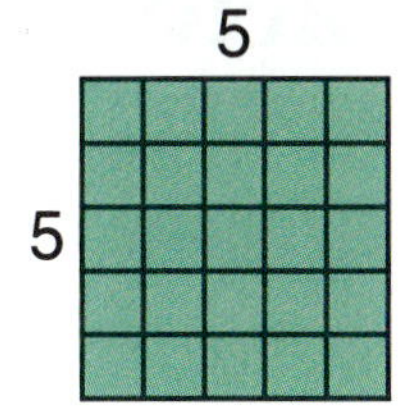

$5 \times 5 = 25$

When both factors are the same, the product is called a **square number.**

3. Name the other square numbers on your workmat.

4. Why do you think these numbers are called square numbers?

5. Complete the workmat for rows of 6, 7, 8, and 9 tiles. Write a multiplication fact for each. What new square numbers have you found?

Summing Up

You can use a calculator to find other square numbers.

6. Try $10 \times 10 = \square$

$11 \times 11 = \square$

$12 \times 12 = \square$

Record the new square numbers.

7. Now try any numbers you wish. Record your new square numbers.

8. Share your square numbers with your classmates. Make a class display showing some of the square numbers you found.

GETTING STARTED

Use mental math. Double each number.
3 9 12 6

6 as a Factor

Ed saw deer. There were 6 groups with 4 deer in each. How many deer were there?

6 groups of **4**

$$\begin{array}{r} 4 \\ \times 6 \\ \hline 24 \end{array}$$

6 × 4 = 24

There were 24 deer.

You can also use doubles.
6 is the double of 3.

Think $3 \times 4 = 12$
So $6 \times 4 = 12 + 12$
$6 \times 4 = 24$

When you multiply with 6, you can think of a multiplication fact with 3 and double it.

What if you want to find 7×6?
You can use the Order Property.

Think $7 \times 6 = 6 \times 7$
Now 6×7 is double 3×7.
$3 \times 7 = 21$
So $6 \times 7 = 21 + 21$
$6 \times 7 = 42$ and $7 \times 6 = 42$

$$\begin{array}{r} 7 \\ \times 6 \\ \hline 42 \end{array} \quad \begin{array}{r} 6 \\ \times 7 \\ \hline 42 \end{array}$$

Check Your Understanding

Find each product.

1. $\begin{array}{r} 5 \\ \times 6 \\ \hline \end{array}$ **2.** $\begin{array}{r} 2 \\ \times 6 \\ \hline \end{array}$ **3.** $\begin{array}{r} 8 \\ \times 6 \\ \hline \end{array}$ **4.** $\begin{array}{r} 6 \\ \times 1 \\ \hline \end{array}$ **5.** $\begin{array}{r} 6 \\ \times 3 \\ \hline \end{array}$ **6.** $\begin{array}{r} 0 \\ \times 6 \\ \hline \end{array}$ **7.** $\begin{array}{r} 7 \\ \times 6 \\ \hline \end{array}$

8. $6 \times 9 = \square$ **9.** $6 \times 6 = \square$ **10.** $4 \times 6 = \square$ **11.** $2 \times 6 = \square$

Share Your Ideas How can knowing $3 \times 8 = 24$ help you find the product of 6×8?

Practice

Find each product.

12. $\begin{array}{r} 0 \\ \times 6 \\ \hline \end{array}$ 13. $\begin{array}{r} 6 \\ \times 3 \\ \hline \end{array}$ 14. $\begin{array}{r} 6 \\ \times 6 \\ \hline \end{array}$ 15. $\begin{array}{r} 6 \\ \times 7 \\ \hline \end{array}$

16. $\begin{array}{r} 6 \\ \times 9 \\ \hline \end{array}$ 17. $\begin{array}{r} 3 \\ \times 7 \\ \hline \end{array}$ 18. $\begin{array}{r} 6 \\ \times 2 \\ \hline \end{array}$ 19. $\begin{array}{r} 1 \\ \times 6 \\ \hline \end{array}$

20. $\begin{array}{r} 5 \\ \times 6 \\ \hline \end{array}$ 21. $\begin{array}{r} 8 \\ \times 6 \\ \hline \end{array}$ 22. $\begin{array}{r} 9 \\ \times 6 \\ \hline \end{array}$ 23. $\begin{array}{r} 6 \\ \times 4 \\ \hline \end{array}$

24. $6 \times 6 = \square$ 25. $0 \times 6 = \square$

26. $\square = 6 \times 1$ 27. $\square = 5 \times 9$

28. $\square = 6 \times 8$ 29. $\square = 6 \times 5$

Complete. Write the rule if not given.

Rule: Multiply by 6.

	Input	Output
30.	3	
31.	1	
32.	7	
33.	9	

Rule:

	Input	Output
	15	30
	21	42
34.		24
35.		48

Think and Apply

36. At the top of the mountain there were 6 groups of 8 wild flowers. Draw a picture to show this grouping. Then write a multiplication fact for your picture.

37. Jan pressed her wild flowers. She put 2 in the first row, 4 in the second row, and 6 in the third row. If she continues this pattern, how many flowers will she put in the fifth row?

Write an exciting story about finding something while climbing a mountain. Use the fact $6 \times 5 = 30$.

SUMMING UP

Mixed Review

1. $\begin{array}{r} 17 \\ +\ 8 \\ \hline \end{array}$ 2. $\begin{array}{r} 25 \\ -\ 9 \\ \hline \end{array}$

3. $\begin{array}{r} 30 \\ +16 \\ \hline \end{array}$ 4. $\begin{array}{r} 46 \\ -23 \\ \hline \end{array}$

5. $\begin{array}{r} 80 \\ -19 \\ \hline \end{array}$ 6. $\begin{array}{r} 50 \\ +48 \\ \hline \end{array}$

7. $\begin{array}{r} 299 \\ +146 \\ \hline \end{array}$ 8. $\begin{array}{r} 350 \\ -298 \\ \hline \end{array}$

9. $\begin{array}{r} 29 \\ 16 \\ +32 \\ \hline \end{array}$ 10. $\begin{array}{r} 47 \\ 63 \\ +85 \\ \hline \end{array}$

11. $\begin{array}{r} 3,642 \\ +9,205 \\ \hline \end{array}$

12. $\begin{array}{r} 2,000 \\ -1,060 \\ \hline \end{array}$

13. $\begin{array}{r} \$95.42 \\ -\ 16.20 \\ \hline \end{array}$

Round to the nearest ten.

14. 82 15. 47

16. 73 17. 59

18. 65 19. 94

Round to the nearest hundred.

20. 113 21. 783

22. 668 23. 524

24. 851 25. 349

GETTING STARTED

How many days are there in one week? in two weeks? in three weeks? Write a multiplication sentence to show each.

7 and 8 as Factors

Some campers picked flowers. Other campers found fossils. How many flowers did they pick? How many fossils did they find?

Flowers

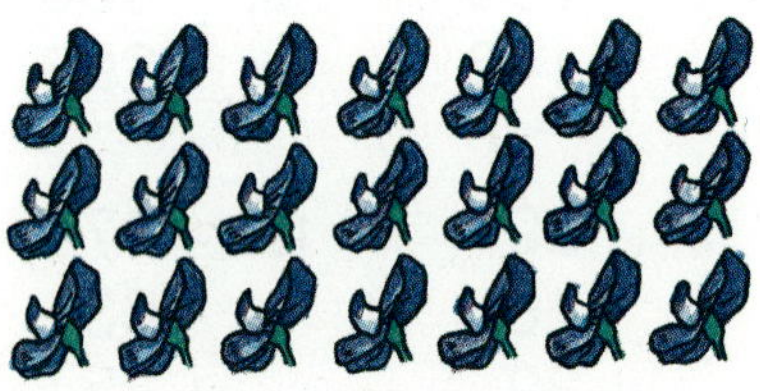

3 rows of **7**

$3 \times 7 = 21$

They picked 21 flowers.

Fossils

2 rows of **8**

$2 \times 8 = 16$

They found 16 fossils.

You already know these facts:

7	7	7	7	7	7	7	
×0	×1	×2	×3	×4	×5	×6	
0	7	14	21	28	35	42	
8	8	8	8	8	8	8	8
×0	×1	×2	×3	×4	×5	×6	×7
0	8	16	24	32	40	48	56

The new facts are:

7	7	7
×7	×8	×9
49	56	63
8	8	
×8	×9	
64	72	

Check Your Understanding

Multiply.

1. 7×2 **2.** 3×7 **3.** 4×8 **4.** 8×5 **5.** 7×4 **6.** 6×7 **7.** 8×6

Share Your Ideas Make a drawing to show 4×8. How can you use the drawing to find 8×8?

Practice

Multiply.

8. 2 ×8	**9.** 4 ×8	**10.** 8 ×3	**11.** 5 ×8	**12.** 8 ×1	**13.** 7 ×5	**14.** 0 ×7
15. 3 ×7	**16.** 6 ×8	**17.** 8 ×7	**18.** 7 ×7	**19.** 6 ×4	**20.** 5 ×7	**21.** 4 ×7
22. 7 ×0	**23.** 3 ×8	**24.** 8 ×9	**25.** 6 ×4	**26.** 0 ×8	**27.** 7 ×2	**28.** 8 ×4

29. 7 × 6 = □ **30.** 8 × 9 = □ **31.** 7 × 9 = □ **32.** 1 × 5 = □

33. □ = 1 × 7 **34.** □ = 0 × 8 **35.** □ = 7 × 6 **36.** □ = 8 × 8

Compare. Use >, <, or = for ●.

37. 8 × 7 ● 40 **38.** 7 × 6 ● 76 **39.** 7 × 0 ● 7

40. 5 × 8 ● 8 × 6 **41.** 9 × 7 ● 60 + 3 **42.** 8 × 6 ● 9 × 8

Complete the table.

43.

Number of Weeks	0	1	2	3	4	5	6	7	8	9
Number of Days	0	7								

Think and Apply

44. Sal is collecting fossils for 8 kits. He needs 6 different types of fossils to put in each kit. How many fossils does he need?

45. The museum has fossils in bin A for 21¢, in bin B for 28¢, and in bin C for 35¢. If the pattern continues, predict the cost of fossils in bin D.

Logical Thinking

Play pick a pair.

46. Pick a pair of numbers with a product that ends in 6.

47. Pick a pair with a product of 24 and a sum of 11.

Explain how knowing other multiplication facts helps you to know multiplication facts with 7 and 8.

SUMMING UP

GETTING STARTED

How do you already know that 9 × 1 = 9? What other facts with nine do you also know?

9 as a Factor

Justin collects rocks. At the top of the hill, he found shiny rocks for his collection. How many rocks did he put in the box?

3 rows of **9**

3 × 9 = 27

He put 27 rocks in the box.

You already know these facts with 9:

9	9	9	9	9	9	9	9	9
×0	×1	×2	×3	×4	×5	×6	×7	×8
0	9	18	27	36	45	54	63	72

The only new fact is:

$$\begin{array}{r} 9 \\ \times 9 \\ \hline 81 \end{array}$$

What is the sum of the digits in each product? What other patterns do you see?

Check Your Understanding

Find each product.

1. $\begin{array}{r} 9 \\ \times 2 \\ \hline \end{array}$ 2. $\begin{array}{r} 9 \\ \times 5 \\ \hline \end{array}$ 3. $\begin{array}{r} 9 \\ \times 4 \\ \hline \end{array}$ 4. $\begin{array}{r} 9 \\ \times 7 \\ \hline \end{array}$ 5. $\begin{array}{r} 1 \\ \times 9 \\ \hline \end{array}$ 6. $\begin{array}{r} 3 \\ \times 9 \\ \hline \end{array}$ 7. $\begin{array}{r} 8 \\ \times 1 \\ \hline \end{array}$

8. 9 × 9 = □ 9. 0 × 9 = □ 10. 6 × 9 = □ 11. 9 × 8 = □

Share Your Ideas If you know 8 × 9 = 72, how can you find 9 × 9?

Practice

Find each product.

12. $\begin{array}{r} 9 \\ \times 1 \\ \hline \end{array}$	**13.** $\begin{array}{r} 6 \\ \times 0 \\ \hline \end{array}$	**14.** $\begin{array}{r} 2 \\ \times 9 \\ \hline \end{array}$	**15.** $\begin{array}{r} 3 \\ \times 4 \\ \hline \end{array}$	**16.** $\begin{array}{r} 9 \\ \times 3 \\ \hline \end{array}$	**17.** $\begin{array}{r} 5 \\ \times 9 \\ \hline \end{array}$	**18.** $\begin{array}{r} 6 \\ \times 9 \\ \hline \end{array}$
19. $\begin{array}{r} 8 \\ \times 4 \\ \hline \end{array}$	**20.** $\begin{array}{r} 4 \\ \times 9 \\ \hline \end{array}$	**21.** $\begin{array}{r} 9 \\ \times 9 \\ \hline \end{array}$	**22.** $\begin{array}{r} 9 \\ \times 6 \\ \hline \end{array}$	**23.** $\begin{array}{r} 7 \\ \times 5 \\ \hline \end{array}$	**24.** $\begin{array}{r} 9 \\ \times 8 \\ \hline \end{array}$	**25.** $\begin{array}{r} 7 \\ \times 9 \\ \hline \end{array}$
26. $\begin{array}{r} 9 \\ \times 0 \\ \hline \end{array}$	**27.** $\begin{array}{r} 8 \\ \times 3 \\ \hline \end{array}$	**28.** $\begin{array}{r} 3 \\ \times 9 \\ \hline \end{array}$	**29.** $\begin{array}{r} 2 \\ \times 9 \\ \hline \end{array}$	**30.** $\begin{array}{r} 4 \\ \times 7 \\ \hline \end{array}$	**31.** $\begin{array}{r} 9 \\ \times 1 \\ \hline \end{array}$	**32.** $\begin{array}{r} 9 \\ \times 5 \\ \hline \end{array}$

33. $3 \times 7 = \square$ **34.** $9 \times 0 = \square$ **35.** $7 \times 9 = \square$ **36.** $9 \times 6 = \square$

37. $2 \times 9 = \square$ **38.** $5 \times 5 = \square$ **39.** $4 \times 9 = \square$ **40.** $8 \times 9 = \square$

41. $\square = 9 \times 7$ **42.** $\square = 3 \times 9$ **43.** $\square = 6 \times 7$ **44.** $\square = 1 \times 9$

Choose the correct answer.

45. $(3 \times 9) + (3 \times 9) = \square$

a. 27 **b.** 24
c. 60 **d.** 54

46. $(2 \times 9) + (2 \times 9) + (3 \times 9) = \square$

a. 18 **b.** 2×9
c. 7×9 **d.** 6×9

Think and Apply

47. Pam found 9 rocks. Sandy found three times as many rocks as Pam did. How many rocks did Sandy find?

48. Dave collected 38 leaves. Vic had 17 leaves. How many more leaves than Vic did Dave have? How many leaves do they have?

Visual Thinking

These hands show products of facts with 9.

$3 \times 9 =$ 2 7

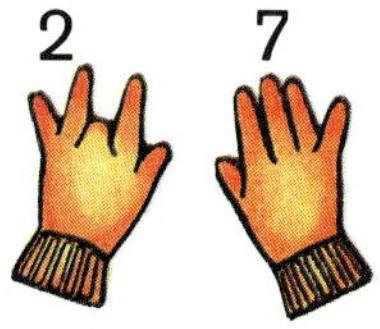

$7 \times 9 =$ 6 3

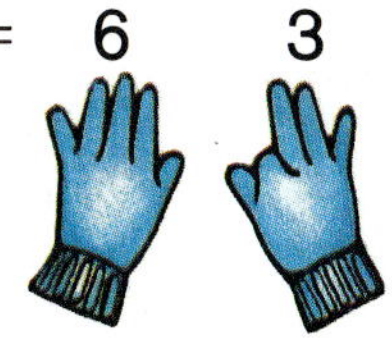

Use your hands to complete each multiplication fact.

49. $4 \times 9 = \square$ **50.** $8 \times 9 = \square$

Explain why you already know most of the facts with 9.

SUMMING UP

Midchapter Review

Find each product. pages 202–209

1. 4×4 **2.** 2×2 **3.** 8×8 **4.** 3×3 **5.** 5×5 **6.** 7×7 **7.** 0×0

8. 1×1 **9.** 6×6 **10.** 9×9 **11.** 6×7 **12.** 9×4 **13.** 6×8 **14.** 9×7

15. 9×8 **16.** 7×8 **17.** 5×7 **18.** 6×9 **19.** 8×9 **20.** 8×7 **21.** 5×8

22. $9 \times 7 = \square$ **23.** $6 \times 5 = \square$ **24.** $6 \times 9 = \square$ **25.** $4 \times 8 = \square$

26. $8 \times 3 = \square$ **27.** $9 \times 4 = \square$ **28.** $6 \times 8 = \square$ **29.** $9 \times 5 = \square$

30. $\square = 3 \times 7$ **31.** $\square = 6 \times 7$ **32.** $\square = 9 \times 3$ **33.** $\square = 7 \times 4$

Choose the correct words to complete each sentence.

Words to Know
square number order product

34. When both factors are the same, the product is a _____.

35. The facts $6 \times 9 = 54$ and $9 \times 6 = 54$ show that you can change the _____ of the factors without changing the _____.

Solve.

36. The mountain climbers climbed in groups of 4. There were 7 groups. How many climbers were there?

37. One mountain climber carried a pack with 8 meals in it. Another climber carried four times as many meals. How many meals were there in all?

Exploring Problem Solving

THINK
EXPLORE
SOLVE
LOOK BACK

Who Wins the Race?

Many people try to climb the mountains on the planet Venta. The mountains are very high and the winds are very strong. On Monday morning, Pat and Juan started a race to see who could reach the top of a Venta mountain first. Each day Pat climbed 2 miles. Then a storm drove her back 1 mile. She camped for the night. Juan climbed 3 miles each day before a storm drove him back 2 miles. Then he camped for the night. The trail to the top of the mountain is 10 miles long.

Thinking Critically

Who won the race? On what day was the race won?

Analyzing and Making Decisions

1. How long is the race? How does a climber win the race?
2. How far did each person climb by the end of the first day? Where were they before the storm drove them back?
3. What happened the second day? Did one person climb higher? Where were they after the storm?
4. Who reached the top of the mountain first? Explain. On what day was the race won?

Look Back What if they raced 20 miles? Who would win? Explain.

Problem Solving Strategies

Using Patterns

Gregor wants to climb Mount Evermost. He wants to be very strong, so he has started to do push-ups. On Monday, he did 20 push-ups. On Tuesday, he did 26. "That's wonderful!" he thought. "I'm sure I'll be able to do 50 push-ups by next Monday." On Wednesday, Gregor did 31 push-ups, and on Thursday he did 35 push-ups. "Hmmm," thought Gregor, "there's a pattern here."

What was the pattern? Did Gregor do 50 push-ups the next Monday if he followed the pattern?

The sequence of numbers in some problems often shows a pattern. Finding the pattern can help you solve the problem.

Solving the Problem

Think What are the questions? What are the facts? One way to show these facts it to make a table.

Day	Push-ups
Mon.	20
Tues.	26
Wed.	31
Thurs.	35
Fri.	

Explore How many push-ups did Gregor do in each of the first four days? How does the number of push-ups change from day to day? What is the pattern?

Solve How many push-ups will he do on Friday, Saturday, Sunday, and Monday? Will he be able to do 50 by Monday? Explain.

Look Back Did you find the answer? When did you know what the answer would be?

Share Your Ideas

1. **What if** Gregor had done 30 push-ups on Monday? How many would he have been able to do by next Monday?

THINK
EXPLORE
SOLVE
LOOK BACK

Practice

Solve. Use a calculator where appropriate.
Use this information to solve 2 and 3.

CHOICES

Edmund and Hillary are preparing to climb a mountain. They plan to reach a main camping area every third day.

2. On which day will they reach the third main camping area?

3. They plan to reach the top of the mountain two days after they reach their fourth main camping area. On what day will they reach the top of the mountain?

Use this information to solve 4 and 5.

Halley's comet appears every 76 years. It appeared in 1986.

4. In what year before 1986 did it last appear?

5. In what year will Halley's comet appear next?

Mixed Strategy Review

6. Marie wants to climb these mountains: Mt. Russell, 14,086 ft; Mt. Castle, 14,269 ft; Mt. Eiger 13,025 ft; and Mt. Meru, 14,979 ft. If she climbs them from the shortest to the tallest, in what order does she climb them?

Make a drawing to help you solve 7.

7. Pat and Sue hike 2 miles an hour. Steve and Jamie hike 3 miles an hour. They are all hiking on the same trail. Pat and Sue leave one hour before Steve and Jamie. How much time will go by before they meet?

Create Your Own

Write a problem which has a pattern in it.

GETTING STARTED

Use mental math! Find these products as quickly as you can.

9×3 8×9 6×8 7×7

Using Mental Math in Multiplication

Astronomers at Mt. Palomar sometimes need calculators and computers in their study of space.

When do you need a calculator? When is mental math a faster way to find an answer?

Multiplication Patterns and Strategies

When 0 is a factor, the product is 0.	When 1 is a factor, the product is the other factor.
When 2, 4, 6, or 8 are factors, think of doubles.	For some factors, think of the order property. $3 \times 9 = 9 \times 3$

Try this experiment.

- Predict whether you can do 20 multiplication facts faster by using mental math or by using a calculator.
- Then carry out an experiment to find out.
- Discuss your results with your classmates.

Check Your Understanding

Use mental math to find each product.

1. 6×0 **2.** 1×7 **3.** 9×2 **4.** 5×4 **5.** 3×6 **6.** 4×8 **7.** 6×9

Share Your Ideas Which patterns or strategies could be used for **1–4**?

Practice

Use mental math to find each product.

8.	9.	10.	11.	12.	13.	14.
$\begin{array}{r} 8 \\ \times 1 \\ \hline \end{array}$	$\begin{array}{r} 0 \\ \times 7 \\ \hline \end{array}$	$\begin{array}{r} 6 \\ \times 2 \\ \hline \end{array}$	$\begin{array}{r} 7 \\ \times 4 \\ \hline \end{array}$	$\begin{array}{r} 9 \\ \times 6 \\ \hline \end{array}$	$\begin{array}{r} 8 \\ \times 8 \\ \hline \end{array}$	$\begin{array}{r} 8 \\ \times 9 \\ \hline \end{array}$

15.	16.	17.	18.	19.	20.	21.
$\begin{array}{r} 6 \\ \times 4 \\ \hline \end{array}$	$\begin{array}{r} 7 \\ \times 6 \\ \hline \end{array}$	$\begin{array}{r} 5 \\ \times 8 \\ \hline \end{array}$	$\begin{array}{r} 7 \\ \times 9 \\ \hline \end{array}$	$\begin{array}{r} 9 \\ \times 0 \\ \hline \end{array}$	$\begin{array}{r} 1 \\ \times 6 \\ \hline \end{array}$	$\begin{array}{r} 8 \\ \times 4 \\ \hline \end{array}$

22. $2 \times 8 = \square$ **23.** $4 \times 8 = \square$ **24.** $9 \times 5 = \square$ **25.** $8 \times 3 = \square$

26. $\square = 0 \times 7$ **27.** $\square = 9 \times 9$ **28.** $\square = 6 \times 3$ **29.** $\square = 4 \times 9$

Decide which has the greater product.

30. 8×6 or 4×6

31. 3×8 or 5×8

32. 4×7 or 9×7

33. 2×9 or 3×9

Use mental math to find each product.

34. $1 \times 9 \times 1 \times 9 \times 1 = \square$

35. $0 \times 8 \times 8 \times 8 \times 8 = \square$

36. $6 \times 6 \times 1 \times 0 \times 6 = \square$

37. $7 \times 1 \times 1 \times 7 \times 1 = \square$

Think and Apply

38. Use a calculator. Start with 0 and add 7 again and again. Record each result. How is adding 7 again and again like multiplying with 7?

DATA **40.** Take a survey. Find out how long it takes each of your classmates to do 20 multiplication facts. What was the most frequent time?

39. Estimate the number of plants in the arrangement below.

Explain how you would multiply 8×9 using mental math and using a calculator.

SUMMING UP

GETTING STARTED

Look for tables in your school books or almanacs. Discuss how the information is organized in the tables you find.

Patterns on the Multiplication Table

The highest point on the Atlantic Coast is Cadillac Mountain. **What if** 7 groups of 3 people picnicked there one day? How many people picnicked?

7 × 3 = ☐

You can use a table to find multiplication facts.

To find 7 × 3 in the table.

- Find the 7 row.
- Find the 3 column.
- The product is where the 7 row and 3 column meet.

21 people picnicked.

A table can help you see multiplication patterns. Which rows have all even products? all odd products? Now find a pattern in the 5 row. What other patterns do you see?

Column ↓ (3); Row → (7)

×	0	1	2	3	4	5	6	7	8	9
0	0	0	0	0	0	0	0	0	0	0
1	0	1	2	3	4	5	6	7	8	9
2	0	2	4	6	8	10	12	14	16	18
3	0	3	6	9	12	15	18	21	24	27
4	0	4	8	12	16	20	24	28	32	36
5	0	5	10	15	20	25	30	35	40	45
6	0	6	12	18	24	30	36	42	48	54
7	0	7	14	21	28	35	42	49	56	63
8	0	8	16	24	32	40	48	56	64	72
9	0	9	18	27	36	45	54	63	72	81

Check Your Understanding

Find each product in the table.

1. 4×3
2. 6×5
3. 2×7
4. 3×9
5. 8×4
6. 9×2
7. 9×9

Share Your Ideas Explain why the product 12 appears more than once on the multiplication table.

Practice

Multiply.

8. 0×3 **9.** 8×6 **10.** 7×2 **11.** 8×4

12. 1×9 **13.** 4×5 **14.** 3×7 **15.** 4×3

16. 9×6 **17.** 7×9 **18.** 6×5 **19.** 7×7

20. $8 \times 9 = \square$ **21.** $9 \times 3 = \square$

22. $\square = 6 \times 6$ **23.** $\square = 8 \times 4$

Use the multiplication table.
Find all the facts that

24. have a product of 18.

25. have a product of 24.

26. have a product of 36.

27. have a product of 48.

28. end in 5.

29. are squares.

Think and Apply

30. At one picnic area there were 3 tables of 8 people. There were 4 tables of 5 people at another area. How many people were there in all?

31. Each day a park ranger visits 8 picnic areas. She keeps a record of her total visits. For which number of days will she write an odd number?

Number of Days	1	2	3	4	5	6	7
Number of Visits	8						

Make a class display to show the patterns you found on the multiplication table. Add to it as you discover new patterns.

SUMMING UP

Mixed Review

1. $9 + 9$ **2.** $8 + 8$

3. $18 - 18$ **4.** $65 - 29$

5. $201 - 145$ **6.** $499 + 580$

7. $975 + 421$ **8.** $300 - 10$

9. $260 + 950$ **10.** $\$.85 - .16$

Write what time it will be.

11. in 3 hours

12. in 15 minutes

13. in 1 hour

14. in 12 hours

Activity

Exploring Multiplying Three Numbers

Does it matter which way numbers are grouped to multiply? What do you think? Explore to find out. Try $3 \times 2 \times 4 = \square$.

Working together

Materials: 50 counters

Record your work in a table like this.

Multiplication Example	Counters in All
$(3 \times 2) \times 4 = \square$	
$3 \times (2 \times 4) = \square$	

A. Numbers are multiplied two at a time. Group the first two factors and multiply them.

$$(3 \times 2) \times 4 = \square$$
$$\underbrace{3 \times 2}_{6}$$

Show 6 groups of 4 counters to find 6×4.

B. Now group the factors a different way. Group the second two factors and multiply them.

$$3 \times (2 \times 4) = \square$$
$$\underbrace{2 \times 4}_{8}$$

Show 3 groups of 8 counters to find 3×8.

C. Try $3 \times 2 \times 2 = \square$. Group the factors in two different ways. Show the product with counters each time.

Sharing Your Results

Look back at your table.

1. How do the two products for $3 \times 2 \times 4$ compare? How do the two products for $3 \times 2 \times 2$ compare?

2. Does it matter which way three numbers are grouped when you multiply? Explain.

Extending the Activity

> **The Grouping Property**
> The way in which numbers are grouped does not change the product.

Use this property as you find these products. Group the factors in any way.

3. $2 \times 3 \times 2 = \square$
4. $5 \times 1 \times 2 = \square$
5. $1 \times 3 \times 3 = \square$
6. $5 \times 1 \times 3 = \square$
7. $2 \times 3 \times 3 = \square$
8. $2 \times 1 \times 4 = \square$

Summing Up

9. **What if** you have to multiply four factors?

$$3 \times 1 \times 2 \times 4 = \square$$

How can the grouping property be used to solve a multiplication like this? Explore to find out.

10. How is the grouping property for multiplication like the grouping property for addition? How is it different?

11. Create your own multiplications with four factors. Find each product. Use a calculator if you like.

Using Problem Solving

Interview: Calculators in Conservation

Pat Kane works for the New Jersey Audubon Society. This is a conservation organization. Pat said, "I help protect the wildlife in the mountains and forests. I gather and record data about the wildlife. The calculator helps me to find totals and to plan."

A. What if you helped to protect wildlife? During a long, harsh winter, you had to set up feeding stations. On each of 3 days, you had to set up 6 stations at 8 locations. How many stations would you have set up?

Use a calculator to multiply:

3 × 6 × 8 = 144

You would have set up 144 feeding stations.

B. Explain how you would use a calculator to multiply $8 \times 9 \times 7$.

Sharing Your Ideas

Pretend you are helping Pat Kane.

Use a calculator to help you solve each problem.

1. You collected data about the deer population. You traveled 9 miles a day. You did this 5 days a week for 4 weeks. How many miles did you travel?

2. You photographed the deer as you traveled. You used 3 boxes of film. Each box had 4 rolls with 12 pictures each. How many pictures did you take?

THINK
EXPLORE
SOLVE
LOOK BACK

Practice

Find each product. Use a calculator to help you.

3. $6 \times 7 \times 8 = \square$	**4.** $9 \times 2 \times 8 = \square$
5. $7 \times 7 \times 9 = \square$	**6.** $6 \times 8 \times 9 = \square$
7. $9 \times 7 \times 5 = \square$	**8.** $8 \times 7 \times 9 = \square$
9. $8 \times 8 \times 6 = \square$	**10.** $6 \times 8 \times 8 = \square$

Solve.

11. You need to find the number of deer in a region covering 6 square miles. Your data shows that there are 32 deer for each square mile. What is the total number of deer in the region? Explain how pressing [6] [×] [32] [=] will give the answer.

12. When a ground squirrel hibernates for the winter, its heartbeat slows down to 20 beats per minute. How many times does its heart beat in 1 hour? in 1 day? Explain why pressing [60] [×] [20] [=] and [60] [×] [20] [×] [24] [=] will give the answers.

13. The normal heartbeat of a ground squirrel is about 200 beats per minute. How many beats is that per hour? per day? Explain how to find the answer on a calculator.

14. You are organizing a tree planting project. **What if** 7 helpers can each plant 10 trees on each of 3 days. How many trees can they plant?

Summing Up

15. How many times does your heart beat in one day? Design an experiment to find out. Use a calculator to help you.

16. Look back at **9** and **10**. What do you notice about the products? Does the order in which numbers are multiplied matter? Explain. Use a calculator. Explore with greater numbers.

Chapter Review

Find each product. pages 202–209, 214–217

1. 4×7	2. 6×6	3. 9×4	4. 9×8	5. 5×5	6. 9×6	7. 5×8
8. 2×2	9. 7×6	10. 1×1	11. 5×7	12. 8×9	13. 8×8	14. 9×3
15. 7×8	16. 6×9	17. 4×4	18. 7×9	19. 9×5	20. 3×3	21. 4×8
22. 9×9	23. 9×7	24. 4×6	25. 7×7	26. 0×0	27. 8×7	28. 6×8

29. $5 \times 8 = \square$ 30. $8 \times 7 = \square$ 31. $9 \times 9 = \square$ 32. $7 \times 9 = \square$

33. $\square = 7 \times 5$ 34. $\square = 6 \times 6$ 35. $\square = 5 \times 9$ 36. $\square = 8 \times 9$

Multiply. pages 218–219

37. $3 \times 3 \times 3 = \square$ 38. $2 \times 3 \times 2 = \square$ 39. $3 \times 2 \times 4 = \square$

Choose the correct word to complete each sentence.

In $5 \times 8 = 40$,

40. 5 is a ____.

41. 8 is a ____.

42. 40 is the ____.

Words to Know
factor product

Solve. pages 211–213

43. Six scientists climbed Mount Washington. Each one brought back 9 samples of plants. How many samples did they bring back altogether?

44. The hikers climbed for 5 minutes. Then they climbed for 8 minutes. Later they climbed for 11 minutes. If they continue this pattern, how many minutes will they climb the next time?

Chapter Test

Multiply.

1. $\begin{array}{r} 6 \\ \times 3 \\ \hline \end{array}$ 2. $\begin{array}{r} 8 \\ \times 8 \\ \hline \end{array}$ 3. $\begin{array}{r} 5 \\ \times 9 \\ \hline \end{array}$ 4. $\begin{array}{r} 4 \\ \times 4 \\ \hline \end{array}$ 5. $\begin{array}{r} 7 \\ \times 8 \\ \hline \end{array}$

6. $\begin{array}{r} 8 \\ \times 4 \\ \hline \end{array}$ 7. $\begin{array}{r} 0 \\ \times 0 \\ \hline \end{array}$ 8. $\begin{array}{r} 6 \\ \times 6 \\ \hline \end{array}$ 9. $\begin{array}{r} 6 \\ \times 7 \\ \hline \end{array}$ 10. $\begin{array}{r} 9 \\ \times 6 \\ \hline \end{array}$

11. $\begin{array}{r} 7 \\ \times 7 \\ \hline \end{array}$ 12. $\begin{array}{r} 7 \\ \times 9 \\ \hline \end{array}$ 13. $\begin{array}{r} 1 \\ \times 1 \\ \hline \end{array}$ 14. $\begin{array}{r} 3 \\ \times 3 \\ \hline \end{array}$ 15. $\begin{array}{r} 7 \\ \times 6 \\ \hline \end{array}$

16. $\begin{array}{r} 6 \\ \times 8 \\ \hline \end{array}$ 17. $\begin{array}{r} 6 \\ \times 1 \\ \hline \end{array}$ 18. $\begin{array}{r} 8 \\ \times 6 \\ \hline \end{array}$ 19. $\begin{array}{r} 5 \\ \times 5 \\ \hline \end{array}$ 20. $\begin{array}{r} 2 \\ \times 2 \\ \hline \end{array}$

21. $6 \times 5 = \square$ 22. $8 \times 9 = \square$ 23. $7 \times 7 = \square$

24. $9 \times 6 = \square$ 25. $7 \times 6 = \square$ 26. $7 \times 9 = \square$

27. $\square = 9 \times 8$ 28. $\square = 8 \times 7$ 29. $\square = 9 \times 9$

30. $4 \times 2 \times 3 = \square$ 31. $3 \times 2 \times 2 = \square$

Solve.

32. The hikers did 8 sit-ups on the first day, 12 sit-ups on the second day, and 16 sit-ups on the third day. If they continue this pattern, how many sit-ups will they do on the fourth day?

33. The cog railway at Pikes Peak has 3 cars. There are 9 people in each car. How many people are there in all?

THINK The sample box is filled. How many rocks are in the box?

EXTENSION

Odd and Even Patterns

What if you multiply two even numbers? Will you get a product that is odd or even? Make a prediction.

Now experiment to find out. Record your work.

1. Find products of two even numbers.

2. Find products of two odd numbers.

3. Find products of one odd number and one even number.

4. What patterns do you see? Complete each sentence in the chart.

Multiplication Patterns for Odd and Even Numbers
The product of two even numbers is _____.
The product of two odd numbers is _____.
The product of an odd number and an even number is _____.

5. How do these patterns compare with the patterns for addition?

Remember the addition patterns for odd and even numbers.
- The sum of two even numbers is even.
- The sum of two odd numbers is even.
- The sum of an odd number and an even number is odd.

Family Math

In Chapters **5–7** your child has been learning about multiplication facts, time, and graphing. Now he or she can help to plan some fun for your family in this project.

Together plan and schedule a day for your family to do some things you can all enjoy.

A Day for Your Family

Cumulative Review

Choose the correct answer. Write A, B, C, or D.

1. What time is it?

A 11:36 **C** 10:36
B 9:36 **D** not given

2. What time will it be in 14 minutes?

A 4:12 **C** 4:01
B 4:11 **D** not given

Use the grid to solve 3–4.

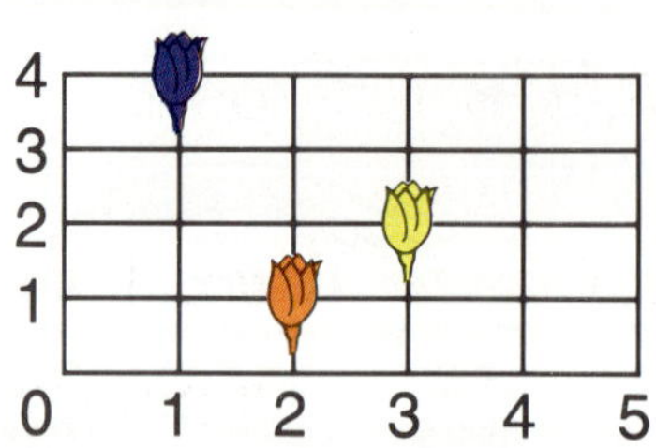

3. Which ordered pair names the location of the orange flower?

A (2, 1) **C** (2, 2)
B (1, 2) **D** not given

4. Which color flower is at (3, 2)?

A red **C** orange
B yellow **D** not given

5. $8 \times 2 = \square$

A 16 **C** 20
B 18 **D** not given

6. $\begin{array}{r} 5 \\ \times 3 \\ \hline \end{array}$

A 15 **C** 20
B 16 **D** not given

7. $9 \times 3 = \square$

A 17 **C** 27
B 28 **D** not given

8. $\begin{array}{r} 4 \\ \times 0 \\ \hline \end{array}$

A 4 **C** 8
B 1 **D** not given

9. $4 \times 7 = \square$

A 28 **C** 24
B 30 **D** not given

10. $5 \times 8 = \square$

A 13 **C** 32
B 40 **D** not given

11. $\begin{array}{r} 5 \\ \times 1 \\ \hline \end{array}$

A 1 **C** 0
B 5 **D** not given

Choose the correct answer. Write A, B, C, or D.

12. $6 \times 7 = \square$

A 38
B 46
C 42
D not given

13. $\begin{array}{r} 9 \\ \times 6 \\ \hline \end{array}$

A 54
B 56
C 15
D not given

14. $4 \times 8 = \square$

A 32
B 36
C 24
D not given

15. $7 \times 7 = \square$

A 42
B 49
C 50
D not given

16. $\begin{array}{r} 7 \\ \times 9 \\ \hline \end{array}$

A 72
B 61
C 63
D not given

17. $2 \times 3 \times 2 = \square$

A 7
B 6
C 10
D not given

18. $4 \times 3 \times 3 = \square$

A 24
B 36
C 21
D not given

Solve.

19. Third-graders play softball and soccer at recess. Twenty-seven children play softball. How many of the 63 children play soccer?

A 36
B 27
C 90
D not enough information

20. Twelve pencils in the box need to be sharpened. How many pencils are in the box?

A 24
B 7
C 12
D not enough information

Complete each pattern.

21. A, A, B, B, C, C, D, D, _____, _____

A D, D
B D, E
C E, E
D not given

22. A, A, B, _____, _____, C, D, D

A A, A
B B, B
C B, C
D not given

23. 5¢, 10¢, _____, 20¢, 25¢

A 11¢
B 15¢
C 19¢
D not given

8 Division Facts

THEME People: Caring and Sharing

Sharing What You Know

Look at the picture. What might the boy and his grandfather be talking about? What might they be sharing? Could they have just finished doing chores together? What do you share with other people? Do you ever share chores?

Using Language

If you wash the dishes and your sister dries them, how are you **dividing** the chores? When you **divide** a chore you share it. You do part and the other person does part. In mathematics, **division** means separating things into equal groups. How is dividing chores like dividing in mathematics? How is it different?

Words to Know: dividend, divisor, quotient, division, division fact, multiplication fact, fact family

Be a Problem Solver

Jim and his grandfather share cleaning chores. Jim says, "I will clean the kitchen if you will clean the living room." His grandfather says, "It usually takes me an hour to clean both. I'll clean for a half hour, then you can finish the job." How would you divide this chore?

List all of the chores that need to be done in your classroom every day. Make a weekly "Get the Job Done" schedule for your class.

Activity

Exploring Division

You have 12 crayons. How many children can have 2 crayons?

Working together

Materials: 12 crayons or counters, Workmat 10

Record your work.

A. Make as many groups of 2 as you can. How many children will get 2 crayons?

B. **What if** you give each child 3 crayons? How many groups of 3 are there?

C. Now make groups of 1, groups of 4, groups of 6, groups of 12.

D. Could you make groups of 7 with no counters left over? Explain.

In All	Number in Each Group	Number of Groups

Sharing Your Results

Look back at your record.

1. When you divided 12 crayons into groups of 2, you made 6 groups. **What if** you started with 6 groups of 2 crayons? How many crayons would you have in all? Explain how multiplication helps you to know.
2. Look at other rows of your table. What multiplication facts can you write?

Extending the Activity

When you make equal groups, you divide. You found that 12 can be divided into groups of 1, 2, 3, 4, 6, and 12 with no crayons left.

Work in a small group.

Materials: 20 crayons or counters

3. Use 7 crayons. Make groups of 2, 3, and 4. What happens? Now try groups of 1, then groups of 7.
4. Now use 13 crayons to make equal groups. Try to make groups of 1, 2, 3, 4 and so on. What happens?

Summing Up

5. What equal groups could you make with 7 crayons? with 13 crayons? When did you have crayons left over?

Use what you know about multiplication.

6. What are some numbers of crayons that can be divided into groups of 2? of 3? of 4?
7. Write a sentence that tells what happens when you divide.

Activity

Understanding Division

What if 2 children wanted to share 12 crayons? How many crayons would each get?

Working together

In All	Number of Groups	Number in Each Group

Materials: 18 crayons or counters, Workmat 11

Record your work.

A. Put 12 crayons into 2 equal groups. How many are in each group?

B. Show how 3 children could share 12 crayons. Make 3 equal groups. Now share the crayons among 4 children. 6 children. Could 5 children share 12 crayons equally? Explain.

C. What if you have 17 crayons? Can 2 children share 17 crayons equally? How many equal groups can you make with 17 crayons?

D. Try other numbers of crayons.

Sharing Your Results

1. Look at your record. Which numbers of crayons could you share only 2 ways?

DATA **Compare this activity with the activity on page 230.**

2. How are they alike? How are they different?
3. What do you notice about the numbers in the tables for each lesson?
4. How can this multiplication sentence help you to divide 12 into 2 equal groups?

$$2 \times \square = 12$$

Extending the Activity

▸ We can divide to find how many groups.

12	÷	2	=	6
↑		↑		↑
in all		in each group		groups

$$2\overline{)12}$$

6 ← groups
2)12 ← in all
↑
in each group

Read Twelve divided by two equals six.

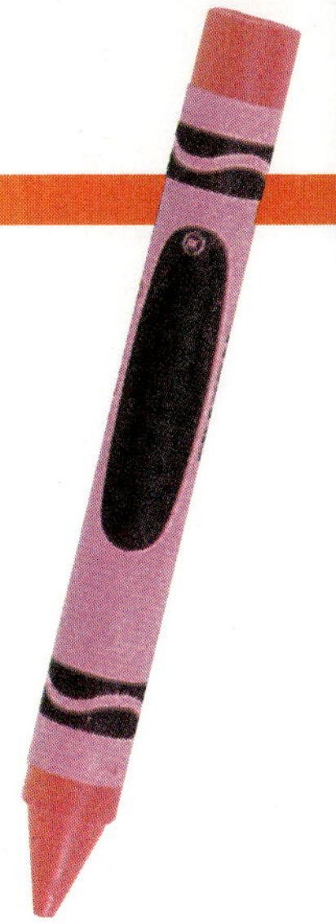

▸ We can divide to find how many in each group.

12	÷	2	=	6
↑		↑		↑
in all		groups		in each group

6 ← in each group
2)12 ← in all
↑
groups

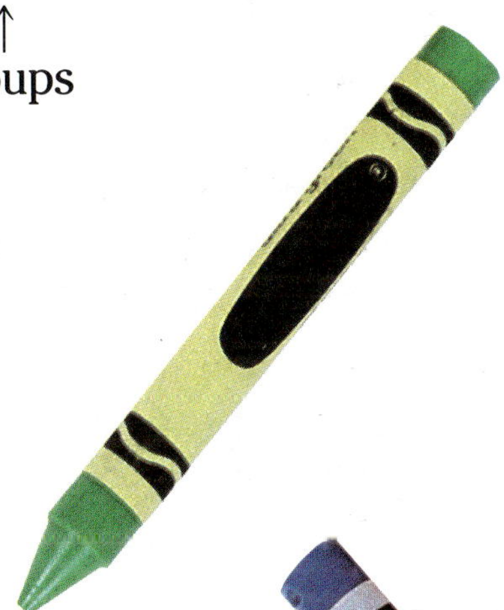

So, 12 ÷ 2 = 6 can mean make groups of 2, or make two equal groups. Either way, the answer is the same.

5. Write the numbers in each row of your table as a division sentence.

6. Write the numbers in each row of your table on page **230** as a division sentence.

7. Tell what each division sentence means.

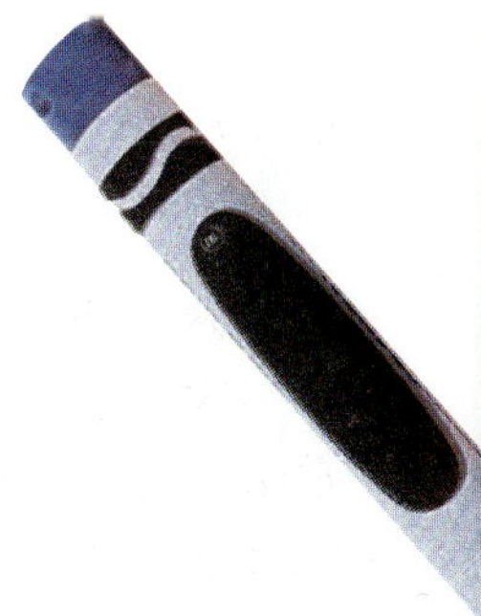

Summing Up

Solve. Then give two meanings for each division sentence.

8. 8 ÷ 4 = ☐ **9.** 6 ÷ 3 = ☐ **10.** 14 ÷ 2 = ☐

11. 10 ÷ 2 = ☐ **12.** 9 ÷ 3 = ☐ **13.** 16 ÷ 4 = ☐

GETTING STARTED

What numbers of pencils can you share equally with a friend? How would you decide how many each person will get?

Dividing by 2

Jan has 8 toy animals on her window seat. She makes 2 equal groups. How many animals are in each group?

Divide to find how many are in each group.

8	÷	2	=	□
animals in all		groups		number in each group

Think 8 animals in 2 equal groups

There are 4 animals in each group.
2 groups of 4 equals 8,
or $2 \times 4 = 8$.

So $8 \div 2 = 4$.

8	÷	2	=	4.
↑ dividend		↑ divisor		↑ quotient

Jan has 4 animals in each group.

Check Your Understanding

Write a division fact for each.

1. 6 animals
   2 equal groups
   How many are in each group?

2. 10 animals
   2 equal groups
   How many are in each group?

Find each quotient.

3. $18 \div 2 = \square$ 4. $16 \div 2 = \square$ 5. $14 \div 2 = \square$ 6. $10 \div 2 = \square$

Share Your Ideas Draw 12 animals in 2 equal groups. How many are in each group?

Practice

Write a division fact for each.

7. 4 animals
   2 equal groups
   How many are in each group?

8. 16 animals
   2 equal groups
   How many are in each group?

Find each quotient.

9. $12 \div 2 = \square$
10. $4 \div 2 = \square$
11. $8 \div 2 = \square$
12. $10 \div 2 = \square$
13. $16 \div 2 = \square$
14. $2 \div 2 = \square$
15. $18 \div 2 = \square$
16. $14 \div 2 = \square$
17. $6 \div 2 = \square$
18. $14 \div 2 = \square$
19. $2 \div 2 = \square$
20. $12 \div 2 = \square$

Complete.

21. Count by twos to 18:
    2, 4, 6, __, __, __, . . .

22. Count backwards by twos:
    18, 16, 14, __, __, __, . . .

Complete. Follow each rule.

Rule: Double the number.

	Input	Output
23.	3	
24.	8	
25.	5	
26.	9	

Rule: Divide in half.

	Input	Output
27.	14	
28.	16	
29.	12	
30.	4	

Think and Apply

31. Jan and her friend shared 10 books equally. How many books did each get?

32. Jan has 14 shoes scattered on the floor. What is the greatest number of pairs there could be?

SUMMING UP

Make a drawing to show 16 books in 2 equal groups. Write a division fact to tell how many are in each group.

GETTING STARTED

Count by threes to 27. How many numbers did you say? How many numbers do you say each time if you count to 18? 21? 30?

Dividing by 3

Lila, Ali, and Kim carry 15 gifts into the house. The gifts are divided equally among the girls. How many gifts is each girl carrying?

15 ÷ 3 = □

Think 15 gifts in 3 equal groups

3 groups of 5 equals 15,
or $3 \times 5 = 15$.

So $15 \div 3 = 5$.

$\overline{)\ }$ and ÷ both mean division.

$$\begin{array}{r} 5 \\ 3\overline{)15} \end{array}$$

5 ← quotient
15 ← dividend
3 ↑ divisor

Each girl is carrying 5 gifts.

Check Your Understanding

Write a division fact for each.

1. 9 gifts
   3 equal groups
   How many are in each group?

2. 12 gifts
   3 equal groups
   How many are in each group?

Divide.

3. 18 ÷ 3 = □
4. 3 ÷ 3 = □
5. 21 ÷ 3 = □
6. 6 ÷ 3 = □
7. $3\overline{)24}$
8. $3\overline{)9}$
9. $3\overline{)21}$
10. $3\overline{)27}$
11. $3\overline{)15}$

Share Your Ideas What if you had 24 gifts? Explain how you would form 3 equal groups.

Practice

Write a division fact for each.

12. 18 gifts
    3 equal groups
    How many are in each group?

13. 21 gifts
    3 equal groups
    How many are in each group?

Divide.

14. $9 \div 3 = \square$
15. $12 \div 3 = \square$
16. $15 \div 3 = \square$
17. $6 \div 3 = \square$
18. $3 \div 3 = \square$
19. $27 \div 3 = \square$
20. $8 \div 2 = \square$
21. $24 \div 3 = \square$
22. $3\overline{)21}$
23. $3\overline{)15}$
24. $3\overline{)3}$
25. $3\overline{)18}$
26. $2\overline{)10}$
27. $3\overline{)24}$
28. $2\overline{)14}$
29. $3\overline{)27}$
30. $3\overline{)6}$
31. $3\overline{)9}$

Compare. Use >, <, or = for ⬬. Choose mental math, paper and pencil, or a calculator. Explain your choices.

CHOICES

32. $12 \div 3$ ⬬ $12 \div 2$
33. $9 \div 3$ ⬬ $18 \div 3$
34. $24 \div 3$ ⬬ $15 \div 3$
35. $6 \div 2$ ⬬ $6 \div 3$
36. $(18 \div 3) - 1$ ⬬ $8 \div 4$
37. $(27 \div 3) + 1$ ⬬ $(14 \div 2) + 3$

Think and Apply

38. Do not count. Estimate about how many gifts there are for 3 children to share equally.

39. Start with 27. How many times must you subtract 3 in order to reach 0? What is $27 \div 3$?

SUMMING UP

Make a drawing to show 21 balloons in 3 equal groups. Write a division fact to tell how many are in each group.

GETTING STARTED

How many groups of 4 counters can you take away from 12 counters? Try it.

Dividing by 4

There are 12 cub scouts going to visit the Seeing Eye Dog School. Only 4 scouts can visit at a time. How many groups of 4 will go?

Divide to find how many groups.

$12 \div 4 = \square$

Think 12 scouts, 4 in a group

There are 3 groups.
3 groups of 4 equals 12.

$3 \times 4 = 12$

So $12 \div 4 = 3$. $\quad 4\overline{)12}$ (quotient 3)

Three groups of 4 cub scouts will go.

Check Your Understanding

Write a divison fact for each.

1. 16 scouts
   4 in each group
   How many groups are there?

2. 24 scouts
   4 in each group
   How many groups are there?

Find each quotient.

3. $8 \div 4 = \square$ 4. $20 \div 4 = \square$ 5. $4 \div 4 = \square$ 6. $32 \div 4 = \square$

Share Your Ideas How would you write $28 \div 4$ using the form $\overline{)}$?

Practice

Write a division fact for each.

7. 8 scouts
   4 in each group
   How many groups are there?

8. 28 scouts
   4 in each group
   How many groups are there?

Find each quotient.

9. $4 \div 4 = \square$
10. $32 \div 4 = \square$
11. $16 \div 4 = \square$
12. $36 \div 4 = \square$
13. $12 \div 4 = \square$
14. $8 \div 4 = \square$
15. $27 \div 3 = \square$
16. $20 \div 4 = \square$
17. $36 \div 4 = \square$
18. $12 \div 3 = \square$
19. $24 \div 4 = \square$
20. $28 \div 4 = \square$

21. $4\overline{)24}$
22. $3\overline{)18}$
23. $4\overline{)28}$
24. $4\overline{)12}$
25. $2\overline{)16}$
26. $4\overline{)16}$
27. $4\overline{)32}$
28. $4\overline{)20}$
29. $3\overline{)9}$
30. $4\overline{)32}$
31. $3\overline{)24}$
32. $4\overline{)12}$
33. $2\overline{)14}$
34. $4\overline{)8}$
35. $4\overline{)36}$

Complete the table.

36.

Number of Cars	1	2	3			6	7
Number of Wheels	4			16	20		

Think and Apply

37. The cub scouts use 3 slices of bread to make each club sandwich for lunch. **What if** the loaf of bread has 25 slices? How many club sandwiches could they make? Explain.

38. Rene can brush 4 dogs in one hour. Can he brush 34 dogs in 8 hours? Explain.

Logical Thinking

Find the mystery numbers.

39. Our sum is 12.
    Our quotient is 2.
    What numbers are we?

40. Our sum is 20.
    Our quotient is 4.
    What numbers are we?

Use the numbers **3**, **4**, and **12** to write two division facts. Explain what each means.

SUMMING UP

GETTING STARTED

Count by fives to 45. How many numbers did you say? Now start at 45. Count backward by fives.

Dividing by 5

At summer camps, children share happy times singing their favorite songs. Sometimes they take turns singing in groups.

Around this campfire, 20 campers are singing in groups of 5. How many groups of 5 are there?

$$20 \div 5 = \square$$

Think 20 campers, 5 in a group
There are 4 groups.

4 groups of 5 equals 20, or $4 \times 5 = 20$.

So $20 \div 5 = 4$. $5\overline{)20}$ with quotient 4

There are 4 groups of campers.

Check Your Understanding

Write a division fact for each.

1. 25 campers
   5 in each group
   How many groups are there?

2. 40 campers
   5 in each group
   How many groups are there?

Divide.

3. $30 \div 5 = \square$
4. $35 \div 5 = \square$
5. $5 \div 5 = \square$
6. $15 \div 5 = \square$
7. $5\overline{)20}$
8. $5\overline{)40}$
9. $5\overline{)10}$
10. $5\overline{)45}$
11. $5\overline{)25}$

Share Your Ideas Write a division fact with 5 as the divisor and 8 as the quotient. What is the dividend? Explain how you know.

Practice

Write a division fact for each.

12. 30 campers
    5 in each group
    How many groups?

13. 45 campers
    5 in each group
    How many groups?

Divide.

14. $15 \div 5 = \square$
15. $10 \div 5 = \square$
16. $5 \div 5 = \square$
17. $20 \div 5 = \square$
18. $40 \div 5 = \square$
19. $24 \div 3 = \square$

20. $5\overline{)35}$
21. $3\overline{)18}$
22. $5\overline{)45}$
23. $4\overline{)36}$
24. $5\overline{)25}$
25. $2\overline{)10}$

Write two division facts for each.

26. 4, 5, 20
27. 3, 5, 15

28. Explore with a calculator.
    - Start with 0.
    - Add 5 again and again until you get to 45.
    - Record each result.
    - Subtract 5 again and again until you get to 0.
    - Record each result.
    - What do you notice about your numbers?

Think and Apply

29. Draw a picture to solve. There are 30 campers in 5 tents. Each tent has the same number of campers. How many campers are in each tent?

Mixed Review

1. $8 + 0$
2. 8×0
3. 6×5
4. $6 + 5$
5. $9 + 4$
6. 9×4
7. $26 + 72$
8. $48 - 25$
9. $65 - 19$
10. $86 + 57$
11. $320 + 655 = \square$
12. $1{,}729 - 98 = \square$
13. $\$2.81 + \$.85 = \square$
14. $\$2.50 - \$.75 = \square$
15. $468 + 392 = \square$
16. $714 - 500 = \square$

Write each time.

17. 

18. 

SUMMING UP

How is counting backwards by fives like dividing by 5?

Midchapter Review

Find each quotient. pages 230–241

1. $14 \div 2 = \square$
2. $12 \div 3 = \square$
3. $8 \div 2 = \square$
4. $16 \div 4 = \square$
5. $2 \div 2 = \square$
6. $24 \div 3 = \square$
7. $20 \div 5 = \square$
8. $28 \div 4 = \square$
9. $10 \div 5 = \square$
10. $32 \div 4 = \square$
11. $18 \div 3 = \square$
12. $35 \div 5 = \square$

Divide. pages 230–241

13. $3\overline{)21}$
14. $2\overline{)18}$
15. $4\overline{)20}$
16. $5\overline{)40}$
17. $3\overline{)27}$
18. $4\overline{)24}$
19. $5\overline{)15}$
20. $2\overline{)16}$
21. $3\overline{)9}$
22. $4\overline{)4}$
23. $5\overline{)25}$
24. $3\overline{)15}$
25. $4\overline{)36}$
26. $5\overline{)20}$
27. $2\overline{)12}$
28. $3\overline{)24}$
29. $4\overline{)28}$
30. $5\overline{)30}$
31. $3\overline{)12}$
32. $4\overline{)8}$

Choose the correct word to complete each sentence.

33. The answer in division is called the ______.
34. The number that we divide by is called the ______.
35. The number that we divide into is called the ______.

Words to Know
divisor dividend quotient

Solve.

36. Mary's job is to walk 10 dogs. She can walk 2 at a time. How many trips will she make?
37. There were 12 campers equally divided among 4 tents. How many campers were in each tent?

Exploring Problem Solving

THINK
EXPLORE
SOLVE
LOOK BACK

Arranging the Flowers

The Farragut family is having a reunion. Jack and Susie are decorating the table. They have 3 vases, 6 roses, 9 carnations, and 12 daisies.

Thinking Critically

What if you were helping Jack and Susie? What are the different ways you could arrange the flowers in the vases?

Analyzing and Making Decisions

1. How many flowers are there altogether? How many vases are there?
2. Try some different arrangements of flowers. Draw a picture or make a table to help you.
3. Can you use all of the daisies and put an equal number of daisies in each of the 3 vases? Can you use all the roses? all the carnations?
4. Suppose you wanted to keep the roses separate from the carnations. What arrangements could you make, using all the flowers?
5. Make a list of at least 5 different arrangements. Which is your favorite? Explain.

Look Back What if you could buy some tulips? How many would you buy? How would you change your favorite arrangement?

Problem Solving Strategies

Two-Step Problems

The children in Mrs. Smith's class are collecting books to share among 4 day-care centers. Sal brought 9 books. Luisa brought 11 books. Mark brought 8 books, and Sally brought 8 books. The class will give the same number of books to each day-care center. How many books will go to each day-care center?

Sometimes you must answer a hidden question before you can solve a problem. You must look for the hidden question first.

Solving the Problem

Think What do you need to find out?

Explore How many books did each child bring to school? How many books in all did the children bring?

Solve How many books will each day-care center receive?

Look Back What was the hidden question? Could you solve the problem without answering the hidden question?

Share Your Ideas

1. **What if** Luisa brought in 15 books? How many books would go to each day-care center?

THINK
EXPLORE
SOLVE
LOOK BACK

Practice

Solve. Use a calculator where appropriate.

CHOICES

2. The Library Club brought 20 books to give to 3 day-care centers. They gave 10 books to the largest center. The remaining books were divided equally between the other two centers. How many did each center receive?

3. Students are collecting bottles to raise money for the library. They have 3 cartons with 6 bottles in each. If each bottle is worth $.10, how much money will they raise?

4. Jim wants to buy a present for his father. The present costs $14. He has saved $4. He makes $2 for every lawn he rakes. How many lawns must he rake in order to have $14?

5. Linda and Ralph are selling lemonade for the Playground Fund. They collected $5 in the morning and $6 in the afternoon. It costs them $3 to make the lemonade. How much money did they raise that day?

Create Your Own

Write a problem in which you must answer a hidden question to answer the problem.

Mixed Strategy Review

Use the information below to solve 6 and 7.

Leon and Patty signed up to bring snacks to the film club meetings every third month.

6. Leon brought snacks in March. What is the next month that he should bring snacks?

7. Patty's first month to bring snacks was February. What are the next two months that she should bring snacks?

GETTING STARTED

Explain how you can find each of these products. 6×1 13×1 80×1 6×0 13×0 80×0

Division with 0 and 1

The Palmer children decorate their holiday cards with flowers. Chris made 5 paper flowers. If he puts 1 flower on each card, how many cards can he decorate?

$5 \div 1 = 5$

Chris can decorate 5 cards.

- Any number divided by 1 is that number.

Susie made 5 flowers. She wants to put 5 flowers on a card. How many cards can she decorate?

$5 \div 5 = 1$

Susie has 1 more card to decorate. She has no flowers left. Can she put 5 flowers on the card? Explain.

$0 \div 5 = 0$

- Any number (except 0) divided by itself is 1.
- Zero divided by any number (except 0) is 0.
- A number cannot be divided by 0.

Check Your Understanding

Use the rules to find each quotient.

1. $8 \div 1 = \square$ **2.** $6 \div 6 = \square$ **3.** $7 \div 7 = \square$ **4.** $9 \div 1 = \square$

5. $0 \div 2 = \square$ **6.** $4 \div 1 = \square$ **7.** $0 \div 6 = \square$ **8.** $9 \div 9 = \square$

Share Your Ideas Look back at **8**. Which rule helped you to find the quotient? Name another exercise that used the same rule.

Practice

Use the rules to find each quotient.

9. 4 ÷ 1 = □	**10.** 7 ÷ 1 = □	**11.** 2 ÷ 1 = □	**12.** 8 ÷ 8 = □
13. 3 ÷ 3 = □	**14.** 1 ÷ 1 = □	**15.** 0 ÷ 9 = □	**16.** 0 ÷ 3 = □
17. 0 ÷ 1 = □	**18.** 2 ÷ 2 = □	**19.** 0 ÷ 4 = □	**20.** 16 ÷ 4 = □
21. $3\overline{)3}$	**22.** $1\overline{)2}$	**23.** $4\overline{)0}$	**24.** $5\overline{)0}$
25. $1\overline{)5}$	**26.** $4\overline{)4}$	**27.** $1\overline{)6}$	**28.** $7\overline{)7}$
29. $8\overline{)8}$	**30.** $6\overline{)0}$	**31.** $2\overline{)12}$	**32.** $1\overline{)8}$

Choose +, −, ×, or ÷ for ●.

33. 7 ● 1 = 8	**34.** 9 ● 9 = 1
35. 4 ● 0 = 0	**36.** 6 ● 1 = 6
37. 8 ● 8 = 0	**38.** 0 ● 7 = 7
39. 3 ● 3 = 1	**40.** 25 ● 0 = 25

Use the patterns you have learned to find each quotient.

41. 49 ÷ 49 = □	**42.** 0 ÷ 17 = □	**43.** 20 ÷ 20 = □
44. 65 ÷ 1 = □	**45.** 0 ÷ 30 = □	**46.** 78 ÷ 1 = □
47. 854 ÷ 1 = □	**48.** 619 ÷ 619 = □	**49.** 0 ÷ 86 = □

Think and Apply

50. Peter has 8 flowers. How many cards can he decorate with 1 flower on each card?

51. Judy has 6 flowers and 6 cards. If she divides the flowers equally among the 6 cards, how many flowers will each card get?

Look back at **47–49**. Explain how you knew each quotient.

SUMMING UP

GETTING STARTED

Name as many multiplication facts as you can with 6 as a factor.

Dividing by 6

The science class collected insects to examine under a microscope. There are 6 microscopes to be shared equally by 24 students. How many students will share each microscope?

24 ÷ 6 = □

Think 6 times what number equals 24?

6 × □ = 24

The missing factor is 4.

6 × 4 = 24

So 24 ÷ 6 = 4

$$6\overline{)24}$$ with 4 above ← quotient; 24 ← dividend; 6 ↑ divisor

Each microscope will be shared by 4 students.

When you divide, you can think of a related multiplication fact.

Check Your Understanding

Find each missing factor. Then find each quotient.

1. 6 × □ = 6
   6 ÷ 6 = □
2. 6 × □ = 30
   30 ÷ 6 = □
3. 6 × □ = 42
   42 ÷ 6 = □
4. 6 × □ = 54
   54 ÷ 6 = □

Find each quotient.

5. $6\overline{)18}$
6. $6\overline{)24}$
7. $6\overline{)48}$
8. $6\overline{)12}$
9. $6\overline{)36}$

Share Your Ideas Look back at **9**. Which multiplication fact could you use to find the quotient?

Practice

Find each missing factor. Then find each quotient.

10. 6 × □ = 24 24 ÷ 6 = □	**11.** 6 × □ = 48 48 ÷ 6 = □	**12.** 6 × □ = 0 0 ÷ 6 = □	**13.** 6 × □ = 18 18 ÷ 6 = □
14. 6 × □ = 36 36 ÷ 6 = □	**15.** 5 × □ = 20 20 ÷ 5 = □	**16.** 6 × □ = 12 12 ÷ 6 = □	**17.** 3 × □ = 24 24 ÷ 3 = □

Solve.

18. $6\overline{)48}$	**19.** $6\overline{)24}$	**20.** $4\overline{)36}$	**21.** $6\overline{)30}$	**22.** $6\overline{)12}$
23. $6\overline{)6}$	**24.** $2\overline{)6}$	**25.** $6\overline{)42}$	**26.** $3\overline{)24}$	**27.** $6\overline{)54}$
28. $3\overline{)18}$	**29.** $6\overline{)0}$	**30.** $4\overline{)32}$	**31.** $6\overline{)18}$	**32.** $6\overline{)36}$

33. 6 × 2 = □ 12 ÷ 6 = □	**34.** 6 × 7 = □ 42 ÷ 6 = □	**35.** 6 × 5 = □ 30 ÷ 6 = □	**36.** 6 × 9 = □ 54 ÷ 6 = □
37. 48 ÷ 6 = □	**38.** 6 ÷ □ = 1	**39.** □ ÷ 6 = 4	**40.** □ ÷ 6 = 3
41. 54 ÷ 6 = □	**42.** 18 ÷ □ = 3	**43.** □ ÷ 6 = 7	**44.** □ ÷ 5 = 6
45. 12 ÷ □ = 2	**46.** 24 ÷ □ = 6	**47.** 6 ÷ □ = 6	**48.** □ ÷ 6 = 6

Choose the correct sentence. Then solve.

49. Jason collected 6 boxes of insects. Each box holds 8 insects. How many insects are there?

a. 6 + 8 = □ **b.** 8 − 6 = □
c. 6 × 8 = □ **d.** 8 + 6 = □

Think and Apply

50. Make a drawing of 30 insects divided equally into 6 boxes. Write a multiplication fact and a division fact for your drawing.

51. Rudy bought 6 new slides for the microscope. He paid 54¢ for them. How much did each slide cost?

SUMMING UP

Explain how you can use a multiplication fact to find the quotient of 48 ÷ 6.

GETTING STARTED

Seven days is how many weeks? Fourteen days is how many weeks? How do you know these answers?

Dividing by 7

Jill cared for animals 56 days in a row last summer. How many weeks is 56 days?

56 ÷ 7 = □

Think □ × 7 = 56

The missing factor is 8.

8 × 7 = 56

So **56 ÷ 7 = 8.** $7\overline{)56}$ with quotient **8**

Jill helped for 8 weeks.

Remember: When you divide, you can think of a related multiplication fact.

What if Jill worked only 35 days? How many weeks is that?

Check Your Understanding

Divide.

1. 7 ÷ 7 = □
2. 42 ÷ 7 = □
3. 28 ÷ 7 = □
4. 14 ÷ 7 = □
5. 21 ÷ 7 = □
6. 49 ÷ 7 = □
7. 35 ÷ 7 = □
8. 56 ÷ 7 = □
9. $7\overline{)63}$
10. $7\overline{)0}$
11. $7\overline{)7}$
12. $7\overline{)14}$
13. $7\overline{)56}$

Share Your Ideas Look back at **5**. Draw a picture to show the division. Name the multiplication fact shown by the picture.

Practice

Divide.

14. 14 ÷ 7 = □ **15.** 21 ÷ 7 = □ **16.** 7 ÷ 7 = □ **17.** 0 ÷ 7 = □

18. 45 ÷ 5 = □ **19.** 28 ÷ 7 = □ **20.** 35 ÷ 7 = □ **21.** 49 ÷ 7 = □

22. 56 ÷ 7 = □ **23.** 63 ÷ 7 = □ **24.** 42 ÷ 6 = □ **25.** 21 ÷ 7 = □

26. $7\overline{)14}$ **27.** $7\overline{)35}$ **28.** $7\overline{)49}$ **29.** $7\overline{)7}$ **30.** $7\overline{)42}$

31. $7\overline{)56}$ **32.** $4\overline{)28}$ **33.** $7\overline{)0}$ **34.** $2\overline{)18}$ **35.** $6\overline{)54}$

36. $7\overline{)21}$ **37.** $6\overline{)48}$ **38.** $7\overline{)35}$ **39.** $5\overline{)40}$ **40.** $7\overline{)63}$

Complete the table.

41.

Number of Weeks	1			4		6			
Number of Days	7	14	21		35		49	56	63

Compare. Use >, <, or = for ●. Choose mental math, paper and pencil, or a calculator. Explain your choices.

CHOICES

42. 48 ÷ 6 ● 63 ÷ 7 **43.** 35 ÷ 7 ● 16 ÷ 4

44. 56 ÷ 7 ● 24 ÷ 3 **45.** 7 × 7 ● 6 × 8

46. 6 × 6 ● 7 × 5 **47.** 42 ÷ 7 ● 48 ÷ 8

48. (7 × 5) + 7 ● (6 × 7) − 2

49. 4 × 7 ● (2 × 7) + (2 × 7)

Think and Apply

50. How many days are there in 5 weeks? 9 weeks? 8 weeks and 2 days? 2 weeks and 5 days?

51. Write a funny story about animals. Use the fact 28 ÷ 7 = 4.

Use the numbers 3, 7, and 21. Write two multiplication sentences and two division sentences.

SUMMING UP

GETTING STARTED

Start with 24. How many times must you subtract 8 in order to reach 0?

Dividing by 8

In a National Science Foundation project, volunteers helped care for penguin chicks.

What if Jessica fed 40 penguin chicks? Eight chicks were in each feeding bin. How many feeding bins were there?

40 ÷ 8 = □

Think **□ × 8 = 40**

The missing factor is 5.

5 × 8 = 40

So **40 ÷ 8 = 5.** $\begin{array}{r}\mathbf{5}\\ \mathbf{8\overline{)40}}\end{array}$

There would be 5 feeding bins.

How could you use subtraction to find the answer?

Check Your Understanding

Find each quotient.

1. 16 ÷ 8 = □ **2.** 8 ÷ 8 = □ **3.** 24 ÷ 8 = □ **4.** 32 ÷ 8 = □

5. 48 ÷ 8 = □ **6.** 40 ÷ 8 = □ **7.** 56 ÷ 8 = □ **8.** 64 ÷ 8 = □

9. $8\overline{)72}$ **10.** $8\overline{)8}$ **11.** $8\overline{)24}$ **12.** $8\overline{)56}$ **13.** $8\overline{)0}$

14. $8\overline{)64}$ **15.** $8\overline{)16}$ **16.** $8\overline{)48}$ **17.** $8\overline{)40}$ **18.** $8\overline{)32}$

Share Your Ideas Look back at **9**. Explain how you found the quotient.

Practice

Find each quotient.

19. 16 ÷ 4 = □ **20.** 0 ÷ 8 = □ **21.** 32 ÷ 8 = □ **22.** 24 ÷ 8 = □

23. 56 ÷ 7 = □ **24.** 48 ÷ 8 = □ **25.** 56 ÷ 8 = □ **26.** 64 ÷ 8 = □

27. $8\overline{)8}$ **28.** $8\overline{)72}$ **29.** $6\overline{)48}$ **30.** $5\overline{)25}$

31. $8\overline{)56}$ **32.** $8\overline{)0}$ **33.** $8\overline{)64}$ **34.** $4\overline{)32}$

Find each missing factor.

35. 8 × □ = 72 **36.** □ × 4 = 24 **37.** 8 × □ = 64 **38.** 5 × □ = 40

39. □ × 6 = 48 **40.** □ × 8 = 56 **41.** 8 × □ = 8 **42.** 8 × □ = 0

Complete each flowchart.

43.
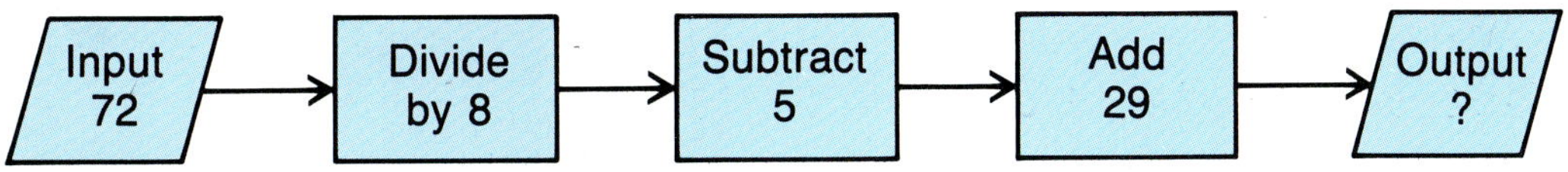

44.
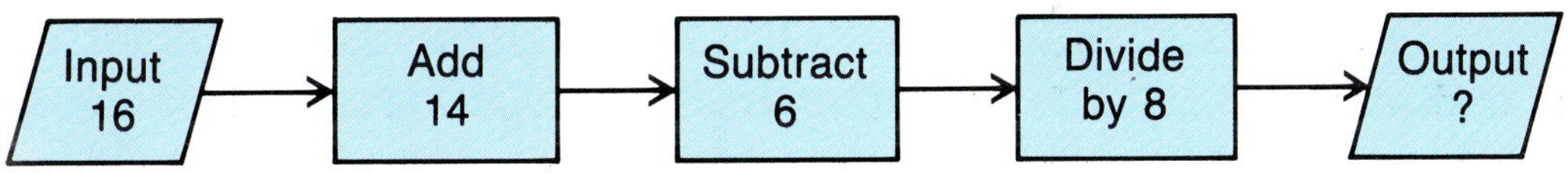

45. Make a flowchart. Have a friend solve it.

Think and Apply

46. Volunteers help 32 penguins into a pool. Suppose 8 penguins were in each group. How many groups were there?

47. Draw 16 penguins waiting in two equal lines to go into a pool. Write a multiplication and division fact for your drawing.

Visual Thinking

48. Estimate how many dots there are. How can you use a fact with 8 to help you?

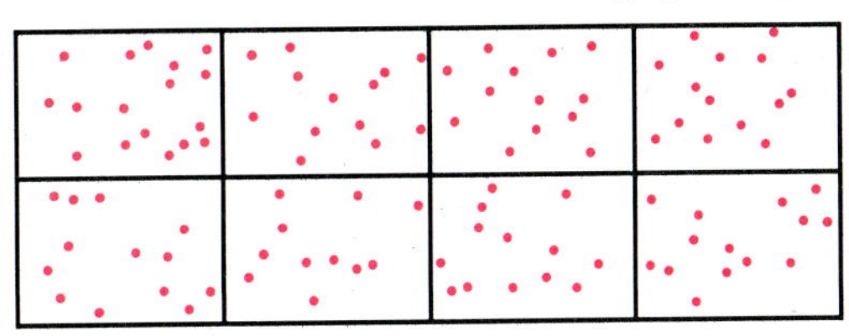

SUMMING UP

Find the quotient for $8\overline{)40}$. Multiply the quotient by the divisor. What number do you get? Tell why.

GETTING STARTED

Name as many multiplication facts with 9 as a factor as you can.

Dividing by 9

Jonathan Chapman, known as Johnny Appleseed, planted a horse load of seeds in Licking Creek, Ohio, in 1801. This became the first apple orchard.

What if Johnny had 63 seeds and he planted them 9 at a time? How many groups of 9 would there be?

63 ÷ 9 = □

Think □ × 9 = 63

7 × 9 = 63

So **63 ÷ 9 = 7.** $9\overline{)63}$ with quotient **7**

There were 7 groups of 9 seeds.

What if Johnny had 81 seeds and he planted them 9 at a time? How many groups would there be?

Check Your Understanding

Divide.

1. 45 ÷ 9 = □ **2.** 18 ÷ 9 = □ **3.** 9 ÷ 9 = □ **4.** 27 ÷ 9 = □

5. 36 ÷ 9 = □ **6.** 81 ÷ 9 = □ **7.** 72 ÷ 9 = □ **8.** 54 ÷ 9 = □

9. $9\overline{)63}$ **10.** $9\overline{)0}$ **11.** $9\overline{)45}$ **12.** $9\overline{)54}$ **13.** $9\overline{)36}$

Share Your Ideas Draw a picture to show 18 ÷ 2. What multiplication fact does your picture show?

Practice

Divide.

14. $45 \div 9 = \square$
15. $21 \div 7 = \square$
16. $36 \div 6 = \square$
17. $72 \div 8 = \square$
18. $27 \div 9 = \square$
19. $54 \div 9 = \square$

20. $6\overline{)12}$ 21. $9\overline{)18}$ 22. $9\overline{)81}$

23. $1\overline{)9}$ 24. $9\overline{)63}$ 25. $8\overline{)64}$

Find each missing number.

26.

27.

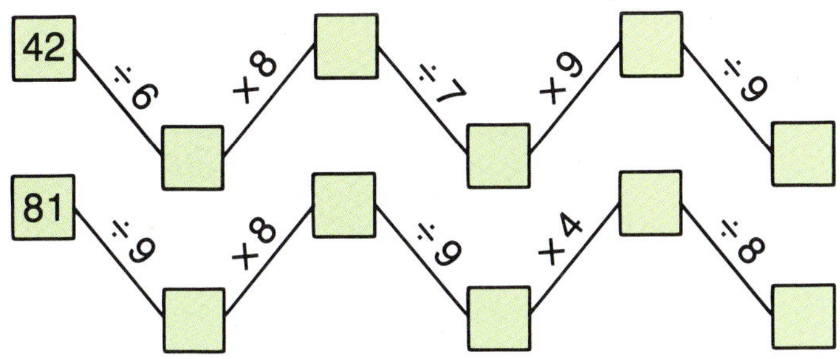

Solve.

28. How many nines are there in 18? 36? 72? How are the quotients related to each other? How are the dividends related to each other?

Think and Apply

29. Johnny had 78 miles to go. He traveled 6 miles. He decided to travel the rest of the miles in 9 days. If he travels the same number of miles each day, how many miles will he travel in one day?

30. Write a story about Johnny Appleseed. Use the fact $27 \div 9 = 3$.

Use the numbers 9, 8, and 72. Write two multiplication sentences and two division sentences.

SUMMING UP

Mixed Review

1. 4×6
2. 8×9
3. 7×6
4. 5×3
5. 4×7
6. 5×0
7. 1×9
8. 6×2
9. 8×7
10. 3×5
11. $1{,}982 - 431$
12. $6{,}549 + 1{,}236$
13. $4{,}382 - 1{,}005$
14. $\$5.28 + 3.91$
15. $\$54.19 - 6.23$
16. $5{,}000 - 999 = \square$
17. $3{,}642 - 642 = \square$
18. $\$6.50 - \$.37 = \square$
19. $\$.91 + \$8.42 = \square$

GETTING STARTED

How are the facts $7 \times 8 = 56$ and $56 \div 7 = 8$ related to each other?

Fact Families

Today many people buy apples that are grown far away.

The number of apples in this box can be described using multiplication and division facts.

How does each fact in the chart tell about the apples?

Fact Family for **3**, **4**, and **12**	
$3 \times 4 = 12$	$12 \div 3 = 4$
$4 \times 3 = 12$	$12 \div 4 = 3$

A fact family uses the same numbers. This family relates multiplication and division.

What is the fact family for the numbers 7, 9, and 63?

Check Your Understanding

Complete each fact family.

1. $3 \times 6 = \square$
$6 \times 3 = \square$
$18 \div 3 = \square$
$18 \div 6 = \square$

2. $4 \times 9 = \square$
$9 \times 4 = \square$
$36 \div 4 = \square$
$36 \div 9 = \square$

3. $5 \times 8 = \square$
$8 \times 5 = \square$
$40 \div 5 = \square$
$40 \div 8 = \square$

4. $2 \times 7 = \square$
$7 \times 2 = \square$
$14 \div 2 = \square$
$14 \div 7 = \square$

Share Your Ideas Explain how to write a fact family for 6, 8, and 48.

Practice

Complete each fact family.

5. $6 \times 7 = \square$
$7 \times 6 = \square$
$42 \div 6 = \square$
$42 \div 7 = \square$

6. $6 + 7 = \square$
$7 + 6 = \square$
$13 - 7 = \square$
$13 - 6 = \square$

7. $3 \times 9 = \square$
$\square \times 3 = 27$
$27 \div 3 = \square$
$27 \div 9 = \square$

8. $3 + 9 = \square$
$\square + 3 = 12$
$12 - 3 = \square$
$12 - \square = 3$

Write the other facts in each family.

9. $4 \times 6 = 24$ **10.** $4 + 6 = 10$ **11.** $6 \times 8 = 48$ **12.** $63 \div 7 = 9$

13. $20 \div 5 = 4$ **14.** $8 \times 4 = 32$ **15.** $18 \div 6 = 3$ **16.** $15 - 9 = 6$

17. $11 - 2 = 9$ **18.** $35 \div 7 = 5$ **19.** $4 \times 6 = 24$ **20.** $72 \div 8 = 9$

Write a multiplication and division fact family for each group.

21. 5, 9, 45 **22.** 8, 9, 72 **23.** 21, 7, 3 **24.** 7, 56, 8

25. 6, 5, 30 **26.** 14, 2, 7 **27.** 8, 8, 64 **28.** 5, 5, 25

Think and Apply

29. What does this mean in most cases? When you know one fact, you really know four facts.

30. Make several drawings that show 36 apples in equal groups. Write as many multiplication and division facts as you can.

Logical Thinking

How many apples?

31. Johnny has less than 10 apples. If they are in rows of 3, none are left over. When put in rows of 4, one is left over.

SUMMING UP

Look back at **27** and **28**. Explain why some fact families have only two facts.

Using Problem Solving

Using Division and Multiplication

The Midland School is collecting food for families who need it. Lynn and Bryant collected 2 cans each from 4 families.

"That's 8 cans of food," said Lynn.

"We could give 2 cans to 4 families," said Bryant.

"I know another way to divide the cans equally," said Lynn.

What other ways can you think of?

Working together

Materials: counters or blocks

Record your work.

A. In what different ways could you distribute the 8 cans equally?

B. Bryant collected 12 cans from families. Each family contributed the same number of cans. How many different families could he have visited? How many cans of food could each family have given?

C. Leslie also collected 12 cans. She and Bryant put their cans together. Each family receiving the food will get the same number of cans. How many families can receive cans? How many cans will each family get?

D. Write a problem about the cans of food. Exchange your problem with another group. Solve the problem.

Sharing Your Ideas

1. Did solving **B** help you to solve **C**? Explain.
2. Lynn cannot remember whether she visited 3 or 4 families. She knows each family gave her the same number of cans. She has 15 cans. How many families did she visit? Explain.

THINK
EXPLORE
SOLVE
LOOK BACK

Extending Your Thinking

3. Lynn collected the same number of cans from the first 3 families that she visited. At the last house, she received 4 cans. She collected 19 cans altogether. How many cans did she collect from each family?

4. Bryant went to some houses on Main Street. Then he went to two houses on Elm Street where he collected 7 cans at each house. He collected 35 cans altogether. How many cans did he collect on Main Street?

5. The students put the cans in two different types of boxes. Type A box held 6 cans. Type B box held 9 cans. One day the students were able to box 51 cans. Each box was full. How many boxes of each type could they have used?

6. After the food drive, the students had a party. There were 28 students. The teacher planned the games so that everyone could play at the same time. They had these games.

Game	Number of Games	Number of People Who Can Play
Rounds	4	2
Buy It	2	2–6
Toss It	3	2–4
Hard Times	1	2–4

Find several ways that show how every student can be taking part in a game at the same time.

Summing Up

7. **Look back** at **6. What if** the teacher needs games for 42 people? She can bring in one more of the games listed. Which one should she bring?

8. Why did you use multiplication and division to solve problems **3–6**?

Chapter Review

Find each quotient. pages 230–241

1. $12 \div 2 = \square$ 2. $15 \div 3 = \square$ 3. $10 \div 5 = \square$
4. $24 \div 4 = \square$ 5. $18 \div 2 = \square$ 6. $36 \div 4 = \square$

7. $5\overline{)30}$ 8. $4\overline{)28}$ 9. $3\overline{)27}$ 10. $2\overline{)14}$ 11. $4\overline{)32}$

12. $5\overline{)40}$ 13. $3\overline{)9}$ 14. $4\overline{)16}$ 15. $5\overline{)15}$ 16. $2\overline{)2}$

Divide. pages 246–255

17. $49 \div 7 = \square$ 18. $48 \div 8 = \square$ 19. $54 \div 9 = \square$
20. $12 \div 6 = \square$ 21. $81 \div 9 = \square$ 22. $32 \div 8 = \square$

23. $7\overline{)21}$ 24. $6\overline{)48}$ 25. $8\overline{)0}$ 26. $7\overline{)42}$ 27. $8\overline{)72}$

28. $9\overline{)9}$ 29. $7\overline{)56}$ 30. $8\overline{)40}$ 31. $6\overline{)30}$ 32. $9\overline{)36}$

Give the other facts in each family. pages 256–257

33. $7 \times 9 = 63$ 34. $7 \times 8 = 56$ 35. $54 \div 9 = 6$

Choose the correct words to complete each sentence.

36. We can use a related _____ to help us find the answer to a _____.

37. A _____ uses the same numbers to build facts related to each other.

Words to Know
fact family division fact multiplication fact

Solve. pages 243–245

38. Third graders took 18 pictures on their class trip. They displayed 6 pictures in each row. How many rows did they have?

39. There are 9 teddy bears and 7 dolls to be shared equally by 8 girls at the slumber party. How many toys will each girl get?

Chapter Test

Find each quotient.

1. $12 \div 3 = \square$
2. $36 \div 4 = \square$
3. $14 \div 2 = \square$
4. $20 \div 5 = \square$
5. $20 \div 4 = \square$
6. $45 \div 5 = \square$
7. $4\overline{)24}$
8. $3\overline{)27}$
9. $5\overline{)30}$
10. $2\overline{)2}$
11. $5\overline{)45}$
12. $3\overline{)24}$
13. $4\overline{)36}$
14. $2\overline{)18}$

Divide.

15. $42 \div 6 = \square$
16. $63 \div 7 = \square$
17. $72 \div 8 = \square$
18. $54 \div 9 = \square$
19. $48 \div 6 = \square$
20. $56 \div 7 = \square$
21. $6\overline{)0}$
22. $7\overline{)28}$
23. $9\overline{)45}$
24. $8\overline{)56}$
25. $7\overline{)7}$
26. $8\overline{)32}$
27. $6\overline{)18}$
28. $9\overline{)27}$

Give the other facts in each fact family.

29. $5 \times 9 = 45$
30. $8 \times 4 = 32$
31. $42 \div 7 = 6$

Solve.

32. There are 56 campers at the campfire.
There are 7 campers on each log.
How many logs are there?

33. **What if** 6 hikers each collected 6 rocks?
They put 4 rocks into each box.
How many boxes did they use?

THINK Try to find 8 different ways to group 24 rocks in equal groups. Make a drawing for each way you find.

EXTENSION

At the Supermarket: Then and Now

When Rod's mother takes him grocery shopping, she sometimes says, "It's the same store, but not the same prices! Groceries cost so much more than they did when I was a child."

How much more does Rod's mother pay now for these groceries?

THEN

5 pounds for 25¢

Step 1 What is the cost per pound?

25¢ ÷ 5 = 5¢ per pound

NOW

69¢ per pound

Step 2 How much more do they cost per pound now?

69¢ − 5¢ = 64¢ more per pound

THEN

8 for 40¢

Step 1 What does one lemon cost?

40¢ ÷ 8 = 5¢ for one

NOW

5 for 45¢

Step 2 What does one lemon cost now?

45¢ ÷ 5 = 9¢ for one now

Step 3 How much more does one lemon cost now?

9¢ − 5¢ = 4¢ more now

Make up a **Then** and **Now** bulletin board. Cut out some supermarket advertisements. Ask adults about how much the items cost when they were children. Write two problems for each item.

Maintaining Skills

Choose the correct answer. Write A, B, C, or D.

1. $2 \times 7 = \square$

   A 9 C 14
   B 16 D not given

2. $\begin{array}{r} 3 \\ \times 8 \\ \hline \end{array}$

   A 32 C 28
   B 24 D not given

3. $\begin{array}{r} 9 \\ \times 4 \\ \hline \end{array}$

   A 24 C 36
   B 32 D not given

4. $5 \times 5 = \square$

   A 10 C 25
   B 15 D not given

5. $\begin{array}{r} 5 \\ \times 6 \\ \hline \end{array}$

   A 30 C 15
   B 35 D not given

6. $\begin{array}{r} 0 \\ \times 9 \\ \hline \end{array}$

   A 9 C 2
   B 1 D not given

7. $6 \times 4 \times 2 = \square$

   A 48 C 40
   B 20 D not given

8. $2\overline{)16}$

   A 7 C 8
   B 6 D not given

9. $32 \div 4 = \square$

   A 8 C 6
   B 9 D not given

10. $7\overline{)49}$

   A 9 C 8
   B 7 D not given

Solve.

11. It snowed every third day at the ski resort. If it snowed on the 5th day of the month, what day did it not snow?

   A 8th C 14th
   B 10th D not given

12. One skier leaves the starting gate every 4 minutes. **What if** the first skier leaves at 9:00 A.M. and the second skier leaves at 9:04 A.M.? What time will the sixth skier leave the gate?

   A 9:20 C 9:25
   B 9:18 D not given

9 Fractions • Probability

THEME Fun at Home

Sharing What You Know

Parties can be fun! Talk about different reasons to have a party. What are some of the things you need to think about? Should you plan an indoor party or an outdoor party? Why might an outdoor party be a good idea? When might an outdoor party be a problem?

Using Language

You decide to have an outdoor party. You hope it doesn't rain. The **probability** of it raining depends on where you live and the season. In mathematics, the chance of something happening is called **probability**. Look out your window. What is the probability that it will rain today?

Words to Know: fraction, equivalent fraction, mixed number, frequency, probability, outcome

Be a Problem Solver

Half of the Taylor children are girls and the other half are boys. Altogether there are six children. Ann, Bruce, and Cathy are in the kitchen baking. Another one of the Taylor children comes into the kitchen. Is this child more likely to be a boy or a girl?

Create a recipe for a new dish for a party. List all the important information that your recipe must tell the chef.

GETTING STARTED

Which shapes are divided into equal parts?

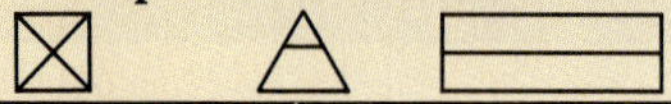

Parts of Regions

Some quilts are made from squares. What part of the square is blue?

You can use a fraction to name part of a region.

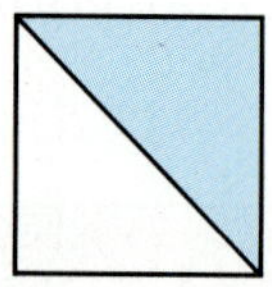

Think How many equal parts in all?
How many parts are blue?
There are **2** equal parts.
1 of **2** equal parts is blue.

write $\frac{1}{2}$

read one half

What part of each region is blue?

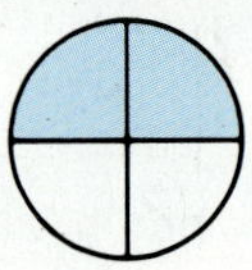

blue parts → $\frac{2}{4}$
equal parts in all →

Two fourths is blue.

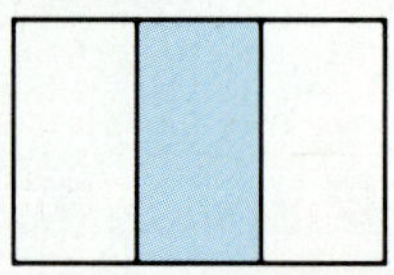

blue part → $\frac{1}{3}$
equal parts in all →

One third is blue.

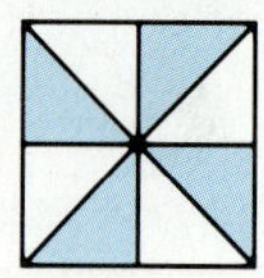

blue part → $\frac{4}{8}$
equal parts in all →

Four eighths is blue.

What part of each region is not blue? Explain.

Check Your Understanding

Write a fraction for the part that is blue.

1. 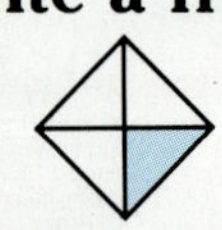

blue parts ☐
equal parts ☐

2. 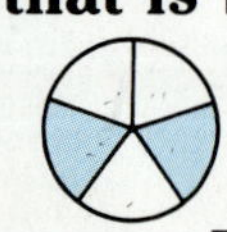

blue parts ☐
equal parts ☐

3. 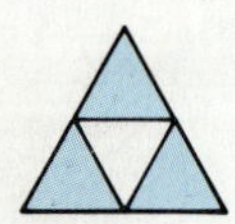

blue parts ☐
equal parts ☐

Share Your Ideas Draw, shade, and label a circle that is $\frac{1}{2}$ blue. Repeat for $\frac{1}{3}$ blue. $\frac{1}{4}$ blue.

Practice

Write a fraction for the part that is blue.

4. 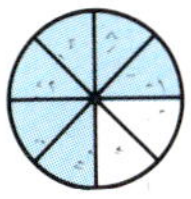

$\frac{\text{blue parts } \square}{\text{equal parts } \square}$

5. 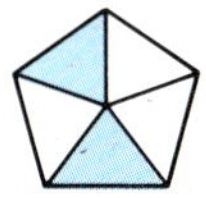

$\frac{\text{blue parts } \square}{\text{equal parts } \square}$

6. 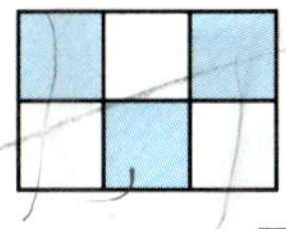

$\frac{\text{blue parts } \square}{\text{equal parts } \square}$

7. 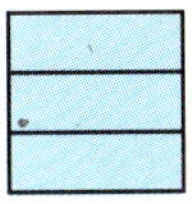

8. 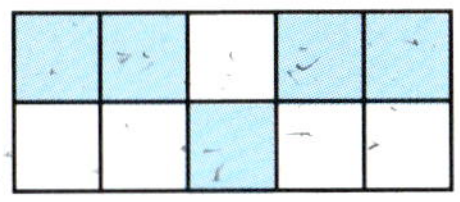

9. 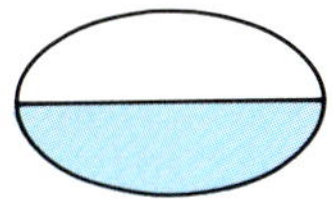

10. 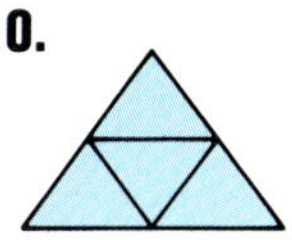

11. 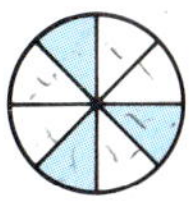

12. 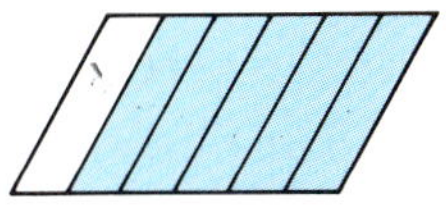

13. 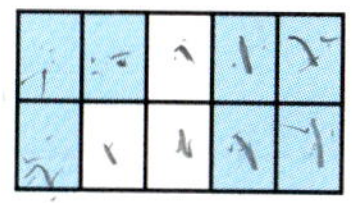

14. 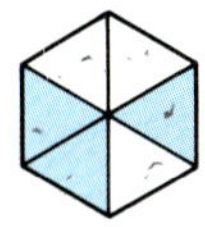

Copy and complete the table.

		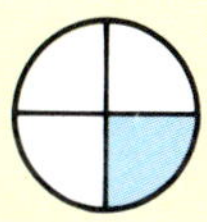	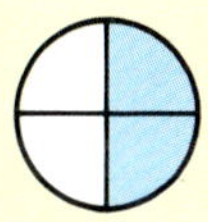	**15.** 	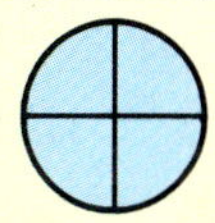
Blue part	**16.**	**17.**	**18.**	$\frac{3}{4}$	**19.**
Part that is NOT blue	$\frac{4}{4}$	**20.**	**21.**	**22.**	$\frac{0}{4}$

Think and Apply

23. Trace this square 4 times. Cut out the squares.

24. Divide each square into 4 equal parts in different ways. Label each part $\frac{1}{4}$.

25. Do all $\frac{1}{4}$ parts look alike? How may ways can a square be divided into fourths?

Is $\frac{1}{3}$ of this shape blue?
Explain your thinking.

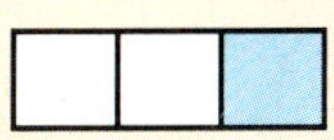

GETTING STARTED

How are this region  and this group  alike? How are they different?

Parts of Groups

A quilting bee was held at Molly's home. Quilters used pieces of fabric like this.

You can use a fraction to name part of a group.

How many equal parts are in the group?
What part of the group is red?
Two of three pieces are red.
Two thirds, or $\frac{2}{3}$, of the group is red.

$\frac{2}{3}$ — 2 red parts, 3 equal parts in all

What part of each group is red?

Think How many equal parts in all?

Two fourths is red. $\frac{2}{4}$

Three eighths is red. $\frac{3}{8}$

What part of each group is blue? yellow? Explain.

Check Your Understanding

Write a fraction for the part that is red.

1. 

2. 

3. 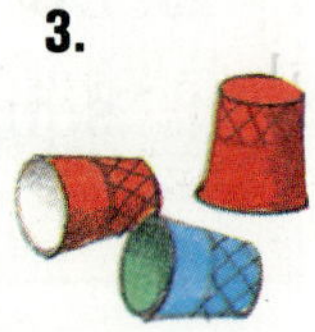

4. 

Ideas Draw, shade, and label uttons that is $\frac{1}{3}$ red. $\frac{3}{4}$ red.

Practice

Write a fraction for the part that is red.

5. 

6. 

7. 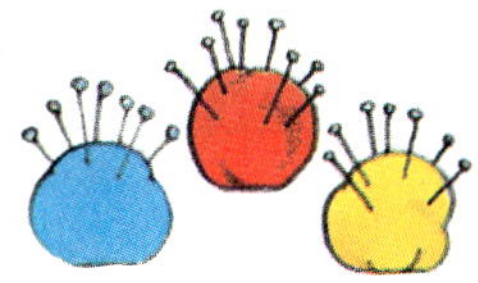

8. 

9. 

10. 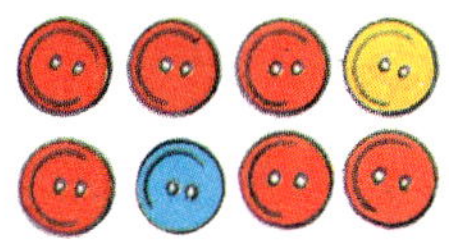

Write a fraction for each.

11. Three of four quilters are sewing.
12. Six of ten needles are on the table.
13. One of six pin cushions is red.
14. Four of four thimbles are silver.
15. Zero of ten squares are finished.

Think and Apply

Complete the table for each handful that was taken from a bag of red and blue buttons.

	Buttons				Fractions	
	Handful	Red	Blue	Total	Red Part	Blue Part
	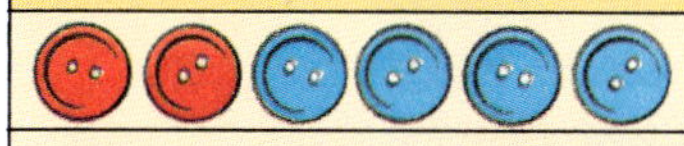	2	4	6	$\frac{2}{6}$	$\frac{4}{6}$
16.						
17.						

18. **Look back** at the table. Guess. Do you think the bag contains more red or blue buttons? Explain.

How is using a fraction to name part of a group like using a fraction to name part of a region? How is it different?

SUMMING UP

GETTING STARTED

Use 6 counters to show 2 equal groups. 3 equal groups.

Fractional Parts

Doreen and her dad bought 8 cans of paint. One half of the cans of paint are blue. How many cans of paint are blue?

$\frac{1}{2}$ of **8** = ☐

Think 8 cans of paint, **2** equal groups

Divide by **2**. **8 ÷ 2 = 4**

$\frac{1}{2}$ of **8 = 4**

Four cans of paint are blue.

Another Example

$\frac{1}{3}$ of 6 = ☐

Think 6 in all, 3 equal groups

Divide by 3. 6 ÷ 3 = 2

$\frac{1}{3}$ of 6 = 2

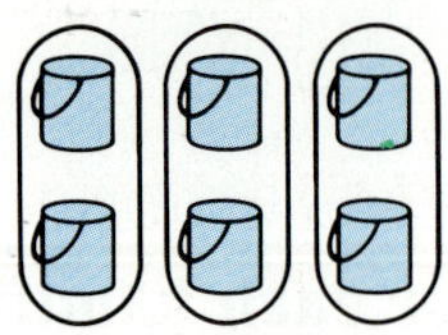

Check Your Understanding

Complete. Use counters or draw a picture if you wish.

1. 

$\frac{1}{2}$ of 6 = ☐

2. 

$\frac{1}{3}$ of 9 = ☐

3. 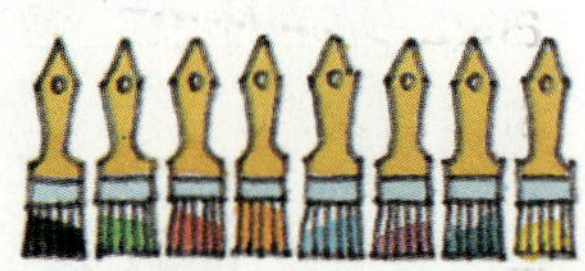

$\frac{1}{4}$ of 8 = ☐

4. $\frac{1}{3}$ of 18 = ☐

5. $\frac{1}{3}$ of 12 = ☐

6. $\frac{1}{8}$ of 16 = ☐

Share Your Ideas Look back at **4**. Explain how you divided to find the fractional part.

Practice

CHOICES **Complete. Use counters if you wish.**

7. 

$\frac{1}{3}$ of 6 = ☐

8. 

$\frac{1}{2}$ of 8 = ☐

9. $\frac{1}{7}$ of 21 = ☐

10. $\frac{1}{9}$ of 54 = ☐

11. $\frac{1}{8}$ of 48 = ☐

12. $\frac{1}{5}$ of 40 = ☐

13. $\frac{1}{4}$ of 24 = ☐

14. $\frac{1}{10}$ of 80 = ☐

Complete. Follow the rule.

Rule: Find $\frac{1}{4}$ of the input.

	Input	Output
15.	4	
16.	16	
17.	20	
18.	36	

Rule: Find $\frac{1}{2}$ of the input.

	Input	Output
19.	18	
20.	14	
21.	6	
22.	16	

Think and Apply

23. Draw 15 paint brushes. Shade $\frac{1}{3}$ of them red and $\frac{1}{5}$ of them green. How many are not shaded?

24. Tom bought 12 cans of paint. Of these, $\frac{1}{2}$ are purple and $\frac{1}{3}$ are red. The rest are green. How many cans of paint are green?

> Explain why you divide by 4 to find $\frac{1}{4}$ of a group.

SUMMING UP

Mixed Review

1. 8×6
2. 9×4
3. 7×3
4. 2×5
5. 8×7
6. 9×8
7. $3\overline{)9}$
8. $4\overline{)28}$
9. $6\overline{)54}$
10. $5\overline{)40}$
11. $2\overline{)16}$
12. $7\overline{)42}$
13. 74 − 36 = ☐
14. \$6.00 − \$.87 = ☐
15. 39 + 23 = ☐
16. 84 + 36 = ☐
17. 5 × 9 = ☐
18. 0 × 6 = ☐
19. 72 ÷ 9 = ☐
20. 9 ÷ 1 = ☐
21. 81 ÷ 9 = ☐
22. 48 ÷ 6 = ☐
23. 56 ÷ 7 = ☐

Activity

Exploring Equivalent Fractions

Ann and Chris use pieces of trim to frame a puppet stage. Ann asked for $\frac{1}{2}$ of a piece to finish her side. Chris said she needed $\frac{2}{4}$ of a piece. Who is correct?

Do this activity with fraction pieces to find out.

Working together

Materials: fraction pieces

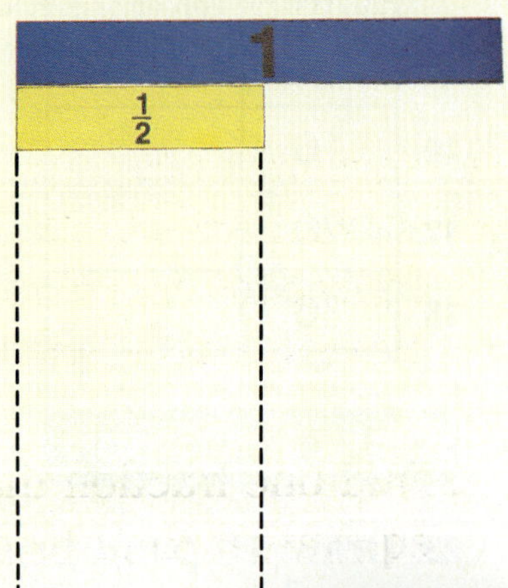

A. Start with **1 whole.** Then show $\frac{1}{2}$ with one fraction piece. Draw and label the fraction pieces each time.

B. Now form $\frac{1}{2}$ with fraction pieces that show $\frac{1}{4}$. that show $\frac{1}{8}$.

C. Explore. Find other fraction pieces of equal length that form $\frac{1}{2}$.

Sharing Your Results

1. Was Chris or Ann correct?
2. Did the length change as you changed the size and number of pieces? Explain.
3. **Look back** at your drawing. What do you notice about the fractions that equal $\frac{1}{2}$?

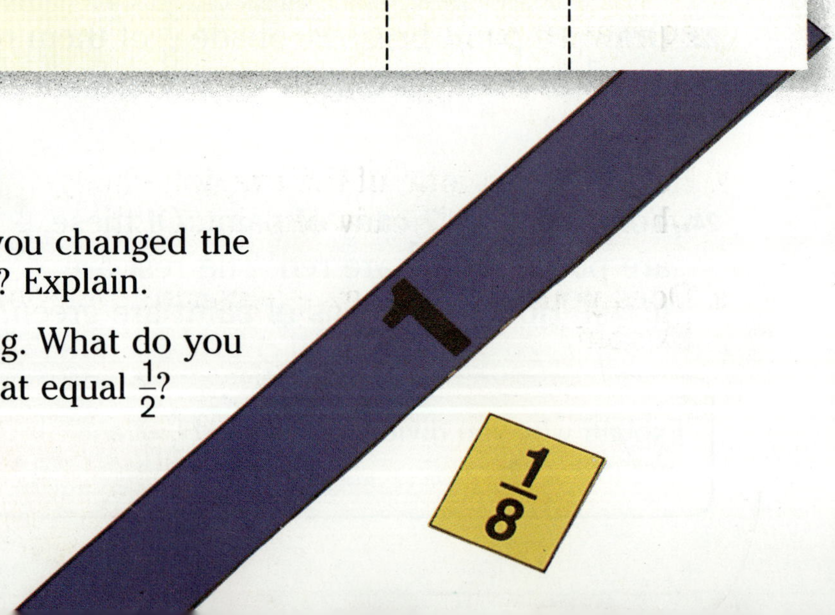

Extending the Activity

All the fractions you recorded name the same part of one whole. These fractions are equivalent fractions.

▸ **Equivalent fractions** name the same amount.

We write: $\frac{1}{2} = \frac{2}{4}$ $\frac{2}{4} = \frac{4}{8}$ $\frac{1}{2} = \frac{4}{8}$

Now do the activity this way. Record your work.

4. Start with the fraction piece that shows $\frac{1}{4}$. Find the equivalent fractions with $\frac{1}{8}$ pieces. with $\frac{1}{12}$ pieces.

5. Now start with the fraction that shows $\frac{1}{3}$. What are the equivalent fractions with $\frac{1}{6}$ pieces? with $\frac{1}{12}$ pieces?

6. Find one fraction that is equivalent to $\frac{1}{5}$.

Summing Up

7. How did the name of the fraction change when you found equivalent fractions?

8. Does your record show equivalent fractions? Explain.

Activity

Understanding Equivalent Fractions

Ann saw that $\frac{1}{2}$ and $\frac{2}{4}$ are equivalent fractions. Now find other equivalent fractions.

Working together

Materials: 6 strips of paper the same length

A. Label a strip 1 whole.

B. Fold another strip of paper into 2 equal parts. Mark the fold. Label each part.

C. Fold other strips into 3, 4, 6, and 8 equal parts. Mark the folds. Label all the parts.

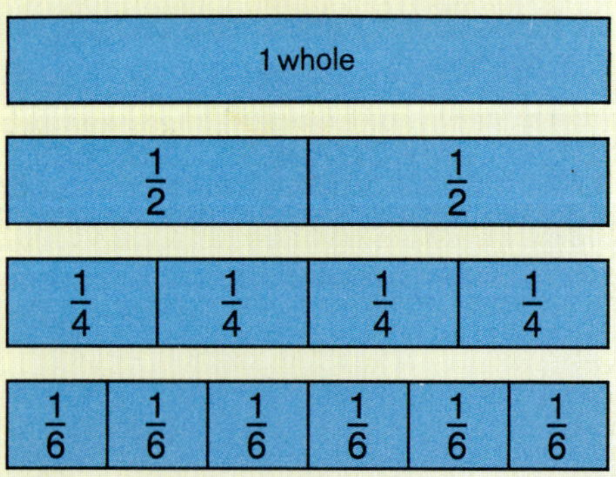

D. Start with any strip. Match other strips to it. Do any of the parts match?

E. Record the parts that match.

Sharing Your Results

1. Compare your findings with those of other groups. List all the equivalent fractions you found.
2. Did any groups find equivalent fractions that you did not find? Why do you think your group missed finding them?

Practice

Compare. Use >, <, or = for ●.

4. 

$\frac{1}{4}$ ● $\frac{3}{4}$

5. 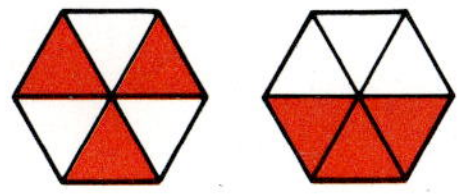

$\frac{3}{6}$ ● $\frac{3}{6}$

6. 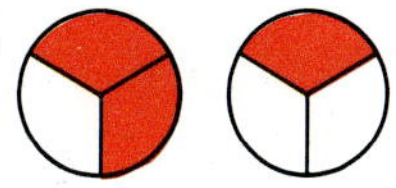

$\frac{2}{3}$ ● $\frac{1}{3}$

7. $\frac{8}{8}$ ● $\frac{2}{8}$

8. $\frac{1}{5}$ ● $\frac{4}{5}$

9. $\frac{1}{2}$ ● $\frac{2}{2}$

Complete each pattern.

10. $\frac{0}{8}$, $\frac{1}{8}$, $\frac{2}{8}$, ___, ___, ___

11. $\frac{5}{5}$, $\frac{4}{5}$, $\frac{3}{5}$, ___, ___, ___

12. $\frac{0}{10}$, $\frac{2}{10}$, $\frac{4}{10}$, ___, ___, ___

Match the fractions to the points on the number line.

13. $\frac{2}{4}$, $\frac{1}{4}$, $\frac{3}{4}$

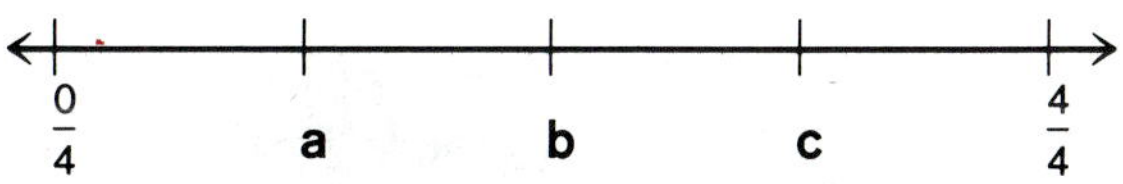

14. $\frac{3}{5}$, $\frac{1}{5}$, $\frac{2}{5}$, $\frac{4}{5}$

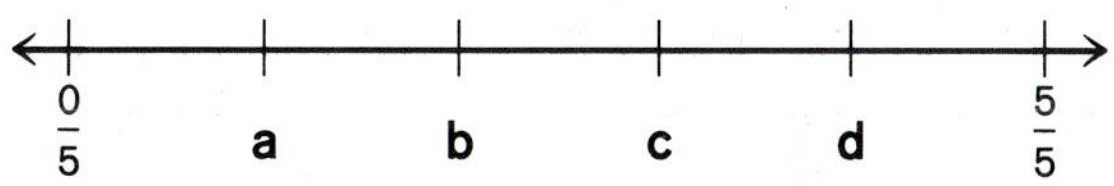

Think and Apply

15. **What if** Dave and Meg need $\frac{4}{8}$ cup of milk and $\frac{2}{8}$ cup of blueberries for their muffin recipe? Do they need more milk or blueberries?

Logical Thinking

16. Four quarters is the same amount as one dollar. One quarter is $\frac{1}{4}$ of a dollar. What part of a dollar is two quarters? three quarters? four quarters?

Write a few sentences to describe how to compare $\frac{3}{8}$ and $\frac{5}{8}$.

SUMMING UP

What part of a dollar is one quarter? What part is one dime? Tell which you would rather have and why.

Comparing Unlike Fractions

You can use $\frac{1}{3}$ or $\frac{1}{4}$ of the ribbon to make a bow. Which will give you the bigger bow?

Use fraction pieces to help you decide.

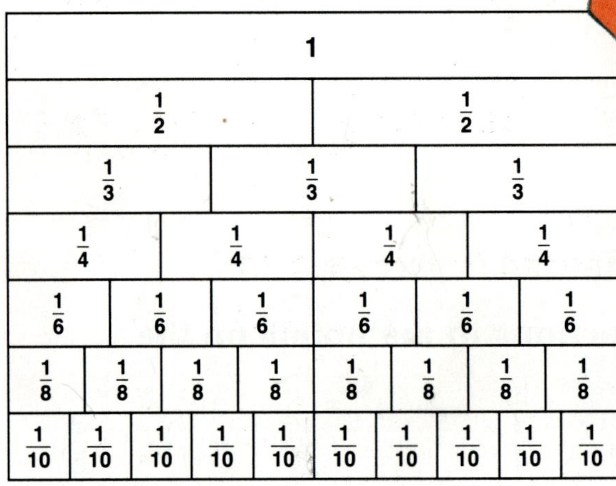

Compare $\frac{1}{3}$ and $\frac{1}{4}$.

$$\frac{1}{3} > \frac{1}{4}$$

Use $\frac{1}{3}$ of the ribbon for the bigger bow.

More Examples

Use fraction pieces to compare.

$\frac{1}{8}$ and $\frac{1}{6}$ $\qquad$ $\frac{3}{4}$ and $\frac{1}{2}$ $\qquad$ $\frac{2}{6}$ and $\frac{2}{3}$

$\frac{1}{8} < \frac{1}{6}$ $\qquad$ $\frac{3}{4} > \frac{1}{2}$ $\qquad$ $\frac{2}{6} < \frac{2}{3}$

Explain how to compare $\frac{1}{10}$ and $\frac{1}{3}$.

Check Your Understanding

Compare. Use >, <, or = for ⬬. Use fraction pieces, if you like.

1. $\frac{2}{3}$ ⬬ $\frac{2}{4}$
2. $\frac{1}{6}$ ⬬ $\frac{1}{4}$
3. $\frac{2}{4}$ ⬬ $\frac{5}{8}$

Share Your Ideas Look back at **2**. Explain how you knew which fraction was greater.

Practice

Compare. Use >, <, or = for ●. Use fraction pieces for 7–14, if you like.

4. 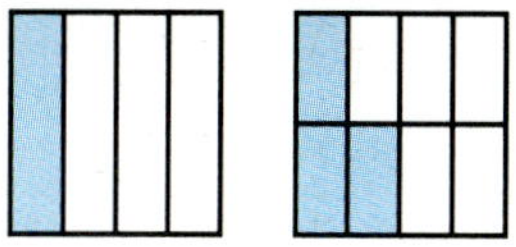

$\frac{1}{4}$ ● $\frac{3}{8}$

5. 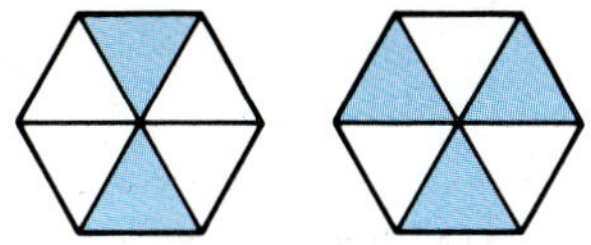

$\frac{2}{6}$ ● $\frac{3}{6}$

6. 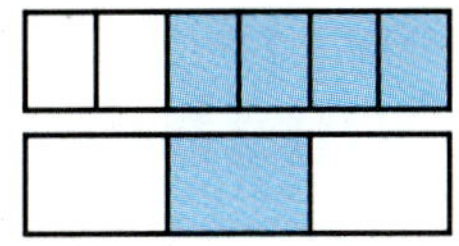

$\frac{4}{6}$ ● $\frac{1}{3}$

7. $\frac{5}{8}$ ● $\frac{4}{10}$

8. $\frac{1}{2}$ ● $\frac{4}{8}$

9. $\frac{1}{8}$ ● $\frac{1}{4}$

10. $\frac{1}{3}$ ● $\frac{1}{10}$

11. $\frac{2}{4}$ ● $\frac{5}{8}$

12. $\frac{3}{8}$ ● $\frac{8}{8}$

13. $\frac{3}{4}$ ● $\frac{2}{3}$

14. $\frac{10}{10}$ ● $\frac{5}{5}$

Choose the correct fraction for each.

15. $\frac{3}{4} < \frac{\square}{\square}$

a. $\frac{1}{2}$ b. $\frac{4}{4}$

c. $\frac{2}{8}$ d. $\frac{2}{4}$

16. $\frac{5}{10} = \frac{\square}{\square}$

a. $\frac{4}{6}$ b. $\frac{1}{2}$

c. $\frac{5}{5}$ d. $\frac{1}{3}$

17. $\frac{2}{3} > \frac{\square}{\square}$

a. $\frac{1}{3}$ b. $\frac{3}{3}$

c. $\frac{4}{6}$ d. $\frac{5}{6}$

Think and Apply

18. Chris used $\frac{2}{3}$ of a sheet of wrapping paper. Ruth used $\frac{1}{2}$ of a sheet. Who used more paper?

19. A box of wrapping paper is $\frac{3}{4}$ full. Another box is $\frac{5}{8}$ full. Which box has more paper? Explain.

Visual Thinking

Name the fraction for each shaded part.

20. 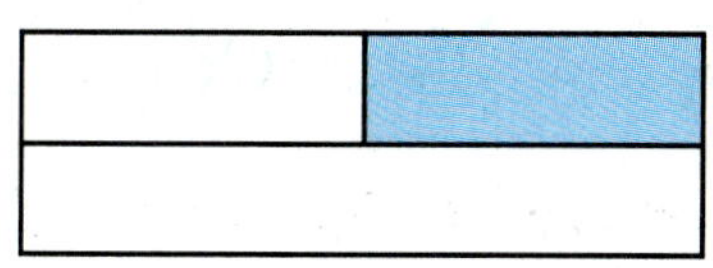

21. 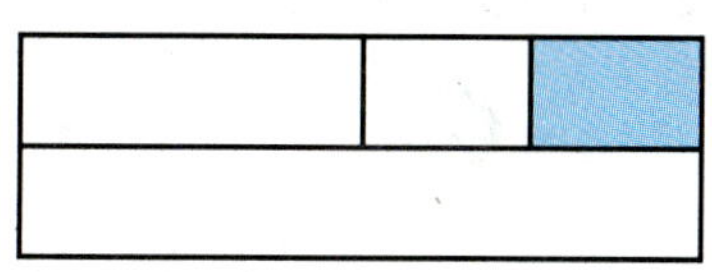

Look back at **7–14**. Which fractions did you compare without using fraction pieces? Explain.

SUMMING UP

GETTING STARTED

Make a drawing of one full glass of milk and one glass of milk half full. How would you describe the total amount of milk?

Mixed Numbers

After school, the Ruiz children have a snack. Only one orange and half an orange are left.

1 and $\frac{1}{2} = 1\frac{1}{2}$

write $1\frac{1}{2}$

read one and one half

$1\frac{1}{2}$ is a **mixed number.** It has a whole number and a fraction.

More Examples

2 and $\frac{1}{3} = 2\frac{1}{3}$

two and one third bananas

3 and $\frac{1}{2} = 3\frac{1}{2}$

three and one half apples

How would you show 3 and $\frac{1}{4}$ pizzas? Explain.

Check Your Understanding

Write a mixed number for the part that is blue.

1. 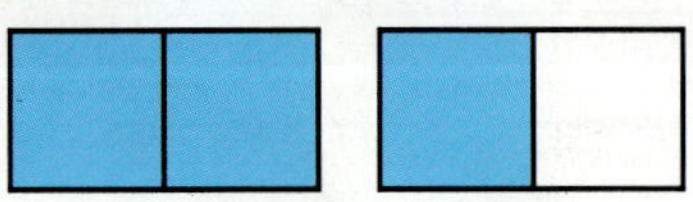

2. 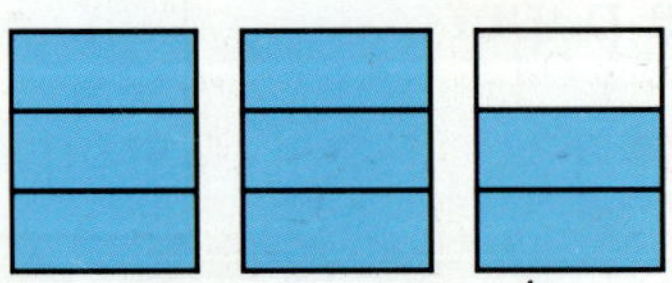

3. 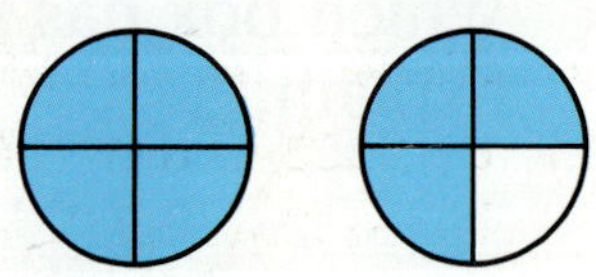

Share Your Ideas Make a drawing to show $4\frac{1}{4}$. Compare your drawing with a classmate's.

Practice

Write a mixed number for the part that is blue.

4. 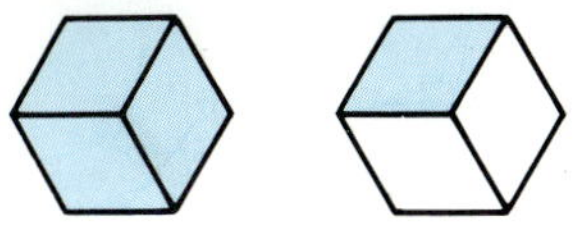

5. 

6. 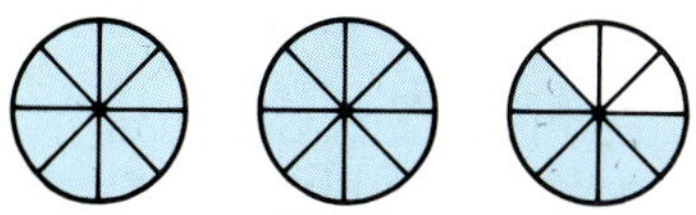

7. 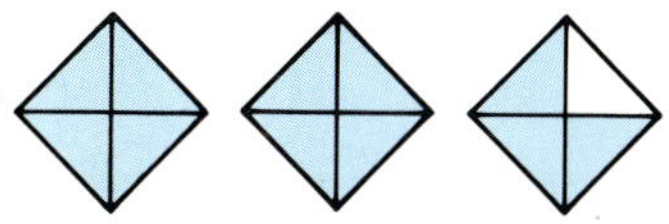

8. 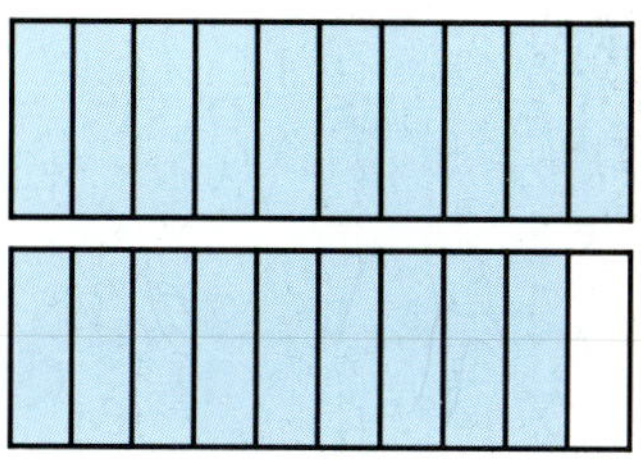

9. 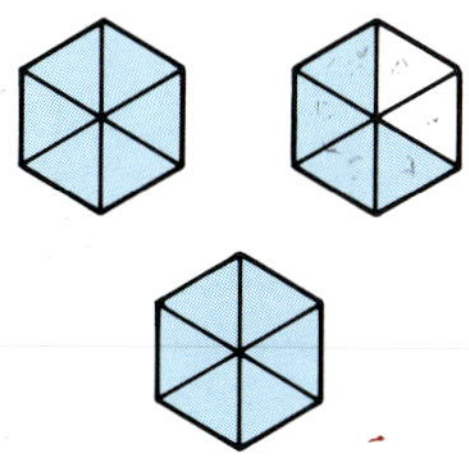

Complete each pattern.

10. 6, $6\frac{1}{2}$, 7, ___, ___, ___

11. 2, $2\frac{1}{3}$, $2\frac{2}{3}$, ___, ___, ___

12. $3\frac{1}{4}$, 3, $2\frac{3}{4}$, ___, ___, ___

Think and Apply

Solve. Draw a picture if you wish.

13. Mrs. Ruiz gave five children half an apple each. Use a mixed number to tell how many apples she gave out.

14. Maria drank 2 glasses of milk at lunch, $\frac{2}{3}$ of a glass after school, and 1 glass at supper. How much milk did she drink in all?

SUMMING UP

Make drawings to show $2\frac{6}{8}$ and $2\frac{3}{4}$. How are your drawings alike? How are they different?

Mixed Review

1. 7×5
2. 6×9
3. 4×3
4. 8×8
5. 1×9
6. 4×0
7. $1\overline{)5}$
8. $2\overline{)6}$
9. $4\overline{)8}$
10. $7\overline{)7}$
11. $6\overline{)12}$
12. $7\overline{)42}$
13. $5 \times 0 = \square$
14. $63 \div 9 = \square$
15. $9 + 6 + 7 = \square$
16. $2 + 15 + 9 = \square$
17. $47 + 8 + 6 = \square$
18. $\$.75 + \$3.12 = \square$
19. $\$9.00 + \$.58 = \square$
20. $\$8.50 - \$.50 = \square$
21. $\$76.99 - \$.25 = \square$
22. $3{,}000 + 40 = \square$
23. $600 - 50 = \square$
24. $480 - 199 = \square$

Midchapter Review

Write a fraction for the part that is blue. pages 266–269

1. 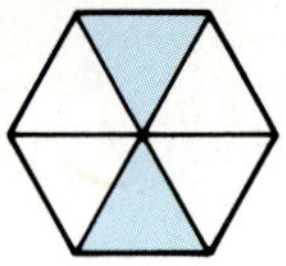
2. 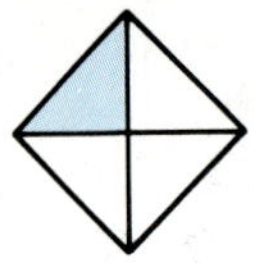
3. 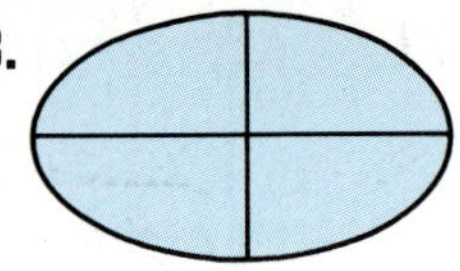
4. 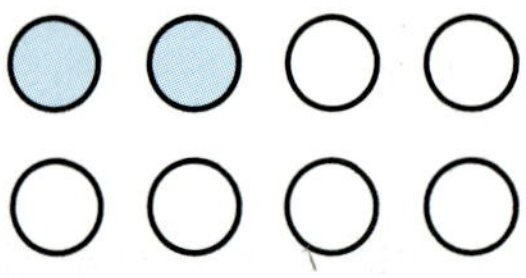
5. 
6. 

Complete. pages 270–271

7. $\frac{1}{8}$ of 32 = □
8. $\frac{1}{6}$ of 48 = □
9. $\frac{1}{3}$ of 24 = □
10. $\frac{1}{4}$ of 36 = □

Compare. Use >, <, or = for ●. pages 276–279

11. $\frac{1}{4}$ ● $\frac{2}{4}$
12. $\frac{4}{8}$ ● $\frac{2}{8}$
13. $\frac{3}{10}$ ● $\frac{7}{10}$
14. $\frac{5}{6}$ ● $\frac{3}{6}$
15. $\frac{1}{2}$ ● $\frac{2}{4}$
16. $\frac{1}{3}$ ● $\frac{1}{2}$
17. $\frac{1}{4}$ ● $\frac{7}{8}$
18. $\frac{2}{3}$ ● $\frac{2}{6}$

Write a mixed number for the part that is blue. pages 280–281

19. 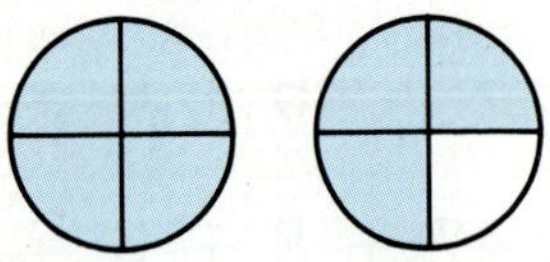
20. 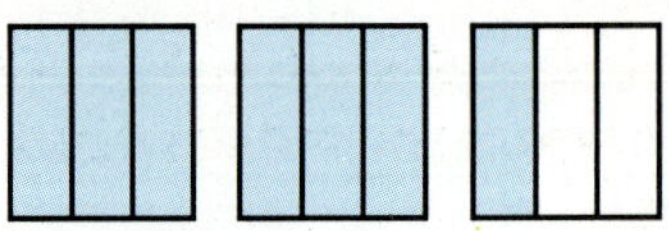
21. 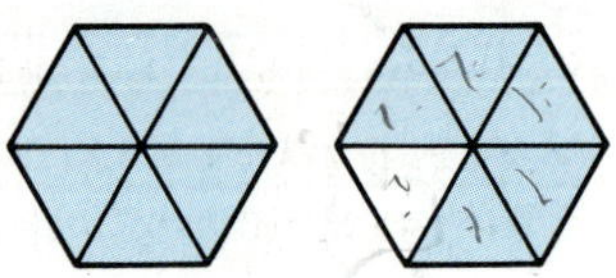

Choose the correct words to complete each sentence.

22. A ______ can name part of a region or group.
23. A ______ has a whole number and a fraction.
24. Different fractions that name the same amount are ______.

Words to Know
equivalent fractions mixed number fraction

Solve.

25. Bill drank $\frac{2}{3}$ of a glass of milk and Ana drank $\frac{3}{4}$ of a glass. Who drank more milk?
26. Al had 8 pears. He gave Sue $\frac{1}{4}$ of them. How many did he give Sue? How many did he have left?

Exploring Problem Solving

THINK
EXPLORE
SOLVE
LOOK BACK

How Do You Cut It?

Mrs. Macaluso is using 2 different-sized pans to make lasagne for 8 people. She wants to be able to cut the lasagne into 12 serving pieces of about equal size.

Thinking Critically

Which two pans should she use? How should she cut the lasagne in the 2 pans so there are 12 servings altogether?

Make drawings of pans that have the same shapes as those shown here, or use Workmat 12.

a.

b.

c.

d.

e.

Analyzing and Making Decisions

1. Look at the pans. Are there any that look like they are the same size? Which ones? Do some look bigger than others? Which ones?
2. **What if** you found two pans that were about the same size? How many servings would you have in each pan?
3. **What if** you found a pan that was twice as big as another? How many servings would you have in each pan?
4. Use your Workmat and try to make 12 equal servings. Show how you would make your servings.
5. **What if** you had to make 15 servings? How would you do it?

Look Back What if you use the circular pan and another pan to make 15 servings? What other pan would you use? Make a drawing to show how your servings would look.

Problem Solving Strategies

Making and Using Tables

Mother had a bad cold and had to stay in bed for 3 days. The children said, "Please let us help do the chores."

Mother asked the children to make the lunches, sweep the floor, and take out the garbage. She said, "Today Becky can choose a chore first, Dan second, and Amy third. Tomorrow Dan can go first, then Amy, and then Becky. On the last day, Amy, Dan, and Becky can go in that order."

How might the children do the chores?

If a problem has many facts, it is wise to write them in a table. A table makes it easier to think about the facts.

Solving the Problem

Think What do you have to find out?

Explore What are the chores? How many days do the children need to do them? How do they decide which chores to do? How would a table like this help you?

	Becky	Dan	Amy
Day 1	Lunches	Sweep	
Day 2			
Day 3			

Solve Show what each child could do each day.

Look Back Look at your chart. Are all three chores done each day? Does each child have to do a different chore each day?

Share Your Ideas

1. Is this the only way that the chores could be done? If not, show another way.

THINK
EXPLORE
SOLVE
LOOK BACK

Practice

Solve. Use a calculator where appropriate.

CHOICES

Use this information to solve 2–5.

During the summer the four Clayton children kept a chart of the books they read.

	Biography	Mystery	Comedy	Sports
Adam	3	2	4	6
Lucy	7	2	2	3
Bruce	0	12	1	1
Mindy	2	3	3	3

2. Who read the most books?
3. Who read the fewest books?
4. Who read the most mystery books?
5. How many more books did Lucy read than Mindy?

Mixed Strategy Review

6. Eva and Jenna are trading stickers. Eva gives Jenna 3 animal stickers for every 5 car stickers. Jenna gives Eva 25 car stickers. How many animal stickers does Eva give Jenna?
7. Channel 5 is showing science-fiction movies all day. A different movie is shown every 2 hours. The first movie begins at 1:30 P.M. When does the fourth movie begin?
8. Bruce and Tanya were in a race with 15 people. Bruce ran the race in 15 minutes. It took Tanya 17 minutes. Who won the race?
9. Tamara bought 5 puzzle books for $3 each. She also bought 4 tapes for $4 each and 3 toy cars for $2 each. How much did she pay for the puzzle books?

Create Your Own

Look back at **8**. Use the information in this problem to write your own problem.

Activity

Exploring Making and Using Frequency Tables

What do you like to collect? Do you collect stamps? baseball cards? rocks? stickers?

What do most students like to collect? Take a survey to find out.

Working together

A. Make a table showing the different things students collect.

B. Students, in turn, tell what they like to collect. Mark each student's choice with a tally (/).

C. Count the tallies to see how many students collect each thing. Each sum is the **frequency** of that choice.

Choice	Tally	Frequency
Stamps	卌 |||	8

Sharing Your Results

1. What did most students collect?
2. How many students collected each item?
3. What is the sum of all the frequencies? Is it the same as the number of students in the survey? Explain.

Extending the Activity

Now try a number cube experiment.

Make another frequency table. This time, record the number that comes up when you toss a number cube.

Work in a small group.

Materials: cube with numbers 1–6

4. Predict how many times you will get each number if you toss the cube 30 times. Record your predictions.

5. Toss the cube 30 times. Make a tally to show what happened each time.

6. Add the tallies to get the frequency of each possible result.

7. Compare the results with your prediction. Did you get the results you expected? How close were you?

Summing Up

8. Discuss the number cube experiment with other groups. Did all the groups get the same results? Why or why not?

9. When you tossed the cube, was there a better chance of one number coming up than another number? Explain.

10. Tell what you think will happen if you toss the cube 60 times. (You may want to do this to test your idea.)

Activity

Exploring Probability

A bag has 3 kinds of fruit snacks. When Joey reaches into the bag, he hopes for pineapple.

What are Joey's chances of getting a pineapple fruit snack? Do this experiment to find out.

Working together

Materials: bag with 3 buttons or counters (1 red, 1 yellow, 1 blue)

A. Predict how many you will get of each color if you draw a button 30 times. Record your prediction.

B. Without looking, draw a button from the bag 30 times. Replace each button before drawing again.

C. Record each result in a frequency table.

D. Count the tallies to get the frequency of each possible result.

E. Compare the results with your prediction. How close was your prediction?

Color	Tally	Frequency
Red		
Blue		
Yellow		

Sharing Your Results

1. Compare your results with those of other groups. Did all groups get the same results? Why do you think this happened?

2. Think about Joey and the bag of snacks. How can you describe Joey's chances of getting the pineapple snack?

Extending the Activity

In the button experiment, there were 3 possible **outcomes:** red, yellow, and blue.

The chance of drawing each color is the same: 1 out of 3. If you draw *many* times, you can expect to get each color about the same number of times.

What if these cubes were in a bag?

3. How many possible outcomes of a draw are there? Name the possible outcomes.

4. What is the chance of drawing black? yellow? green?

5. What is the chance of drawing blue?

Summing Up

Look at the spinner.

6. What are the possible outcomes of a spin?

7. What is the chance of the pointer stopping on each color? Explain your answer.

8. If you spin *many* times, will you get each color about the same number of times? Explain.

Using Problem Solving

A Probability Game

You can play this probability game with one or more partners. The first player to get 12 or more points wins the game.

Working together

Materials: spinner, paper, pencil

Follow these rules.

A. Make a spinner like the one shown.

B. Each player starts with 5 points. If the spinner stops on 1, the player must subtract 1 point. If the spinner stops on 2 or 3, the player adds that many points. Each player keeps a record of the number of spins and the scores.

C. Players take turns.

D. If a player reaches 0, the game is over, and the person with the highest score wins. Otherwise, the player who reaches 12 or more points wins the game.

E. Play at least 3 games.

Spin	Score
Start	5
1	4
2	6
3	9

Sharing Your Ideas

1. How many times did you spin a 1? a 2? a 3? Why do you think this happened?

2. How often did a player's score reach zero? Why do you think this happened?

Extending Your Thinking

3. Suppose you could choose one of these spinners to play this probability game. Which spinner would make the game most exciting?

A

B

C

4. Which spinner would probably help you get 12 points very quickly? Explain.

5. Which spinner would probably make you lose points often? Explain.

6. What would happen with spinner **C**?

7. With a partner, choose one of the designs and make a spinner. Play the game with that spinner.

8. As a joke, Melissa made this design for a spinner. Then she said to her friend, "Please play the game with me. I will even let you go first!" What was Melissa's joke? Why did she let the other person go first?

9. **What if** there was room for only 4 spaces on a spinner? Make a spinner that has 4 spaces and make a set of rules for using the numbers 1, 2, and 3. Then play the game with a friend to see how well it works.

Summing Up

10. **Look back** at 7. Which spinner did you use for your game? Why? How did it work?

11. **Look back** at 8. Would you have gone first against Melissa? Explain.

12. **Look back** at 9. How did your spinner with 4 numbers work? Which spinner made the game the most fun? Why?

Chapter Review

Write a fraction for the part that is blue. pages 266–269

1. 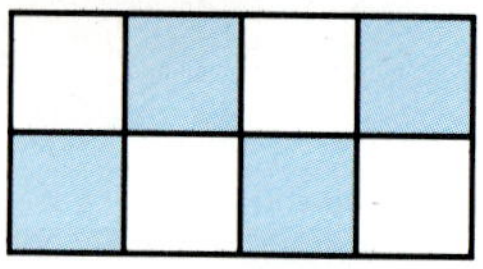

2. 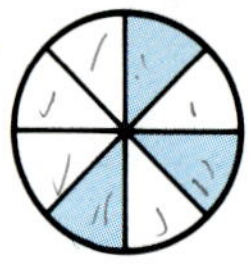

3. 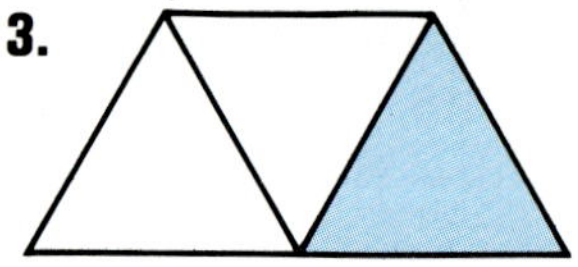

4. 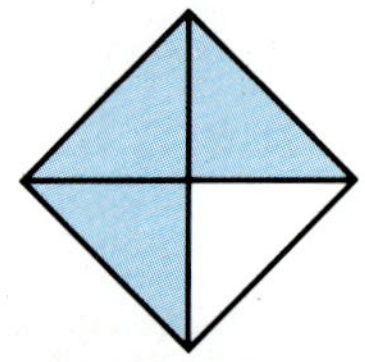

5. 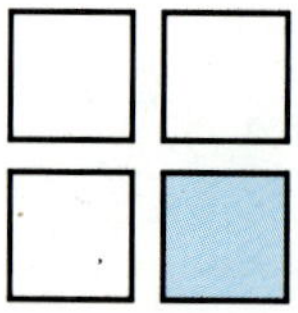

6. 

7. 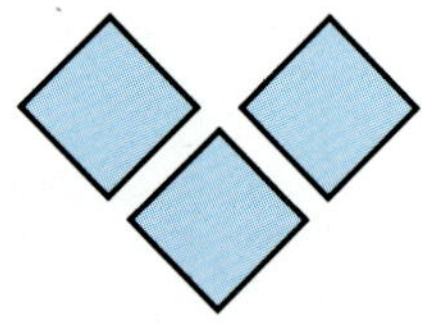

8. 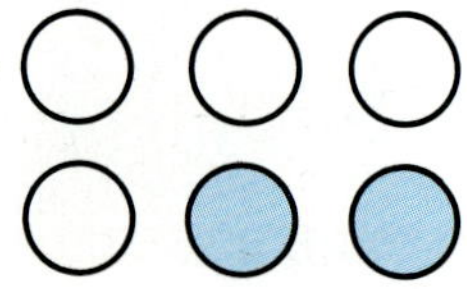

Complete. pages 270–271

9. $\frac{1}{7}$ of 35 = ☐
10. $\frac{1}{5}$ of 45 = ☐
11. $\frac{1}{2}$ of 18 = ☐
12. $\frac{1}{9}$ of 81 = ☐

Compare. Use >, <, or = for ⬬. pages 276–279

13. $\frac{3}{6}$ ⬬ $\frac{2}{6}$
14. $\frac{1}{4}$ ⬬ $\frac{2}{8}$
15. $\frac{4}{6}$ ⬬ $\frac{1}{3}$
16. $\frac{1}{2}$ ⬬ $\frac{4}{6}$

Write a mixed number for the part that is blue. pages 280–281

17. 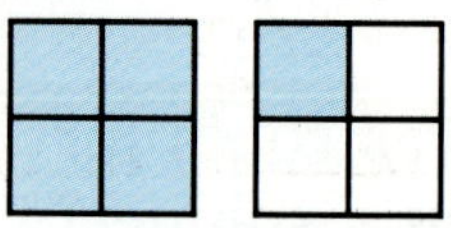

18. 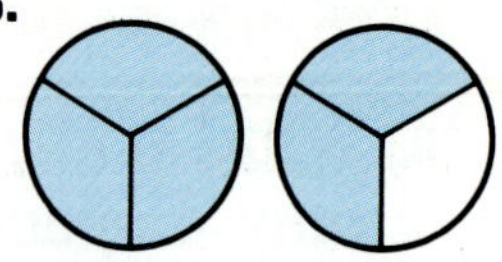

19. 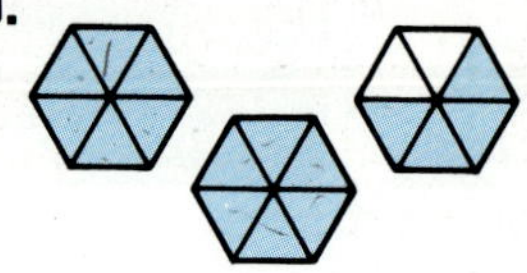

20. 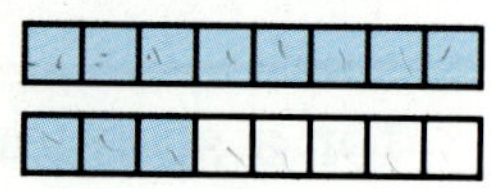

Choose the correct word to complete each sentence.

Words to Know
frequency outcome

21. An _____ is a possible result in an experiment.
22. The sum of the tallies for each outcome tells the _____ of the outcome.

Solve. pages 283–285

23. Third graders kept this record of the fruit they ate on a weekend. Which day did they eat the most fruit?
24. Which fruit was most popular?

Fruit Eaten on a Weekend		
	Sat.	Sun.
Apples	15	5
Pears	2	4
Bananas	21	0
Peaches	8	17

Chapter Test

Write a fraction for the part that is blue.

1. 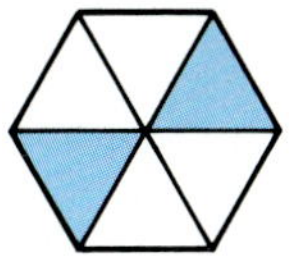

2. 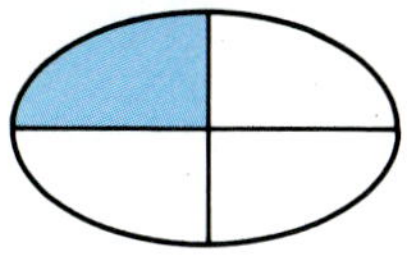

3. 

4. 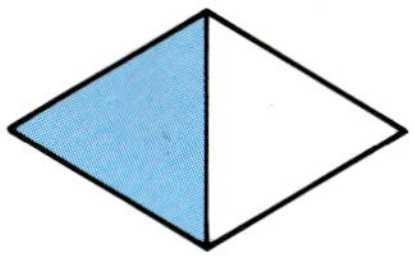

5. 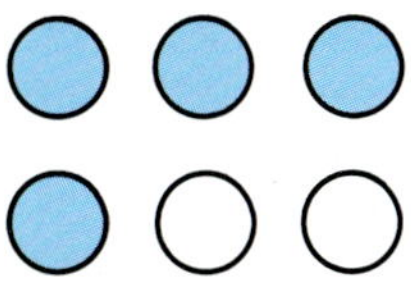

6. 

7. 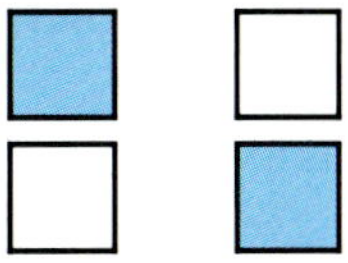

8. 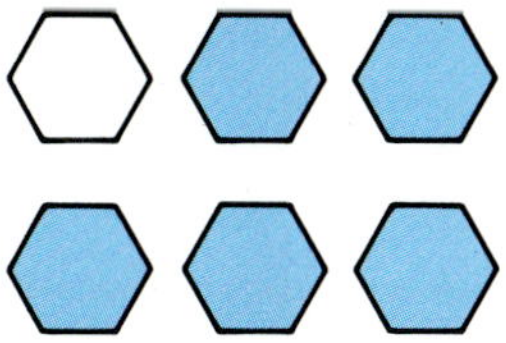

Complete.

9. $\frac{1}{4}$ of 24 $= \square$ 10. $\frac{1}{6}$ of 54 $= \square$ 11. $\frac{1}{8}$ of 40 $= \square$ 12. $\frac{1}{3}$ of 15 $= \square$

Compare. Use $>$, $<$, or $=$ for ●.

13. $\frac{1}{8}$ ● $\frac{6}{8}$ 14. $\frac{2}{4}$ ● $\frac{1}{2}$ 15. $\frac{1}{3}$ ● $\frac{1}{6}$ 16. $\frac{3}{4}$ ● $\frac{2}{3}$ 17. $\frac{3}{5}$ ● $\frac{1}{5}$

Write a mixed number for the part that is blue.

17. 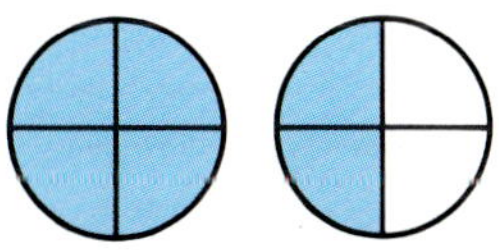

18. 

19. 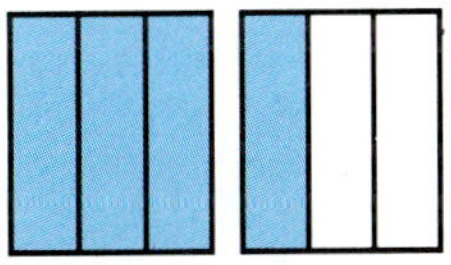

20. 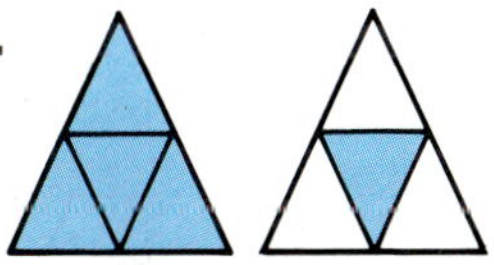

Use the spinner for 22 and 23.

22. What are the possible outcomes of a spin?

23. What is the chance of the pointer stopping on red? on blue? on yellow?

Solve.

24. How many different ways can heads and tails show when a penny, nickel and dime are tossed?

25. One of Chad's ten marbles is red. Three are blue. Write a fraction for the part that is red.

Think A bag has 5 red and 5 blue buttons. In 20 draws would you expect to draw more red or blue buttons? Explain.

Computer Link

Random Numbers

The Logo command **RANDOM** 5 returns a random number from 0 to 4. Do you think that each number 0 to 4 would print if you enter the command 5 times?

random number–a number that is chosen in an unpredictable way.

Materials: Logo

A. Enter the command `PR RANDOM 5` several times. What numbers are shown on the screen?

B. Predict how many times each number from 0 to 4 will print if you print 20 random numbers. Enter the command below. Then record your results with tally marks.

```
REPEAT 20 [PR RANDOM 5]
```

C. Predict how many times each number from 0 to 4 would print out of 100 times. Combine the results of five students to show results for 100 times.

0	1	2	3	4	
					Total

Sharing Your Results

1. Compare your prediction to your results for 20 times. Are they the same? Why or why not?
2. Were the results for 100 times closer to your prediction?

Extending the Activity

Look at how the digits 0 to 9 are printed in a book. Which digits are made of only straight line segments? Which have curves?

3. **What if** you print 100 random digits with the computer. How many times do you expect the computer to print digits with curves? with no curves? Why?

Curves	No Curves

4. Use `RANDOM` commands to print 100 digits from 0 to 9. Enter this command 10 times. Record your results in a chart.

```
REPEAT 10 [PR RANDOM 10]
```

5. Compare the actual results for 100 times with what you expected to happen. Are they the same? Why or why not?

Summing Up

6. Explain how you can make a prediction for
   a. digits from 0 to 4.
   b. curved digits from 0 to 9.

7. What if you use 0 to 9 to print 100 digits randomly? How many would you expect to have no straight parts? Explain.

EXTENSION

Estimating Fractions

How good are you at estimating fractions? Try these activities to find out.

Estimate what fractional part of each shape is blue.

1. 

2. 

3. 

4. 

5. 

6. 

Look at the rectangle *ABCD*. Then draw rectangles that are about:

7. $\frac{1}{2}$ the size of *ABCD*.

8. $\frac{1}{3}$ the size of *ABCD*.

9. $\frac{1}{4}$ the size of *ABCD*.

10. $\frac{3}{4}$ the size of *ABCD*.

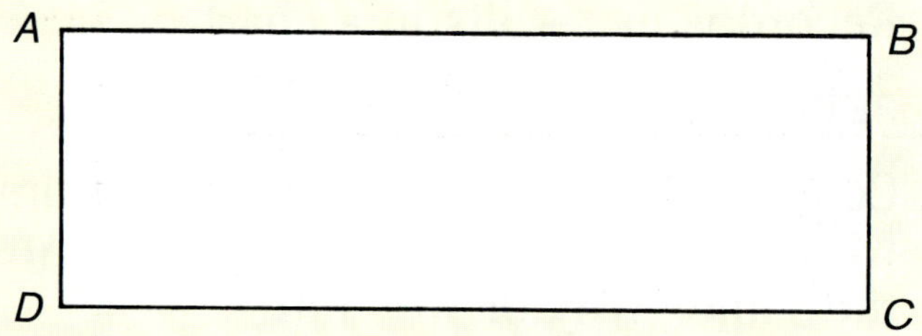

11. Now look at each fraction below. Decide if the fraction is close to 1, close to 0, or close to $\frac{1}{2}$. Then complete a table like the one at the right.

Close to 1	Close to 0	Close to $\frac{1}{2}$

$\frac{3}{8}$ $\frac{5}{8}$ $\frac{7}{8}$ $\frac{9}{10}$ $\frac{1}{8}$ $\frac{1}{6}$ $\frac{4}{10}$

$\frac{1}{3}$ $\frac{2}{10}$ $\frac{5}{6}$ $\frac{6}{10}$ $\frac{4}{6}$ $\frac{2}{6}$ $\frac{2}{3}$

12. Compare your results with a classmate's results. Did you both get the same results? Share your thinking. Did you get better at estimating fractions as you did this lesson? Explain.

Maintaining Skills

Choose the correct answer. Write A, B, C, or D.

1. $9 \times 8 = \square$

A 64 **C** 63

B 72 **D** not given

2. $\begin{array}{r} 9 \\ \times 4 \\ \hline \end{array}$

A 32 **C** 36

B 34 **D** not given

3. $5 \times 2 \times 3 = \square$

A 30 **C** 25

B 21 **D** not given

4. $3\overline{)27}$

A 9 **C** 6

B 7 **D** not given

5. $28 \div 4 = \square$

A 8 **C** 7

B 9 **D** not given

6. What is another fact in this fact family? $9 \times 7 = 63$

A $63 - 7 = 8$ **C** $7 \times 9 = 64$

B $8 \times 7 = 56$ **D** not given

7. What part is shaded?

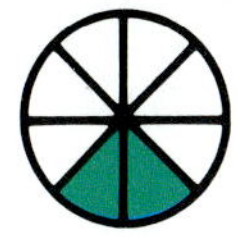

A $\frac{2}{6}$ **C** $\frac{4}{8}$

B $\frac{2}{8}$ **D** not given

8. $\frac{1}{6}$ of $18 = \square$

A 4 **C** 3

B 2 **D** not given

9. Compare. $\frac{1}{5}$ ⬬ $\frac{3}{5}$

A $<$ **C** $=$

B $>$ **D** not given

Solve.

10. Lunch tickets cost $1.25 each. Rita bought three tickets on Monday. How much change did she receive from $5.00?

A $1.25 **C** $2.75

B $2.00 **D** not given

11. Sarah and her 2 sisters bought 2 plants for their grandmother. One cost $5.00, and the other cost $7.00. If they shared the cost evenly, how much did each girl pay?

A $12.00 **C** $6.00

B $5.00 **D** not given

10 Understanding Measurement

THEME The Great Outdoors

Sharing What You Know

Think about the animals shown in the picture. Would you use ounces or pounds to weigh a zebra? Would you use inches or feet to measure a chipmunk? What units of measurement would you use for the other animals?

Using Language

Would you measure a snail in inches or feet? Have you ever seen a snail inch along the sand? To **inch along** means to move very slowly, or little by little. In mathematics, an **inch** is a small unit used to measure length. How are these meanings of **inch** alike? How are they different?

Words to Know: centimeter, decimeter, meter, liter, kilometer, inch, foot, yard, cup, ounce

Be a Problem Solver

A grasshopper is 2 inches long. It can jump 6 times its own length. How can you find out how far it jumps?

List all the animals you can that would think an inch is a long way to go.

GETTING STARTED

Jane says that this book is 6 paper clips wide. Tom says it is 3 toothpicks wide. Why are their measurements different?

Estimating and Measuring Length: Centimeter and Decimeter

- The **centimeter (cm)** is a standard unit of length in the metric system.

The length of the slug is between 1 centimeter and 2 centimeters. Its length to the nearest centimeter is 2 cm.

- The **decimeter (dm)** is used to measure lengths longer than a centimeter.

1 decimeter = 10 centimeters

The earthworm is about 1 decimeter long.

Check Your Understanding

Estimate each length. Then use a ruler to measure.

1. 

2. 

3. 

Share Your Ideas Look around your classroom. What are some objects you could measure in centimeters? decimeters?

Practice

Estimate each length. Then measure to the nearest centimeter.

4. 

5. 

6. 

7. 

8. width of this book

9. length of this book

10. length of a new pencil

11. length of your thumb

Use a ruler to draw a bar each length.

12. 3 centimeters
13. 9 centimeters
14. 1 centimeter
15. 1 decimeter
16. 5 centimeters
17. 2 decimeters

Without using a ruler, draw a bar of each length. Then measure to check.

18. 6 centimeters
19. 4 centimeters
20. 8 centimeters
21. 2 centimeters
22. 1 decimeter
23. 2 decimeters

Think and Apply

DATA **24.** Choose 5 objects in your classroom. Estimate and measure the length of each object. Record your work.

Object	Estimate	Measurement
A.		
B.		
C.		
D.		
E.		

25. Were your estimates usually too high or too low?

26. What objects were less than 10 decimeters? more than 10 decimeters?

27. Make a list of the objects in order from shortest to longest.

SUMMING UP

Look around your classroom. Find 3 objects that you estimate to be shorter than 5 centimeters; 3 objects longer than 1 decimeter.

GETTING STARTED

Which is farther from your classroom: the principal's office or the lunchroom? How could you find out?

Meter and Kilometer

The golden eagle is about 1 meter long. It usually keeps its nest high on a cliff.

▸ We use the **meter (m)** to measure lengths much longer than a centimeter.

1 meter = 100 centimeters
1 meter = 10 decimeters

A doorknob is about 1 meter from the floor.

▸ We use the **kilometer (km)** to measure longer lengths.

1 kilometer = 1,000 meters

The distance from New York City to Washington, D.C. is about 380 kilometers.

Would you use meters or kilometers to measure the length and width of your classroom?

Check Your Understanding

Choose cm, m, or km to complete each sentence.

1. Maps show distances in ________.
2. A shoe measures about 25 ________ long.
3. Your classroom is about 3 ________ high.
4. A paper clip is about 3 ________ long.

Share Your Ideas Would you measure the distance you live from a park in meters or kilometers? Explain.

Practice

Choose cm, m, or km to complete each sentence.

5. It is about 1,325 _____ from Maine to Florida.
6. A flashlight is about 20 _____ long.
7. The distance around a lake is 15 _____.
8. The length of a car is about 4 _____.
9. The height of a flagpole is about 8 _____.

Choose the better measurement for each.

10. width of the classroom door

    a. 1 dm b. 1 m
    c. 10 km d. 10 cm

11. length of pen

    a. 15 m b. 2 dm
    c. 1 km d. 15 cm

Complete.

12. 1 km = _____ m
13. 1 m = _____ cm
14. 1,000 m = _____ km
15. 100 cm = _____ m
16. 1 dm = _____ cm
17. 10 cm = _____ dm
18. 2 m = _____ cm
19. 5 m = _____ cm
20. 2,000 m = _____ km

Think and Apply

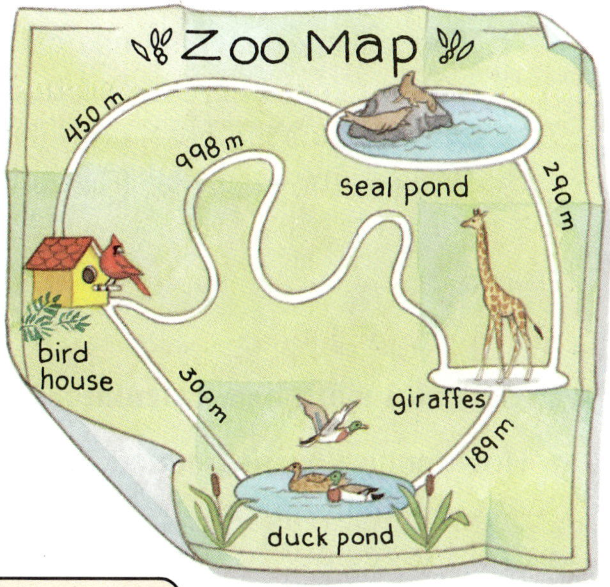

Find the distance

21. from the seal pond to the giraffes and back.
22. from the bird house to the seal pond to the giraffes.
23. of the shortest path from the duck pond to the seal pond.

Name 3 distances you would measure in meters. Then name 3 more you would measure in kilometers.

SUMMING UP

Using Problem Solving

Interview: Calculators in the Bakery

Ron Nardone, a baker at the Somerset Hills Bakery Shop, says, "Calculators help me when I change recipes. Here are some of the ingredients I use for muffins. If I make fewer muffins, I must use a smaller amount of each ingredient. This recipe is for 32 dozen muffins."

20 pounds of flour	12 pounds of water
24 pounds of sugar	18 pounds of eggs

Note: Bakers measure ingredients on a scale, in pounds

A. **What if** a customer orders 8 dozen muffins?
8 is $\frac{1}{4}$ of 32 or $\frac{1}{4}$ of 32 = 8
Find $\frac{1}{4}$ of each ingredient.
Use a calculator to divide each amount by 4.

Think: 8 is one of four equal parts of 32.

Flour: $\frac{20 \text{ pounds}}{4} = 5$ pounds

B. Explain how to find the amount of water needed.

C. The salespeople also use calculators. A customer wants $\frac{1}{2}$ loaf of rye bread that costs 98 cents.
To find $\frac{1}{2}$ of the cost, you divide by 2.
What is the cost of $\frac{1}{2}$ loaf of bread that is 84 cents?

$\frac{98 \text{ cents}}{2} = 49$ cents

Sharing Your Ideas

CHOICES

Use mental math or a calculator.

1. How much flour is needed for $\frac{1}{2}$ of the muffin mix recipe? for $\frac{1}{5}$ of the recipe?

2. The cost for $\frac{1}{2}$ loaf of bread is $\frac{1}{2}$ the cost of a whole loaf at the Pastry Shop. What does $\frac{1}{2}$ loaf of bread cost if one whole loaf costs $.90?

THINK
EXPLORE
SOLVE
LOOK BACK

Practice

Use mental math or a calculator.

CHOICES

3. A recipe calls for 40 pounds of flour and 36 pounds of eggs. Find how much flour and eggs are needed for $\frac{1}{4}$ of the recipe.

4. The bakery has 72 muffins. One fourth of the muffins are oat and one third are bran. The rest are raisin muffins. How many of each kind are there?

5. Mr. Nardone is making 264 rolls. One half have sesame seeds. How many rolls have sesame seeds? How many do not have sesame seeds?

6. One dozen rolls cost $2.50. Explain how you can use a calculator to find the cost of $\frac{1}{2}$ dozen.

7. A baker's helper baked 128 onion rolls. He sold $\frac{1}{4}$ of them while they were still warm. Then he sold $\frac{1}{2}$ of those that were left. How many rolls in all were sold? How many rolls were not sold?

Summing Up

8. Recipes that we use at home usually call for a certain number of eggs. Why do you think a baker would weigh the eggs for a recipe?

9. Write your own bakery problem. Use the numbers 360 and $\frac{1}{6}$. Trade problems with a classmate. Then solve your problem any way you wish.

GETTING STARTED

Some gasoline pumps measure the gasoline in liters. Think of as many products as you can that are measured in liters.

Estimating Capacity: Liter

When Tina plays soccer, she drinks water from a sports bottle. It holds about 1 liter of water.

- We use the **liter (L)** to measure amounts of liquid.

1 L 2 L 5 L

- We use **milliliter (mL)** to measure small amounts of liquid.

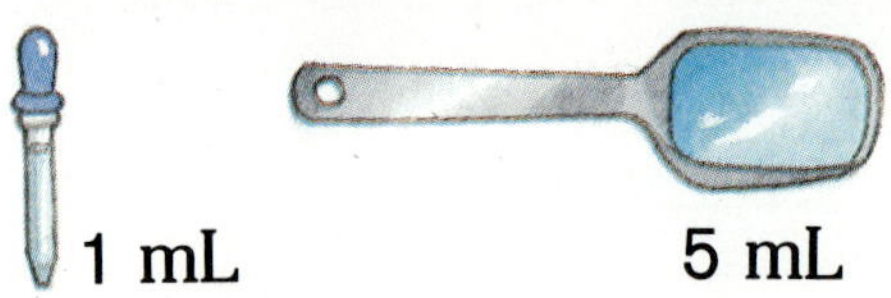

1 mL 5 mL

1 liter = 1,000 milliliters

Would you use liters or milliliters to measure the amount of water in a full bathtub?

Check Your Understanding

Choose liters or milliliters to measure each.

1. medicine in a spoon

2. juice in a glass

3. water in a swimming pool

4. soup in a spoon

5. punch in a punchbowl

6. gasoline in a car

Share Your Ideas Why do you think scientists often use the milliliter unit of measure?

Practice

Choose liters or milliliters to measure each.

7. juice in a spoon
8. water in a fish bowl
9. soup in a bowl
10. water in a water tank
11. paint in a bucket
12. milk in a glass

Choose the best estimate for each.

13. A glass of water contains
 a. 250 L b. 2 mL
 c. 10 L d. 250 mL

14. A pitcher of juice contains
 a. 2 mL b. 25 L
 c. 2 L d. 10 mL

15. A pot of soup contains
 a. 1 mL b. 3 L
 c. 50 L d. 5 mL

16. A cup of tea contains
 a. 150 L b. 1 L
 c. 150 mL d. 1 mL

Compare. Use $>$, $<$, or $=$ for ●.

17. 1 L ● 1 mL
18. 1 L ● 1,000 mL
19. 1 L ● 900 mL
20. 2 L ● 4,000 mL

Think and Apply

21. If Tina uses 100 liters of water in her daily shower, how many liters of water does she use in one week?

22. Five soccer players each drank 200 milliliters of juice. How many milliliters of juice did they drink? How many liters is that?

Logical Thinking

23. **What if** you use water from the pitcher to fill each container? How much water will be left in the pitcher?

A small spoon holds about 5 mL of liquid. How can you use it to estimate how much a glass holds?

SUMMING UP

Activity

Exploring Estimating Weight: Gram

Which weighs more the  or the  ?

Hold one object in each hand. Estimate by comparing the weights you feel. Then use a balance scale to check. Is your estimate correct?

▸ We use the **gram (g)** to measure the weight of light objects.

A paper clip weighs about 1 gram. What other objects weigh about 1 gram?

Working together

Materials: balance scale, 1-gram weight

A. Collect several light objects.

B. To sort the objects, estimate whether each weighs about 1 gram or more than 1 gram.

C. Use the scale to compare each object with the gram weight. Record your results. How do the results compare with your estimates?

About 1 Gram	More Than 1 Gram

Sharing Your Results

1. Make two class lists: one for objects that weigh about 1 gram and one for objects that weigh more than 1 gram.
2. Name any objects on this list that weighed more than, or less than, you expected.

Extending the Activity

Work as a class.

3. Find packages whose contents are labeled in grams. Think about how you order numbers. Now arrange the packages in order from lightest to heaviest.
4. **What if** the labels did not have the packages' weights? Discuss how you could order them from lightest to heaviest.
5. Use your ideas to order ten objects.
6. How could this activity help you know more about the weight of light objects?

Summing Up

Choose the better estimate of weight for each.

7. an apple
   a. 10 g b. 250 g
8. a pen
   a. 500 g b. 5 g
9. a book
   a. 850 g b. 8 g
10. a crayon
    a. 6 g b. 600 g
11. a piece of paper
    a. 150 g b. 1 g
12. a shoe
    a. 2 g b. 200 g
13. an eraser
    a. 5 g b. 500 g
14. a quarter
    a. 10 g b. 950 g
15. eight toothpicks
    a. 1 g b. 180 g

Solve.

16. There are 3 items that weigh 645 grams altogether. Two items weigh 200 grams each. How much does the third item weigh?
17. **What if** you have a nickel that weighs 5 grams? How could you use it to help you find other objects that weigh about 5 grams?

GETTING STARTED

Which weighs more: your math book or your reading book? How could you find out?

Kilogram

The Outdoor Club is taking a river trip. Can 3 people weighing 85, 55, and 35 kilograms share one canoe? The leader said yes after she knew their combined weight. Why?

▸ We use the **kilogram (kg)** to measure the weight of heavy objects.

1 kilogram = 1,000 grams

This book weighs about 1 kilogram.

This sack of potatoes weighs about 10 kilograms.

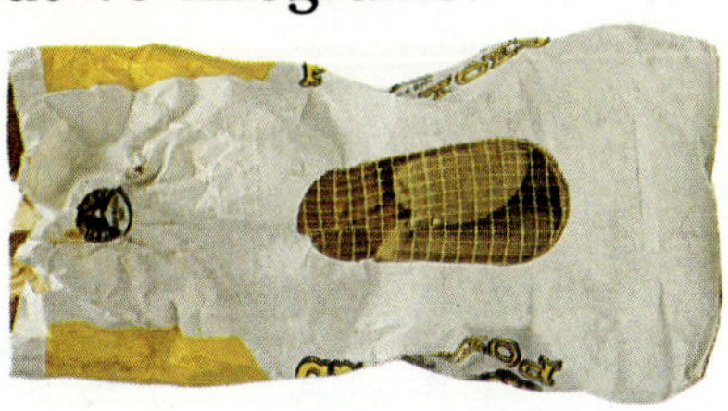

Check Your Understanding

Choose grams or kilograms to measure the weight of each.

1. dime **2.** table **3.** note pad

4. fishing boots **5.** button **6.** piano

7. canoe **8.** fish hook **9.** full knapsack

Share Your Ideas Look back at **1–9**. Estimate which object weighs the least. the most.

Practice

Choose grams or kilograms to measure the weight of each.

10. feather
11. spoon
12. camp stove
13. leaf
14. bag of charcoal
15. tent
16. cast iron pot
17. dollar bill
18. hat
19. fishing rod
20. jeep
21. sleeping bag

Choose the better estimate for each weight.

22. horse **a.** 450 kg **b.** 4 kg
23. bicycle **a.** 200 kg **b.** 13 kg
24. automobile **a.** 5 kg **b.** 900 kg
25. baseball **a.** 1 kg **b.** 400 kg

Compare. Use $>$, $<$, or $=$ for ●.

26. 1 gram ● 1 kilogram
27. 1,000 grams ● 1 kilogram
28. 4,000 grams ● 2 kilograms
29. 5,000 grams ● 5 kilograms

Think and Apply

Work in a group. Use a balance scale with weights.

30. Collect 5 objects and weigh them.
31. Record the results.
32. Make a pictograph of the results.
33. Write the weights in order from lightest to heaviest.
34. Make up two questions that can be answered from your pictograph.

Exchange pictographs with other groups. How could this activity help you estimate the weight of other objects?

SUMMING UP

GETTING STARTED

When was the last time you looked at a thermometer? Why did you need to know the temperature?

Degree Celsius

Inside the cave the temperature is cool. It is 10 degrees Celsius.

Outside the cave the temperature is warm. It is 30°C.

▸ We use the **degree Celsius** to measure temperature.
We write **°C** for degree Celsius.

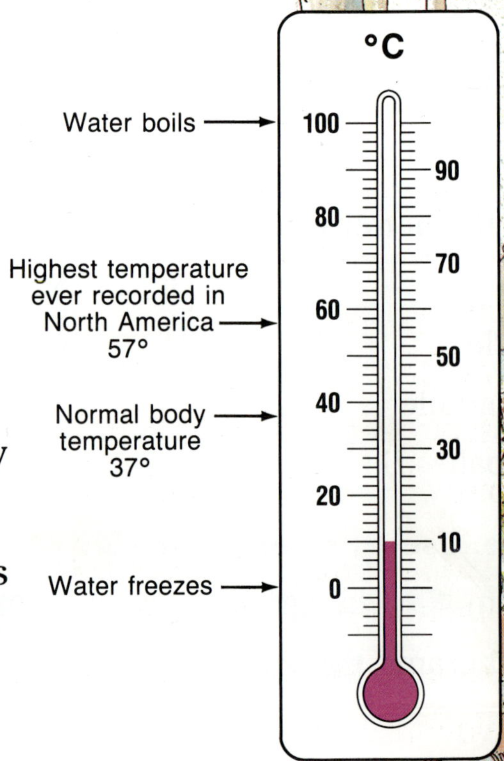

Look at the thermometer.
What is the temperature shown by the red column?

How would you read temperatures that are shown below the freezing point of water?

Check Your Understanding

Use the thermometer to answer each question.

1. What is the normal body temperature?
2. What is the freezing point of water?
3. What is the boiling point of water?
4. What is the highest temperature ever recorded in North America?
5. Is it a hot or cold day when the temperature is 35°C? 2°C?

Share Your Ideas What kind of clothing would you wear when it is 35°C outside? What type of activity could you do at 0°C?

Practice

Write each Celsius temperature shown.

6. 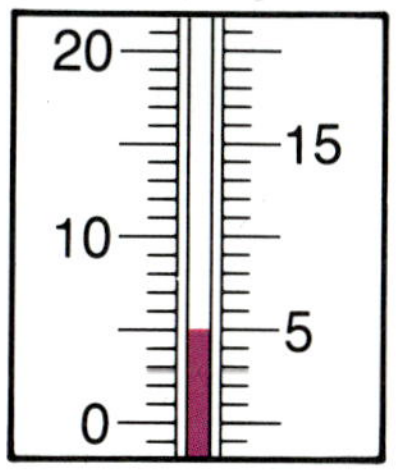

7. 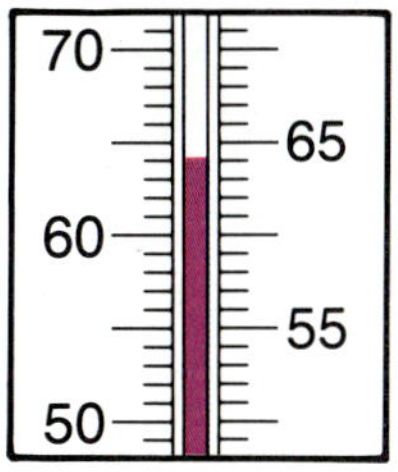

8. 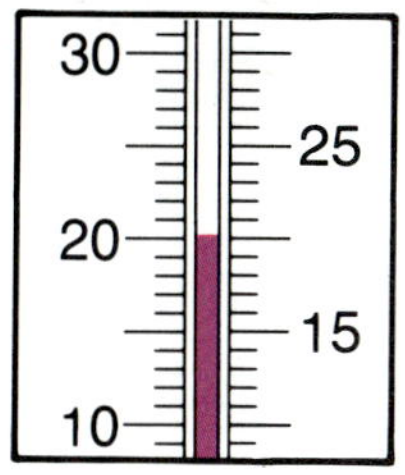

Match. Choose the most reasonable temperature.

9.	hot summer day	**a.**	5°C
10.	cold winter day	**b.**	19°C
11.	room temperature	**c.**	80°C
12.	hot soup	**d.**	40°C

Use the thermometer on page 312. Write the temperature for each.

13. 10° above normal body temperature

14. 10° below the boiling point

Think and Apply

Our body temperature is 37°C. Animals have different temperatures. Use the table to find **15** and **16**.

ANIMAL TEMPERATURES	
Blue jay	43°C
Sparrow	41°C
Goat	40°C
Horned lizard	35°C
Yellow perch	16°C

15. the highest and the lowest body temperature

16. body temperatures higher than ours

> Estimate the Celsius temperature in your room. Compare your estimate to those of others. How can you check your estimate?

Mixed Review

1. $317 - 26$

2. $703 - 79$

3. $689 + 54$

4. $538 + 126$

5. 1×8

6. 5×0

7. 8×9

8. 6×7

9. $9\overline{)45}$

10. $6\overline{)54}$

11. $3\overline{)9}$

12. $4\overline{)0}$

13. $1{,}054 - 643$

14. $1{,}276 + 95$

15. $\square + 391 = 538$

16. $472 - \square = 109$

17. $8 \times \square = 48$

18. $\square \times 6 = 36$

19. $7 \times \square = 49$

20. $\square \times 9 = 81$

21. $63 \div 7 = \square$

22. $16 \div 2 = \square$

23. $\square \div 7 = 8$

SUMMING UP

Midchapter Review

Measure each to the nearest centimeter. pages 300–301

1. 

2. 

Choose cm, m, or km to measure each. pages 300–303

3. length of a crayon
4. distance to the zoo
5. height of a house

Choose liters or milliliters to measure each. pages 306–307

6. bowl of soup
7. tropical fish tank
8. glass of juice

Choose grams or kilograms to measure each. pages 308–311

9. dog
10. safety pin
11. box of apples

Write each Celsius temperature. pages 312–313

12. 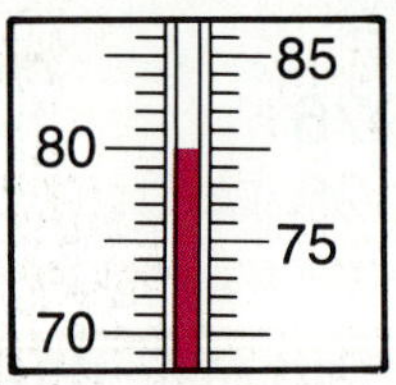

13. 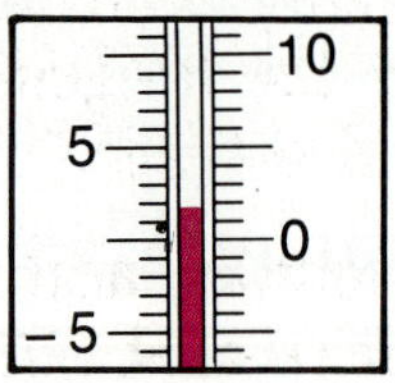

14. 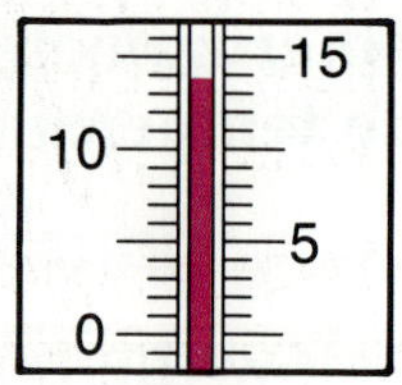

Choose the correct word to complete each sentence.

15. We use the _____ to measure lengths.
16. We use the _____ to measure the weight of light objects.
17. We use the _____ to measure the weight of heavy objects.

Words to Know
gram
centimeter
kilogram

Solve.

18. Jan had 1 liter of milk. She used 300 milliliters of milk to make muffins. She used 500 milliliters to make pancakes. How much milk was left?

19. Steve hiked 220 meters to the pond and then 595 meters to the lodge. How far did he hike? How much more or less than a kilometer was that?

Exploring Problem Solving

THINK
EXPLORE
SOLVE
LOOK BACK

How Far Would You Hike?

The Harris family is planning to hike some trails in the park. The map lists the distances between different starting points marked by the [x]. It also lists the height above sea level for each starting point.

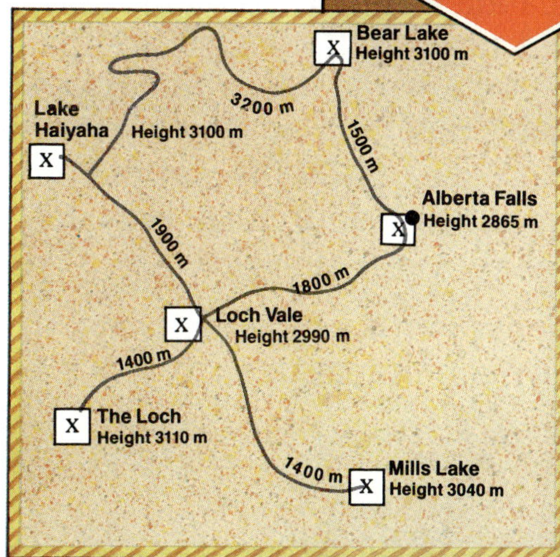

Thinking Critically

Which trails look like they would be easier? Which trails seem harder to walk?

Analyzing and Making Decisions

1. If you hiked from one starting point to the next, which hike would be the longest? Which trail would be the shortest?
2. Look at each trail again. Beside each starting point is the height of that point. **What if** each trail were either all uphill or all downhill? Would you be going uphill or downhill on each trail? How far uphill or downhill is it?
3. What is the longest hike you could take by going from one point to another point? How far uphill or downhill would it be?
4. Which trail do you think would be the easiest? Explain. Which trail do you think would be the hardest to hike? Explain.

Look Back What if you hiked to one point and back to where you started? Which trail would you take? Explain.

Problem Solving Strategies

Alternate Solutions

Amanda wants to go hiking. She does not want to walk more than 3,500 m. Is the following hike less than 3,500 m: from The Loch to Loch Vale (1,400 m) and then from Loch Vale to Lake Haiyaha (1,900 m)?

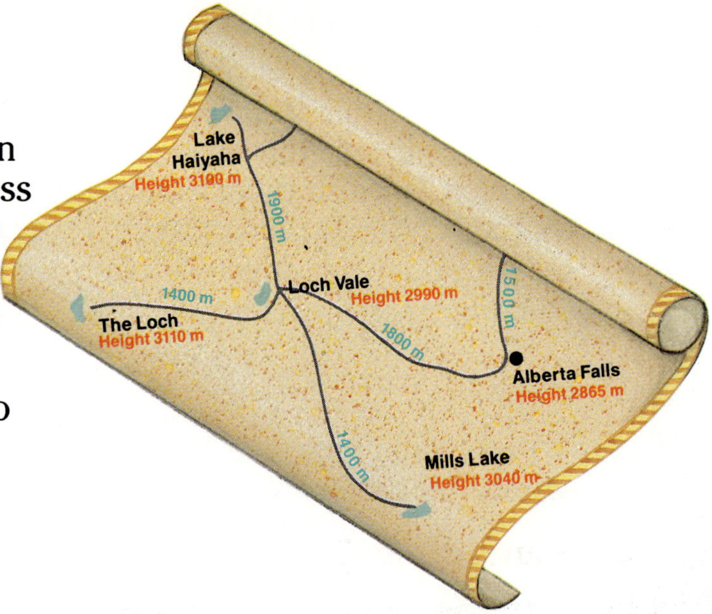

There often are different ways to solve a problem.

Solving the Problem

Think What is the question?

Explore What is the longest distance that Amanda wants to hike? How long is each hike? How can you tell if the total distance is more than 3,500 m by adding the lengths of the two trails? What could you tell by subtracting 1,400 m from 3,500 m?

Solve Is the hike less than 3,500 m?

Look Back How did doing the problem two ways help you check your answer?

Share Your Ideas

1. Which method would you use to solve the problem? Explain.

THINK
EXPLORE
SOLVE
LOOK BACK

Practice

Solve. Use a calculator where appropriate.

CHOICES

2. Mitchell is doing some landscaping work at the park. Beginning at one end of the park, he plants a shrub every 5 feet along a 30-foot walk. How many shrubs does he plant?

3. Ms. Molenta is riding her bike on a trail. She rides 2 km every 5 minutes. If she rides at this speed for 20 minutes, how far will she go?

Use this information to solve 4–5.

Each canoe holds 3 people.

4. Can 12 people ride in 4 canoes? Explain.

5. Can 16 people ride in 5 canoes? Explain.

Create Your Own

Write a problem of your own which you can solve in more than one way.

Mixed Strategy Review

6. The 30 members of the hiking club are going camping. They have 5 tents. Are there enough tents?

7. Each of the 30 campers is put into one of 10 equal groups for chores. How many campers are in each group?

Use this information to solve problems 8–9.

Art Room A has 23 pairs of scissors, 24 rulers, and 15 compasses. Art Room B has 28 pairs of scissors, 19 rulers, and 12 compasses. Art Room C has 16 rulers and 18 compasses.

8. How many rulers are there in all? How many scissors are there? How many compasses are there?

9. Which room has the most of these supplies?

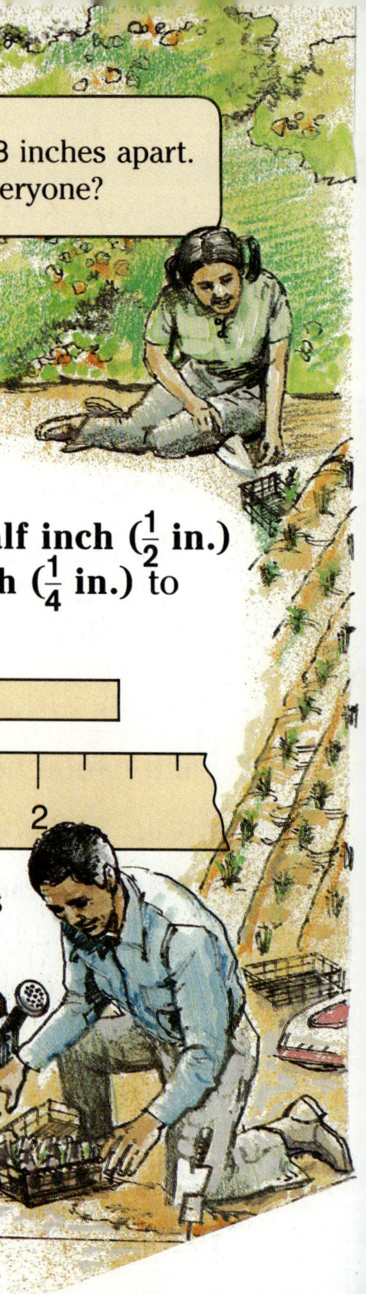

GETTING STARTED

Try to hold your hands about 3 inches apart. Is the distance the same for everyone?

Inch and Fractions of an Inch

Everyday Susie works in the garden with Grandpa. Today they are pulling carrot seedlings so that the remaining plants will be 2 inches apart.

▸ We use the **inch (in.)** as a standard unit in the **customary system** to measure length.

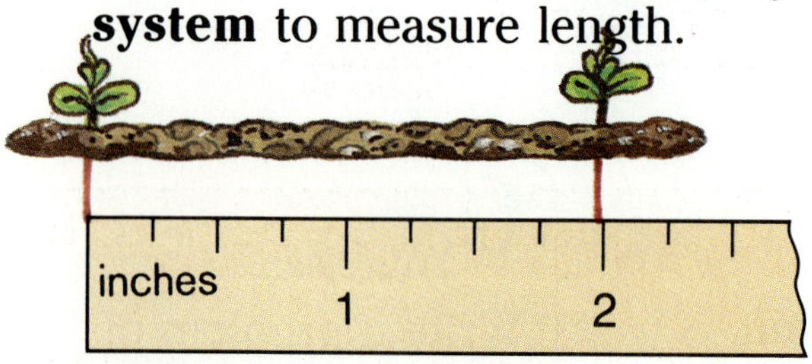

The distance is between 1 inch and 2 inches. It is closer to 2 inches. The distance is 2 inches to the nearest inch.

▸ We also use the **half inch ($\frac{1}{2}$ in.)** and the **quarter inch ($\frac{1}{4}$ in.)** to measure length.

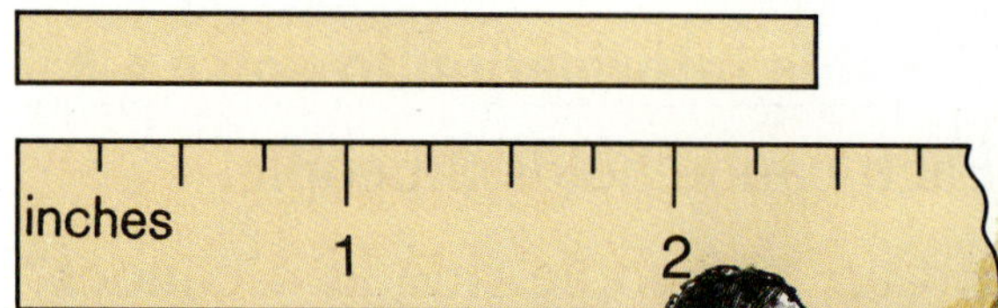

This bar is $2\frac{1}{2}$ inches long to the nearest $\frac{1}{2}$ inch.

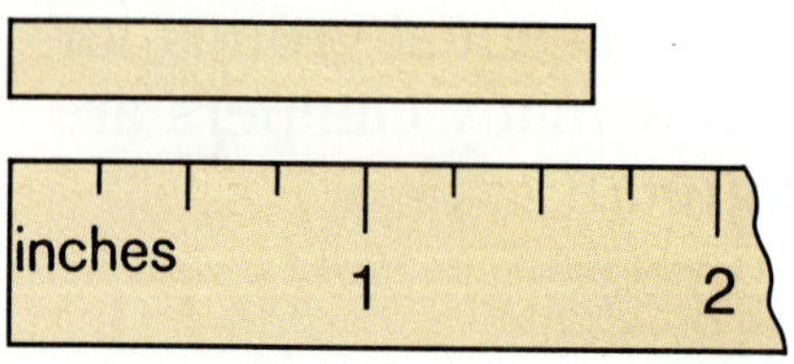

This bar is $1\frac{3}{4}$ inches long to the nearest $\frac{1}{4}$ inch.

Check Your Understanding

Estimate each length. Then measure to the nearest inch.

1. 

2. 

Share Your Ideas Look around the classroom. Find 3 objects that are about 1 inch long.

Practice

Estimate each length. Then measure to the nearest inch.

3. 

4. 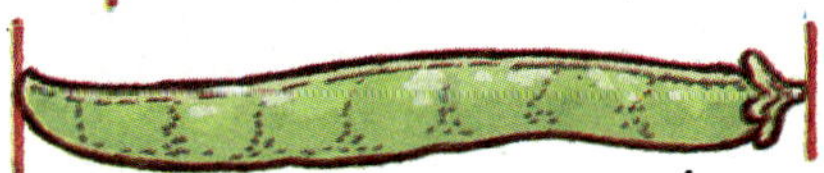

5. 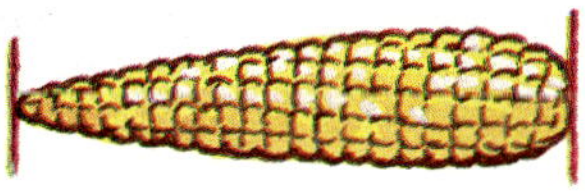

Measure to the nearest $\frac{1}{2}$ inch.

6. 

7. 

8. 

9. 

Measure to the nearest $\frac{1}{4}$ inch.

10. 

11. 

Use a ruler to draw a bar of each length.

12. $1\frac{1}{2}$ in.
13. $4\frac{1}{4}$ in.
14. $5\frac{1}{4}$ in.
15. $3\frac{3}{4}$ in.

Without using a ruler, draw a bar of each length. Then measure to check.

16. 8 in.
17. $3\frac{1}{2}$ in.
18. $6\frac{1}{4}$ in.
19. $1\frac{3}{4}$ in.

Think and Apply

DATA 20. Choose 5 objects. Measure each to the nearest inch. Record your data in a table like this.

Object	Measured Length	Estimated Length

Ask your classmates to estimate the lengths of the objects without looking at your table.

Visual Thinking

21. Decide without measuring: Which bar is the shortest? the longest? Which one is about 1 inch long? 2 inches long?

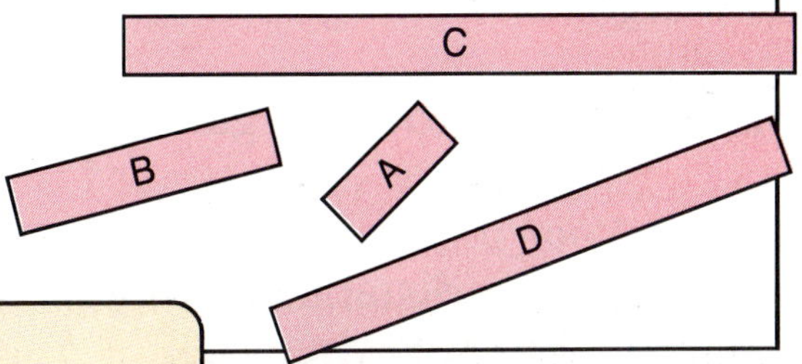

SUMMING UP

Check your classmates' estimates in **20**. Were they closer to the actual length for the shorter objects or for the longer objects?

GETTING STARTED

When people say that someone is 3 feet 2 inches tall, what do they mean? How tall are you?

Foot, Yard, Mile

Jake and his friends held an outdoor paper plate toss. Jake tossed the plate 7 feet. Janice tossed the plate 3 feet. Jake's cousin Harry tossed the plate 2 feet.

- We use the **foot (ft)** to measure length.

12 inches = 1 foot

- We use the **yard (yd)** to measure longer lengths.

36 inches = 1 yard
3 feet = 1 yard

- We use the **mile (mi)** to measure very long distances.

5,280 feet = 1 mile
1,760 yards = 1 mile

Which player tossed the plate 1 yard? more than 1 yard? less than 1 yard?

Check Your Understanding

Choose foot, yard, or mile to measure each.

1. height of a swing

2. length of a jogging track

3. a pass in a football game

4. length of a basketball court

Share Your Ideas Would you use inches to measure the distance from your home to school? Why or why not?

Practice

Choose foot, yard, or mile to measure each.

5. length of a room
6. height of a door
7. width of a library table
8. distance across a lake
9. length of an ice skating rink
10. height of a basketball hoop

Estimate. Then measure to the nearest foot.

11. length of your desk
12. width of your desk
13. height of a classmate
14. length of the chalkboard

Complete.

15. 12 inches = _____ foot
16. 1 foot = _____ inches
17. 36 inches = _____ yard
18. 1 yard = _____ feet
19. 5,280 feet = _____ mile
20. 1 yard = _____ inches
21. 2 yards = _____ feet
22. 15 feet = _____ yards

Think and Apply

23. When the wind was blowing, Jake was able to toss the paper plate 9 feet. How many yards is that?
24. Plan your own paper plate toss.
    - Where would it be?
    - Would you have a starting line?
    - What materials would you need?
    - How would you measure the distances the plates were tossed?

Mathematics and History

Units of measure came from the length of a person's hands, arms, and feet.

25. How wide is the chalkboard in cubits? How long is the classroom in paces?
26. Compare your measurements with those of others. Explain any differences.

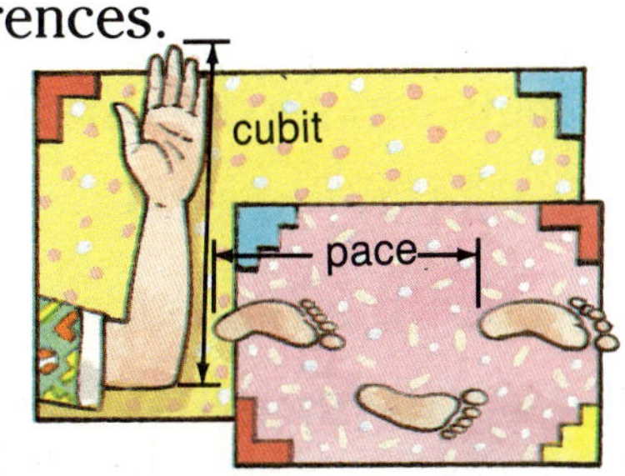

SUMMING UP

You can choose any unit to measure the length of an object. Why are some units better than others?

GETTING STARTED What size milk container does your school have in the lunchroom? Read the label to find out how much milk it contains.

Cup, Pint, Quart, Gallon

Jane's recipe for fruit punch calls for 1 cup of orange juice per serving. How much orange juice does she need for 16 servings?

2 cups are needed for 2 servings.

2 cups (c) = 1 pint (pt)

4 cups, or 2 pints, are needed for 4 servings.

2 pints (pt) = 1 quart (qt)

16 cups, or 4 quarts, are needed for 16 servings.

4 quarts (qt) = 1 gallon (gal)

Jane needs 16 cups, or 1 gallon, of orange juice.

How much orange juice would Jane need for 24 servings?

Check Your Understanding

Choose cup, pint, quart, or gallon to measure each.

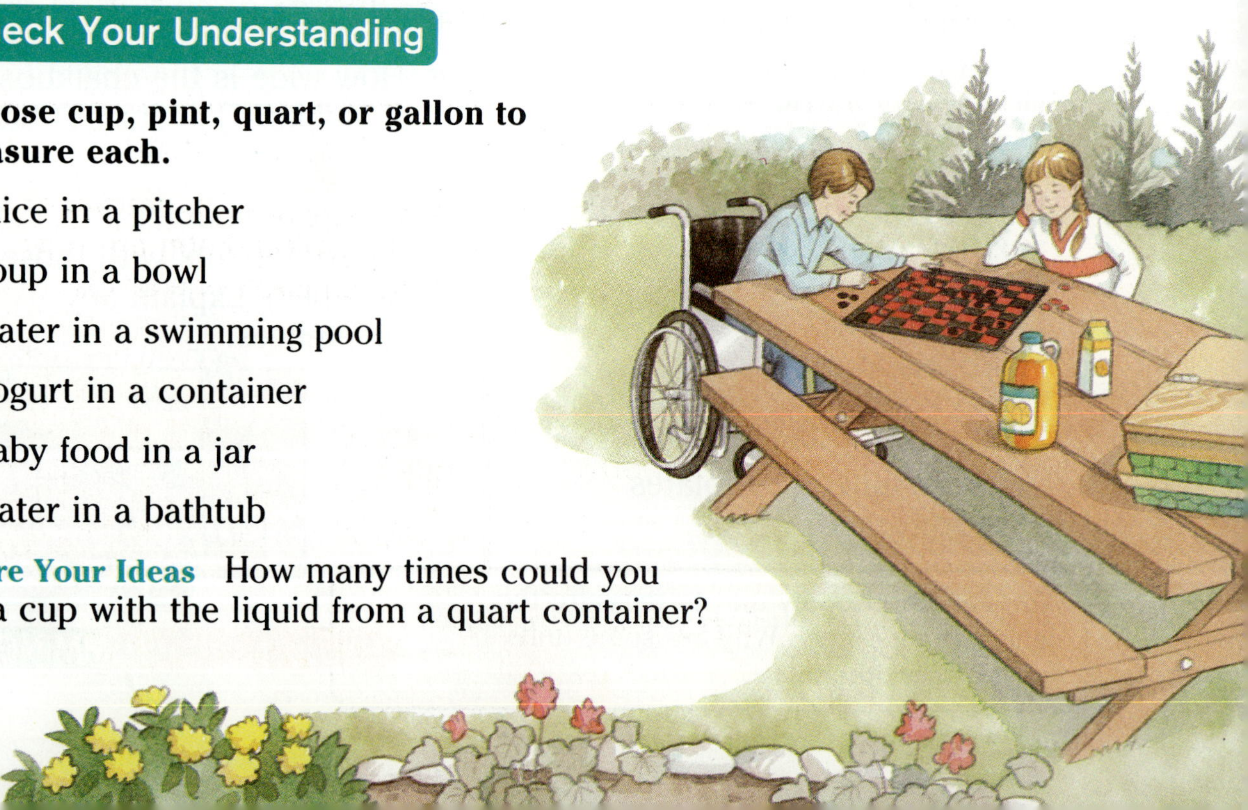

1. juice in a pitcher
2. soup in a bowl
3. water in a swimming pool
4. yogurt in a container
5. baby food in a jar
6. water in a bathtub

Share Your Ideas How many times could you fill a cup with the liquid from a quart container?

Practice

Choose cup, pint, quart, or gallon to measure each.

7. juice in a glass
8. soup in a can
9. gasoline in a car
10. soup in a soup pot
11. water in a bucket
12. water in a storage tower

Choose the best estimate for each.

13.

a. 1 pint b. 3 quarts
c. 1 gallon d. 8 cups

14.

a. 1 cup b. 5 gallons
c. 2 quarts d. 15 pints

Complete.

15. 2 cups = _____ pint
16. 1 pint = _____ cups
17. 2 pints = _____ quart
18. 1 quart = _____ pints
19. 4 quarts = _____ gallon
20. 1 gallon = _____ quarts
21. 2 gallons = _____ quarts
22. 4 cups = _____ pints

Think and Apply

23. Jane used 1 gallon of orange juice for the punch. She used 2 quarts of pineapple juice. How much more orange juice than pineapple juice did she use?

24. Which is more—6 cups of soup or 4 pints of soup? How much more?

Visual Thinking

25. Compare each container to the gallon. Estimate how much each one holds.

Make a class list of things you use that come in quarts and gallons.

SUMMING UP

Activity

Exploring
Estimating Weight: Ounce

The Post Office weighs mail by the ounce. Jerry's letter weighs about 1 ounce.

▸ We use the **ounce (oz)** to measure the weight of light objects.

Take turns holding a 1-ounce weight. How many classroom objects can you find that weigh about 1 ounce?

Working together

Materials: balance scale, 1-ounce weight

A. Collect objects that you estimate to weigh about 1 ounce.

B. Use the scale to compare each object with the 1-ounce weight. List the objects that weigh 1 ounce or less. List the objects that weigh more than 1 ounce.

DATA Sharing Your Results

1. Compare your lists with those of other groups. Did you list some of the same objects?

2. Challenge your classmates to tell whether your objects weigh 1 ounce or less, or more than 1 ounce.

Extending the Activity

Estimate weight with your own hands.

Work in a group.

Materials: 3 packages with weight labeled in ounces, several unlabeled objects

3. Hold a labeled package in one hand. Hold an unlabeled object in the other hand.
4. Estimate whether the object weighs more than, about the same as, or less than, the package. Record your results.

Package	More	Same	Less
3 oz raisins	grapefruit	apple	feather

Summing Up

5. **Look back** at your list. How could you decide which object weighed the most?

Choose the better estimate for each weight.

6. fishing lure
   a. 1 oz b. 50 oz
7. box of cereal
   a. 1 oz b. 12 oz
8. tube of toothpaste
   a. 100 oz b. 9 oz
9. chalkboard eraser
   a. 3 oz b. 50 oz
10. beach ball
    a. 10 oz b. 100 oz
11. bar of soap
    a. 100 oz b. 6 oz
12. orange
    a. 6 oz b. 600 oz
13. hammer
    a. 1 oz b. 10 oz
14. egg
    a. 40 oz b. 4 oz

How much does each object weigh?

15.

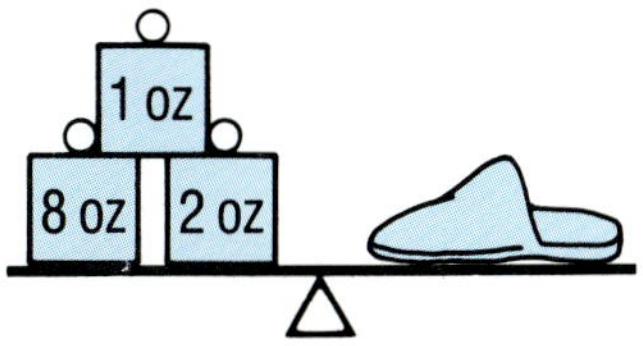

16.

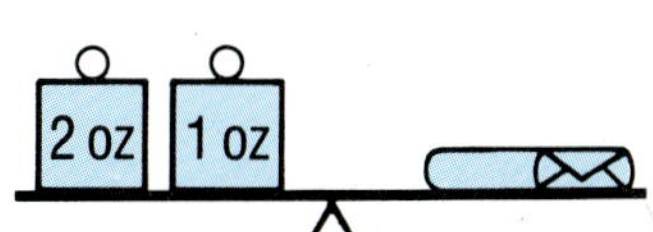

17.

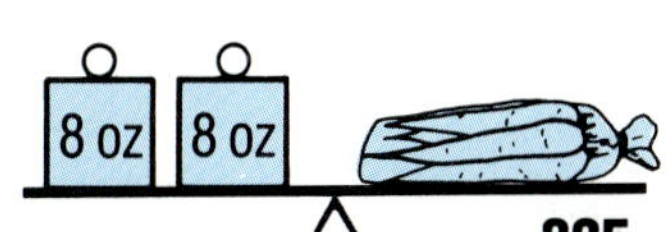

GETTING STARTED

Have you ever tried to carry a 5-pound bag of flour? a 20-pound bag of potatoes? What is the heaviest load you were able to carry?

Pound

Batter up! It's time for the baseball game and Clem is bringing the bag of bats. "It's so heavy," he says. "It weighs over 20 pounds."

▸ We use the **pound (lb)** to measure the weight of heavier objects.

16 ounces = 1 pound

A baseball glove weighs about 1 pound.

A baseball bat weighs about 2 pounds.

How much do you think each base weighs?

How much does a baseball uniform weigh?

Check Your Understanding

Choose ounce or pound to measure the weight of each.

1. bench in the dugout
2. baseball cap
3. pair of socks
4. box of new baseballs
5. catcher's equipment
6. baseball player

Share Your Ideas The weight of a newborn baby is usually measured in pounds and ounces. The weight of an adult is usually measured in pounds. Why do you think this is so?

Practice

Choose ounce or pound to measure the weight of each.

7. team water jug
8. pair of shoes
9. camera
10. score book
11. team shirt
12. folding chair
13. team first-aid kit
14. pair of shoelaces
15. rake

Choose the better estimate for each weight.

16. baseball
 a. 6 oz **b.** 6 lb
17. football
 a. 1 lb **b.** 1 oz
18. bowling ball
 a. 10 lb **b.** 10 oz
19. telephone
 a. 5 oz **b.** 5 lb
20. 2 baseball cards
 a. 1 oz **b.** 1 lb
21. watermelon
 a. 4 lb **b.** 4 oz

Compare. Use $>$, $<$, or $=$ for ●.

22. 1 oz ● 1 lb
23. 1 lb ● 16 oz
24. 2 lb ● 32 oz
25. 3 lb ● 48 oz
26. 7 lb ● 7 oz
27. 1 lb ● 20 oz
28. 40 oz ● 2 lb
29. $\frac{1}{2}$ lb ● 8 oz
30. 4 oz ● $\frac{3}{4}$ lb

Think and Apply

Work in a group. Use a balance scale with weights.

31. Collect 5 objects and estimate their weights.
32. Weigh them and record the results.
33. Which object is the lightest? the heaviest?
34. How much do the objects weigh in all?
35. Make up two more questions that can be answered from your results.

Make a list of three situations in which you estimate weight. How does estimating weight help you?

SUMMING UP

GETTING STARTED

How can you estimate the temperature without looking at a thermometer? Explain.

Degree Fahrenheit

In the Sonoran desert the daytime temperature can be higher than 110 degrees Fahrenheit. The nighttime temperature can be 60°F.

▸ We use the **degree Fahrenheit** to measure temperature. We write **°F** for degree Fahrenheit.

Look at the thermometer.
What is the temperature shown by the red column?

What do you think the temperature in your classroom is right now?

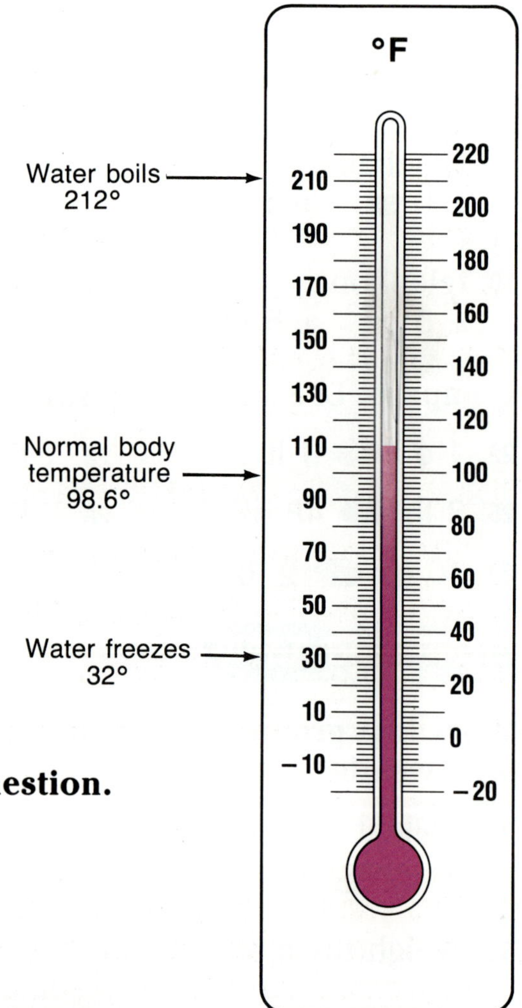

Check Your Understanding

Use the thermometer to answer each question.

1. What is the freezing point of water?
2. What is the boiling point of water?
3. What is normal body temperature?
4. Is it a hot or cold day when the temperature is 92°F?
5. Is it a hot or cold day when the temperature is 10°F?

Share Your Ideas What kind of clothing would you wear when it is 20°F outside? What type of activity would you do at 90°F?

Practice

Write each Fahrenheit temperature shown.

6. 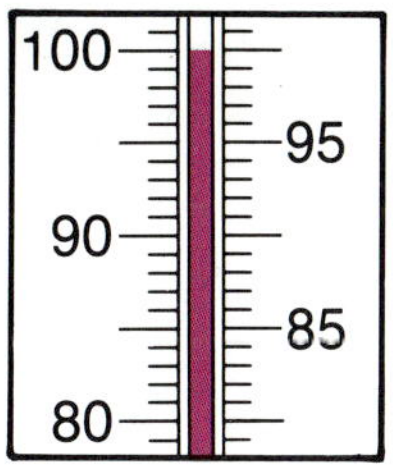

7. 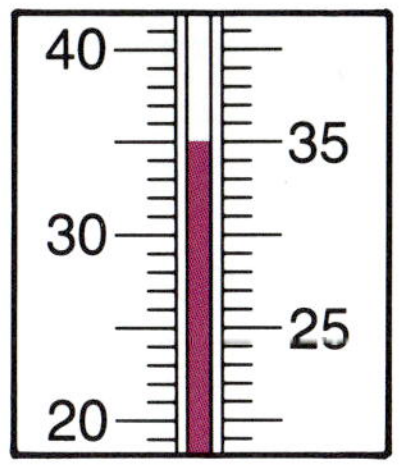

8. 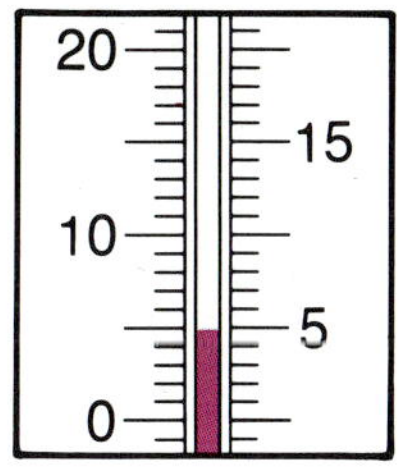

Match. Choose the most reasonable temperature.

9.	cold winter day	**a.**	10°F
10.	hot summer day	**b.**	32°F
11.	room temperature	**c.**	95°F
12.	ice cubes	**d.**	70°F

Write the temperature for each.

13. 20° below the boiling point

14. 30° above the freezing point

15. 40° below the boiling point

16. 3° above normal body temperature

Think and Apply

17. At 3 P.M. the temperature outside was 15°F. By 1 A.M. the temperature dropped 12 degrees. What was the temperature then?

18. The temperature was 60°F. It fell 10 degrees and then rose 15 degrees. What was the temperature then?

What if you recorded the temperature each hour from 9 A.M. until 9 P.M. in one day? What do you think the record would look like?

SUMMING UP

Mixed Review

Use mental math.

1. 80 + 60 = □

2. 90 − 40 = □

3. 60 + 30 = □

4. 50 − 20 = □

5. 600 + 200 = □

6. 800 − 300 = □

7. 900 − 700 = □

8. 1,000 + 400 = □

Estimate.

9. 51 − 39 = □

10. 29 + 28 = □

11. 79 − 59 = □

12. 71 − 38 = □

13. 59 + 12 = □

14. 39 + 49 = □

15. 603 − 499 = □

16. 721 + 285 = □

You have 1 white and 2 red buttons in a bag.

17. What is the chance of drawing a white button?

18. What is the chance of drawing a red button?

Chapter Review

Measure each to the nearest centimeter. pages 300–301

1. 

2. 

3. 

Choose the better estimate for each. pages 300–303, 306–313

4. height of a house
a. 10 cm **b.** 10 m

5. capacity of a sink
a. 200 mL **b.** 200 L

6. weight of a cow
a. 500 kg **b.** 500 g

7. temperature on a warm day
a. 45°C **b.** 10°C

8. temperature on a cold day
a. 50°C **b.** 0°C

Measure each to the nearest inch. pages 318–319

9. 

10. 

Complete. pages 320–327

11. 1 pt = ____ c

12. ____ oz = 1 lb

13. ____ in. = 2 ft

Write each Fahrenheit temperature. pages 328–329

14. 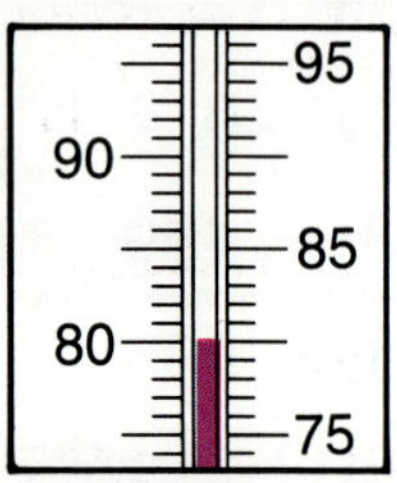

15. 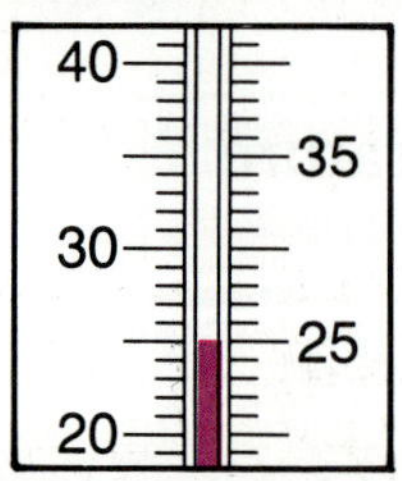

16. 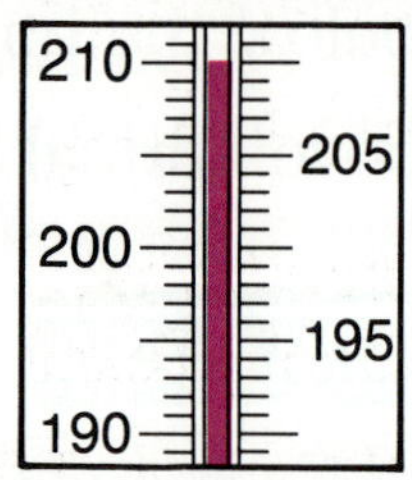

Choose the correct word to complete each sentence.

17. We use the ____ to measure length.

18. We use the ____ to measure weight.

Words to Know
ounce inch

Solve. pages 315–317

19. Jan needs 2 gallons of juice. How many quarts should she buy?

20. Pete had 3 yards of plastic to cover a table. Jan had 6 feet of plastic. Who had more?

Chapter Test

Measure each to the nearest centimeter.

1. 

2. 

3. 

Complete.

4. 100 cm = ______ m
5. ______ mL = 1 L
6. 3 kg = ______ g

Write each Celsius temperature.

7. 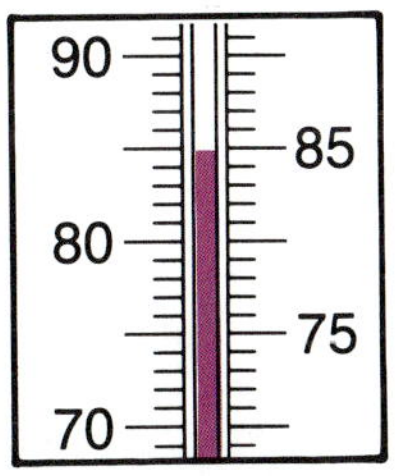

8. 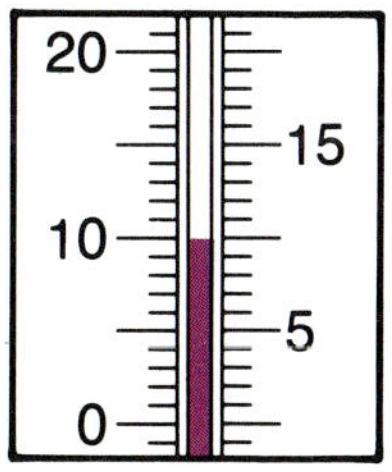

9. 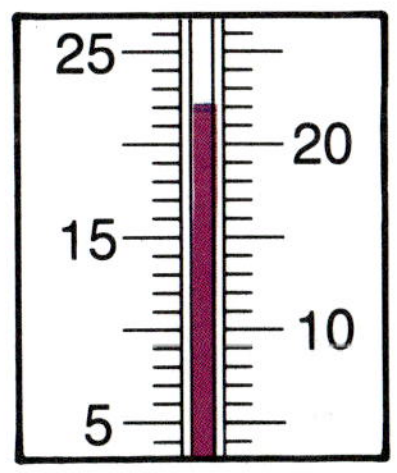

Measure each to the nearest inch.

10. 

11. 

12. 

Choose the better estimate for each.

13. length of a desk
   a. 5 ft b. 5 yd

14. capacity of a mug
   a. 2 qt b. 2 c

15. weight of a pear
   a. 2 oz b. 2 lb

16. distance from Canada to Texas
   a. 2,000 mi b. 2,000 yd

17. temperature on a cold day
   a. 20°F b. 70°F

18. temperature on a hot day
   a. 40°F b. 90°F

Solve.

19. The cook has 2 pounds of meat. He uses 8 ounces for each meat loaf. How many loaves can he make? How many loaves can be made from $2\frac{1}{2}$ pounds of meat?

20. Choose the better way to find how much water is in a jug.
   a. Pour water into another jug and measure it.
   b. Pour water into cups and count.

THINK At 8 A.M. the temperature was 48°F. It rose 2 degrees every hour until 10 A.M. What is the temperature if it drops 10 degrees?

EXTENSION

The Warming Trend: Collecting and Analyzing Data

Scientists often have questions about what they see in nature. To find answers they keep a record of what they see. In this experiment you will use the skills of a scientist and a mathematician to measure temperature.

Work in a small group.

Materials: small jar, thermometer

A. Pour very cold water into a small jar until the jar is half full.

B. Measure and record the time and temperature. Then record the temperature every 15 minutes for 2 hours.

C. Discuss different ways your results could be displayed to show what happened. Then decide on one way to display your results.

D. Write about what happened.

Try these.

1. Discuss your results with other groups. Do you think the size and shape of the jar affected the results? Explain.
2. Compare the different ways the results were displayed. Did some ways show what happened better than other ways? Why?
3. Design another experiment called The Cooling Trend! This time start with hot water.

Family Math

Your child has been working with division, fractions, probability, and measurement in Chapters 8–10. Now you can help him or her use those skills.

Work together to build a mobile that is perfectly balanced. It is called a seesaw mobile. Here is how to build it.

Seesaw Mobile

Materials: 6 matching buttons or other ornaments, 5 poles cut from coat hanger wire, ruler, scissors, string

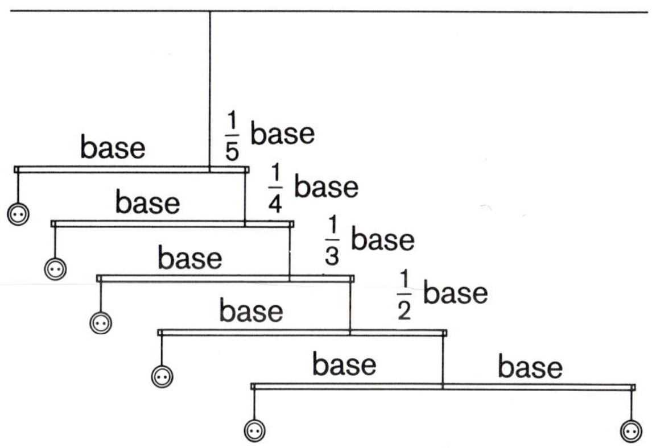

1. The longest pole will be on the bottom. Half of its length is called the base.

   What if the bottom pole is 24 in. long? What would the base length be?

   Think $\frac{1}{2} \times 24$ in. $= 12$ in.

   The base would be 12 in.

   What would the length of the pole next to the bottom be?

   Think 12 in. $+ (\frac{1}{2} \times 12$ in.$) = \square$

   12 in. + 6 in. = 18 in.

   The next pole will be 18 in. long.

2. Choose the length of the bottom pole. Then find the lengths, as shown, of the other poles. Cut the poles.

3. Cut ten 6-in. lengths of string. Tie each string to a button.

4. Hang the buttons 3 in. below the poles. Try to keep the string lengths the same.

5. Cut four 5-in. lengths of string. Tie the strings to the poles as shown. There should be 3 in. between the poles.

6. Hang your seesaw mobile from the ceiling.

Cumulative Review

Choose the correct answer. Write A, B, C, or D.

1. $6 \div 2 = \square$

A 4 **C** 6

B 3 **D** not given

2. $3\overline{)21}$

A 9 **C** 7

B 8 **D** not given

3. $45 \div 5 = \square$

A 9 **C** 5

B 8 **D** not given

4. $6\overline{)42}$

A 7 **C** 8

B 6 **D** not given

5. $6\overline{)30}$

A 4 **C** 6

B 5 **D** not given

6. $48 \div 8 = \square$

A 8 **C** 6

B 7 **D** not given

7. What is another fact in this fact family? $9 \times 8 = 72$

A $7 \times 9 = 64$ **C** $72 \div 9 = 6$

B $72 \div 8 = 9$ **D** not given

8. What fraction is the shaded part?

A $\frac{4}{2}$ **C** $\frac{5}{6}$

B $\frac{6}{4}$ **D** not given

9. What part of the group is red?

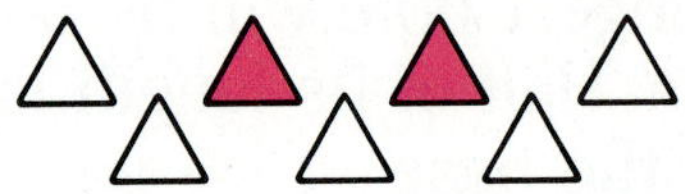

A $\frac{2}{7}$ **C** $\frac{2}{5}$

B $\frac{4}{7}$ **D** not given

10. $\frac{1}{7}$ of $49 = \square$

A 9 **C** 7

B 8 **D** not given

11. Compare. $\frac{1}{2}$ ⬬ $\frac{3}{6}$

A $<$ **C** $=$

B $>$ **D** not given

12. What mixed number names the blue part?

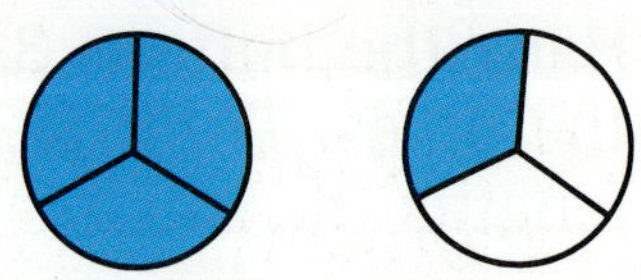

A 1 **C** $\frac{1}{3}$

B $1\frac{1}{3}$ **D** not given

Choose the correct answer. Write A, B, C, or D.

13. Which unit would you use to measure a room?

A meter **C** kilometer

B centimeter **D** not given

14. About how much does a nickel weigh?

A 85 g **C** 5 g

B 110 g **D** not given

15. Compare. 1 g ● 1 kg

A $<$ **C** $=$

B $>$ **D** not given

16. How long is the bar to the nearest $\frac{1}{2}$ inch?

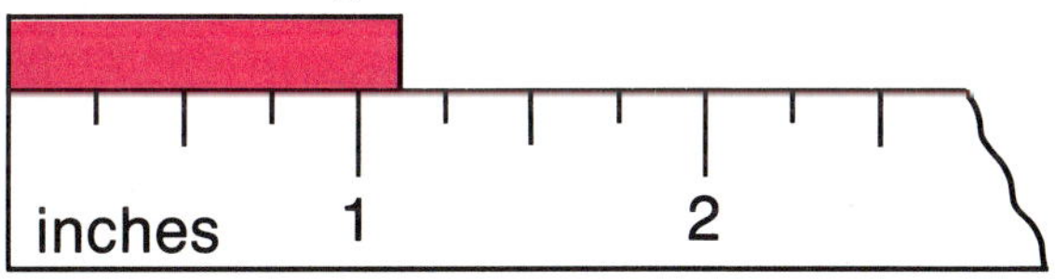

A 1 in. **C** $2\frac{1}{2}$ in.

B 2 in. **D** not given

17. About how much does a nail weigh?

A 25 oz **C** 40 oz

B 1 oz **D** not given

18. What could be the temperature of a room?

A 125°F **C** 72°F

B 16°F **D** not given

Complete the pattern.

19. 10, 20, ______, ______, 50, 60

A 30, 40 **C** 30, 35

B 25, 30 **D** not given

Solve.

20. Mr. Covino raises roses. He planted them in a row using a pattern of 3 red and 2 white roses. How many white roses does he have if there are 15 plants?

A 7 **C** 6

B 9 **D** not given

Use the table to solve **21–22.**

CHILDREN'S BOOKS SOLD		
	Fiction	**Nonfiction**
March	126	52
April	111	48
May	135	50

21. How many books were sold in April?

A 63 **C** 157

B 159 **D** not given

22. How many more fiction books were sold in March than nonfiction books?

A 74 **C** 78

B 64 **D** not given

11 Geometry

THEME Patterns in Our World

Sharing What You Know

Stonehenge is silent and mysterious. Who built it? Why was it built? How was it used? No one knows for sure. Scientists do know that it was built even before people invented the wheel. Look at the picture of Stonehenge. What shapes did its builders use?

Using Language

Some scientists think Stonehenge was used as an ancient calendar. Notice the shallow ditch that circles Stonehenge. The world **circle** means to go around. In mathematics, a **circle** is a closed curve. All the points on the curve are the same distance from the center. What else can you think of that has the shape of a circle?

Words to Know: cylinder, rectangle, square, triangle, circle, line segment, line, ray, congruent, symmetry, perimeter, volume

Be a Problem Solver

How would you and a group of friends arrange yourselves at a campfire? at dinner? in math class? at a football game? When might you want a group of people to sit in a circle? When might you not want to sit in a circle?

List the times when you and other people formed a circle this week. Write the reason for forming each circle.

GETTING STARTED

Describe the shape of your teacher's desk. a globe. How are the shapes alike? different?

Space Figures

When Jill went to the museum, she saw some very old dollhouses. The tiny furniture was made to look just like real furniture.

Which objects in the picture are shaped like each of these?

Rectangular prism Cube Cylinder

Sphere Cone

Some space figures can be described by their **faces, edges,** and **corners.**

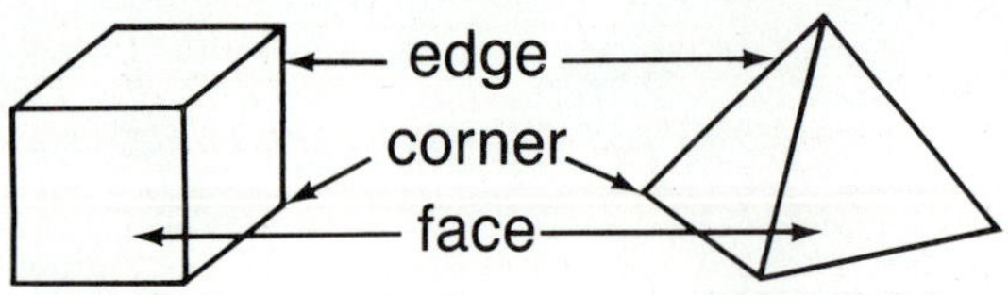

Can all the figures above be described this way? Explain.

Check Your Understanding

Name the space figure that best describes each object.

1. 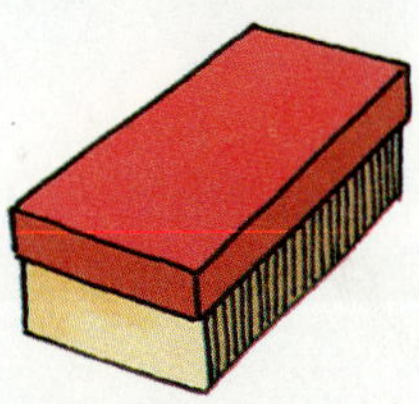 **2.**  **3.**  **4.** 

Share Your Ideas Look around. Name an object that is shaped like each of the space figures.

Extending Your Thinking

Use models of space figures like the ones in **4–9**. Draw each figure as you would see it from the top, the side, the front, and the bottom.

4. cube

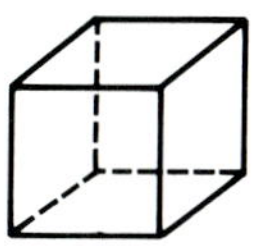

5. cylinder

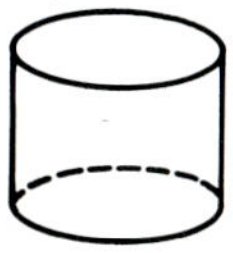

6. cone

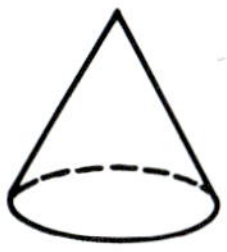

7. sphere

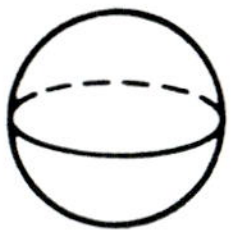

8. pyramid

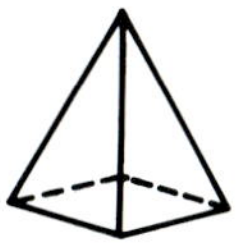

9. triangular prism

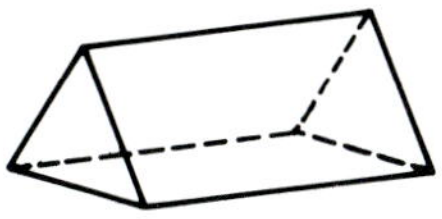

Sometimes you see edges when you look at a figure.

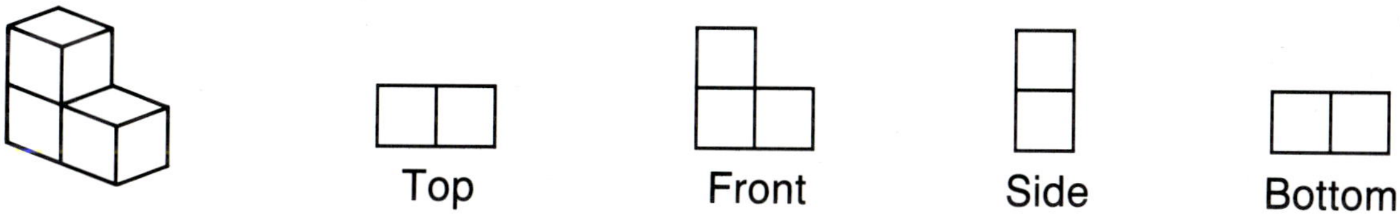

10. Use cubes to make some figures. Draw the top, the front, the bottom and the side view.

Summing Up

11. Look at these drawings. What does this figure look like? Make it with cubes.

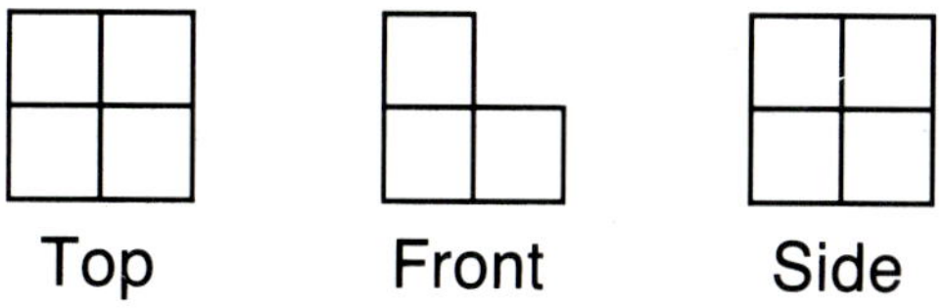

12. **Look back** at your drawings for **4–9**. Which figures look the same from the top, front, and side? For which figures do all your drawings look the same?

GETTING STARTED

What is the shape of this page? of a phonograph record?

Plane Figures

Some famous artists have used simple shapes to create their paintings. Look at the painting by Mondrian. The outline of each shape is a **plane figure.**

Look at these plane figures. Which ones do you see in the painting?

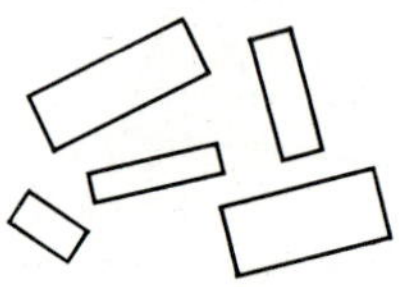

rectangles

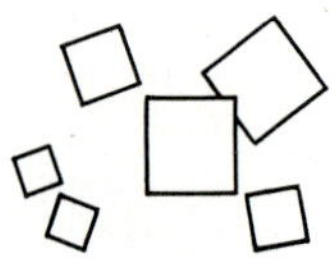

squares

triangles

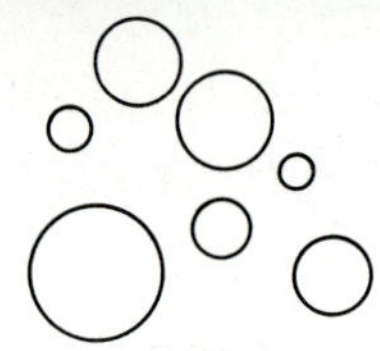

circles

To which set of figures does each of these belong? How do you know?

a. 

b. 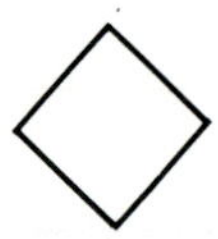

c. 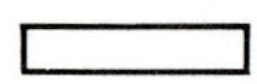

d. 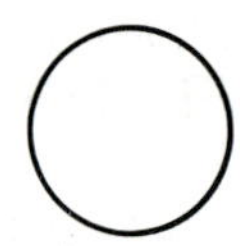

Check Your Understanding

What is the shape of each?

1. 

2. 

3. 

4. 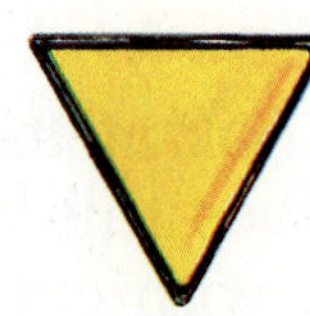

Look back at the figures in a–d. Which figure has

5. 4 equal sides and 4 corners?

6. no straight sides and no corners?

Share Your Ideas Compare the shapes in **1** and **3**. How are they alike? How are they different?

Name the first figure. Which figures are the same kind as the first one in the row?

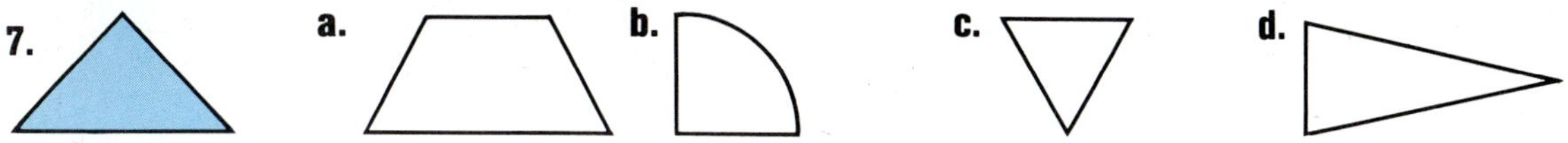

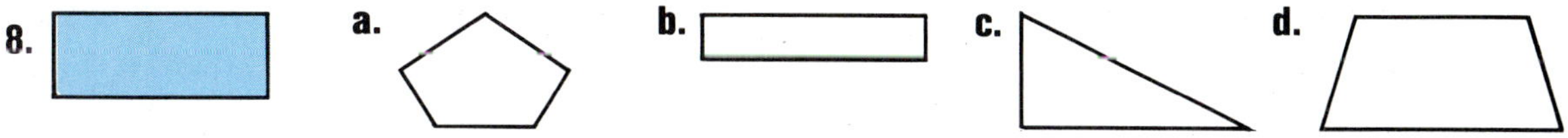

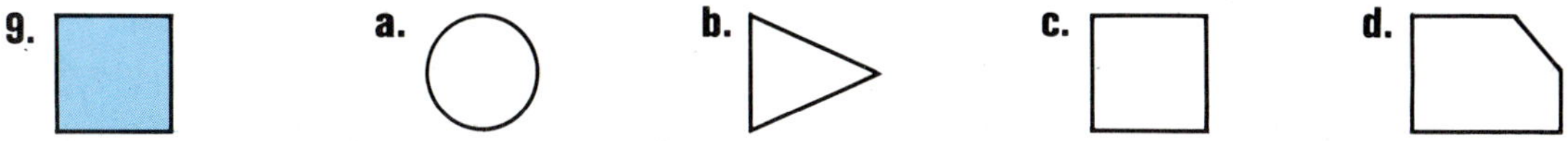

10. Copy and then complete the table. Write the letter of each figure in the correct place.

Rectangles	Squares	Triangles	Circles

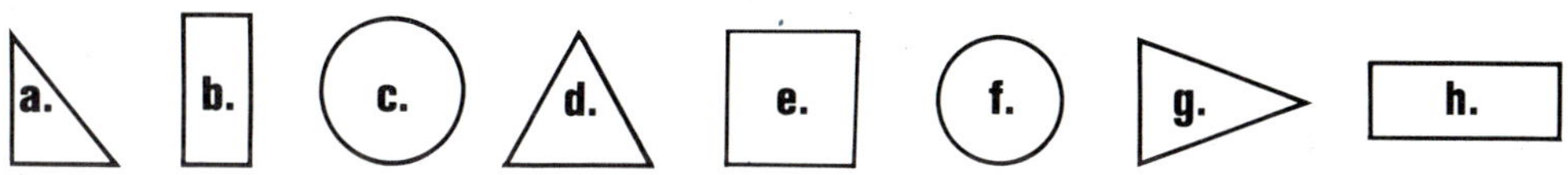

Think and Apply

11. Use Logo to draw a picture or design made of squares. First define the procedure SQUARE. SQUARE 10 draws a square that has sides of 10 turtle steps. Then choose any numbers for the size. Use PENUP(PU) to move without drawing and PENDOWN(PD) to draw again.

```
TO SQUARE :SIZE
REPEAT 4[FD :SIZE RT 90]
END
```

How many different squares can you find?

12. 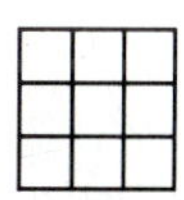

13. 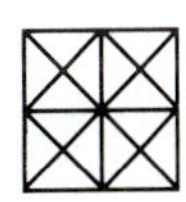

How many different triangles can you find?

14. 

15. 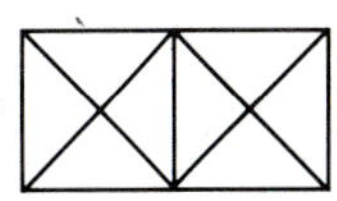

Look back at **12–15**. Make your own plane figure puzzle. Cut out the puzzle pieces. Ask a friend to put them together.

SUMMING UP

GETTING STARTED

What is the shortest distance from point *A* to point *B*? Explain.

A•   •*B*

Lines, Line Segments, and Rays

People have woven rugs and other things for hundreds of years. Look closely at the rugs in the picture. Some parts of the rugs remind us of line segments.

- A **line segment** is straight and has two endpoints. A line segment is part of a line.

line segment ***AB***

- A **line** is straight and has no endpoints. It goes on and on in both directions.

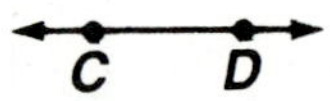

line ***CD***

- A **ray** is part of a line. It has one endpoint and goes on and on in one direction.

ray ***EF***

Is line segment *EF* part of ray *EF?* Explain.

Check Your Understanding

Name each figure.

1. 

2. 

3. 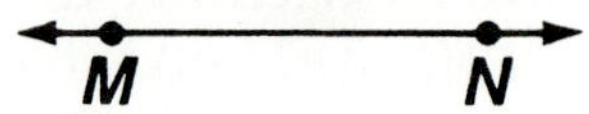

4. 

5. 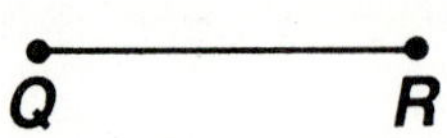

6. 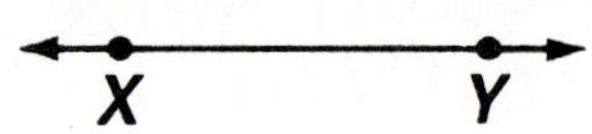

Share Your Ideas Do you think you could measure the length of a line segment? a line? a ray? Explain.

Practice

Name each figure.

7. 

8. 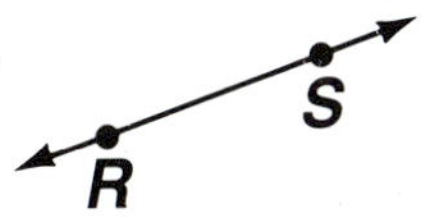

9. 

Study the drawing. Then name one of each.

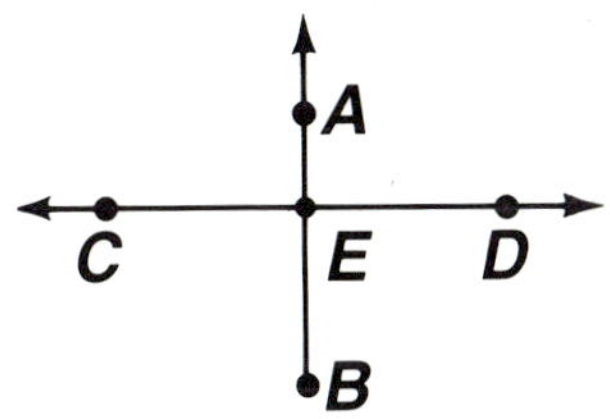

10. a ray
11. a line segment
12. an endpoint

How many line segments are there in each figure? Name each line segment.

13. 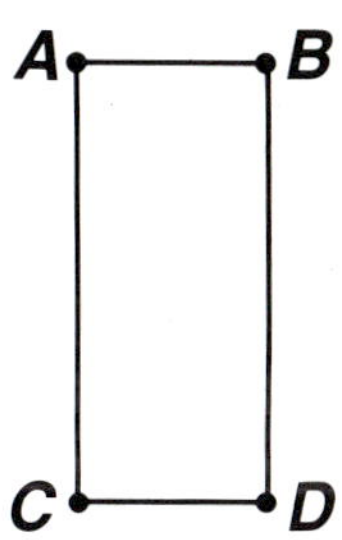

14. 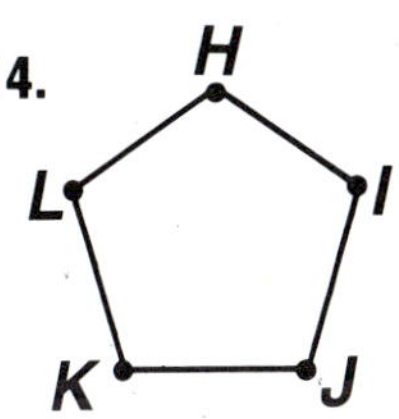

15. 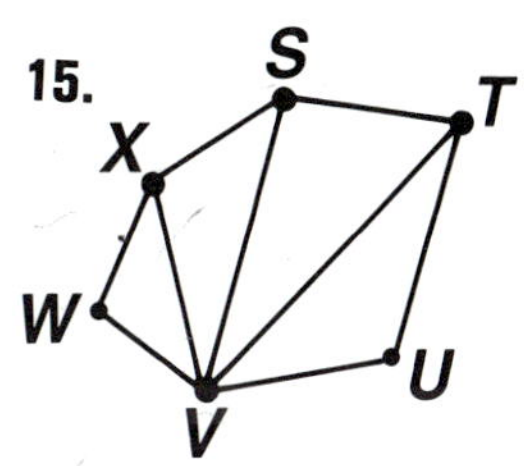

Think and Apply

Finding distances on a map is like finding lengths of line segments. Find the lengths of these line segments on the map.

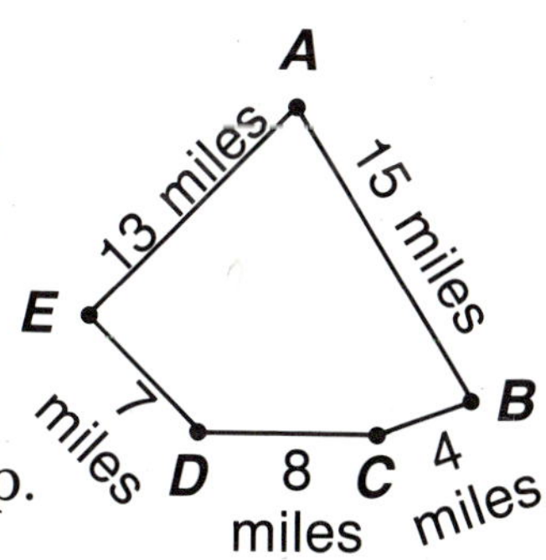

16. line segment *AB*
17. the sum of line segments *AB* and *BC*
18. the sum of all line segments in *ABCDE*

Use line segments to make up a pattern you might weave into a rug.

Mixed Review

1. $\begin{array}{r} 291 \\ +735 \\ \hline \end{array}$

2. $\begin{array}{r} 706 \\ -225 \\ \hline \end{array}$

3. $182 + 97 = \square$
4. $739 + 991 = \square$
5. $800 - 406 = \square$
6. $723 - 88 = \square$

7. $\begin{array}{r} 6 \\ \times 7 \\ \hline \end{array}$

8. $\begin{array}{r} 9 \\ \times 7 \\ \hline \end{array}$

9. $4 \times 9 = \square$
10. $8 \times 5 = \square$
11. $\frac{1}{3}$ of $21 = \square$
12. $\frac{1}{2}$ of $18 = \square$
13. $\frac{1}{5}$ of $30 = \square$
14. $\frac{1}{7}$ of $49 = \square$
15. $\frac{1}{9}$ of $63 = \square$

Complete.

16. 16 oz = ______ lb
17. 2 c = ______ pt
18. 2 pt = ______ qt
19. 12 in. = ______ ft
20. ______ ft = 1 yd

SUMMING UP

Activity

Exploring Angles

The Geodesic Dome is a new type of building. It does not have the usual angles used in other buildings.

You can explore angles of different sizes.

Working together

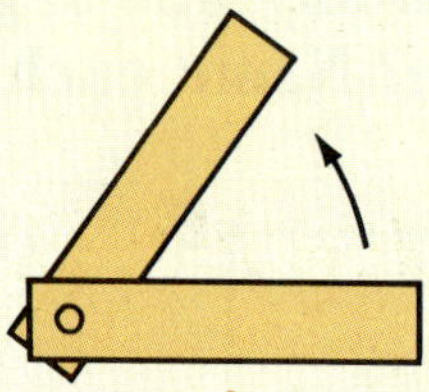

Materials: two 6-inch strips of cardboard and one brass fastener for each student

A. Fasten the two strips at one end to make a model of an angle.

B. Hold one strip still. Move the other.

C. Does the length of the strips change?

D. Does the opening between strips change?

E. Move the model to make an opening about the same as this right angle, less than this right angle, greater than this right angle.

F. Trace some angles on paper. Tell if each angle is the same as a right angle, less than a right angle, or greater than a right angle.

▶ A **right angle** has the same shape as a corner of this page.

Sharing Your Results

1. All right angles are the same size. Are all the right angles made in your group the same size?
2. Look at the angles you traced that are not right angles. How can you tell which is the least angle? the greatest angle?

Extending the Activity

Use your model to form a right angle.

3. Look for places or objects in your classroom that seem to form right angles. Make a list to show what you found.
4. Look for places or objects that seem to form angles less than a right angle and greater than a right angle. Make a list of what you found.

Summing Up

5. Compare your lists with those of your classmates. Which kind of angle did you find most often? Why do you think so?

Tell whether each angle is a right angle, less than a right angle, or greater than a right angle.

6. 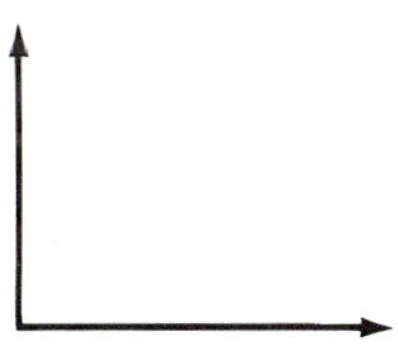

7. 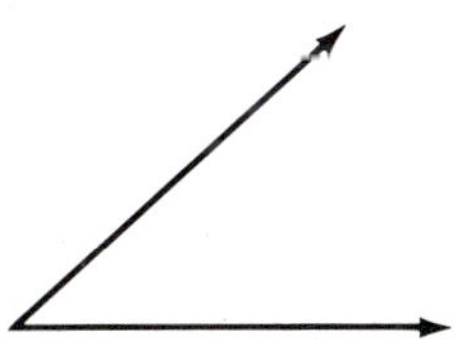

8. 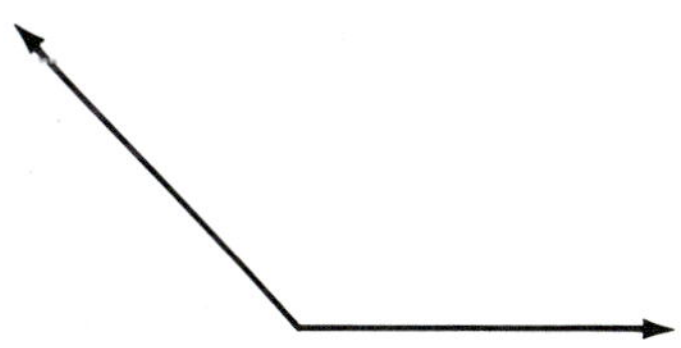

9. 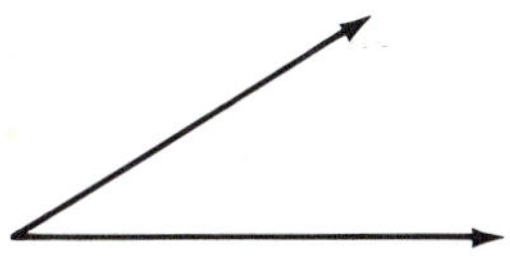

10. 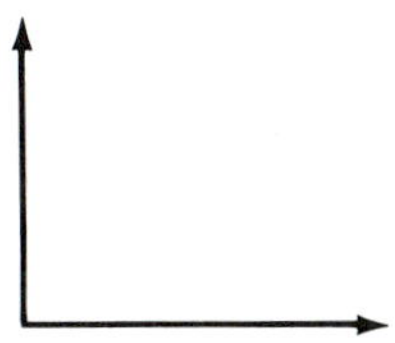

11. 

Midchapter Review

Name each figure. pages 338–339, 342–347

1. 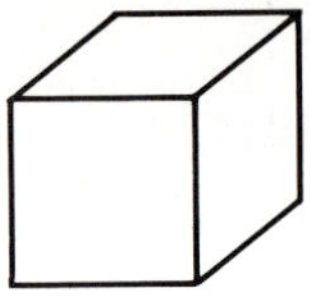

2. 

3. 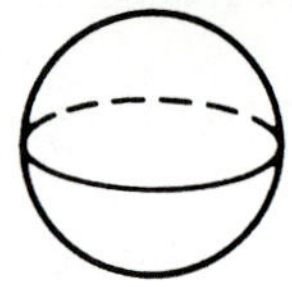

4. 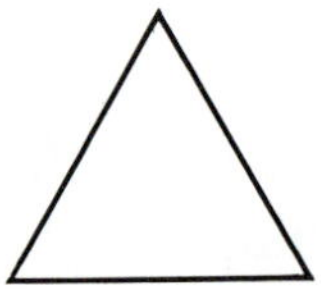

5. 

6. 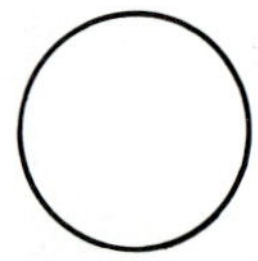

7. 

8. 

9. 

10. 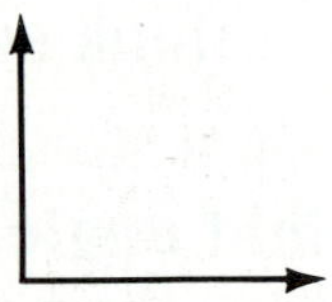

11. 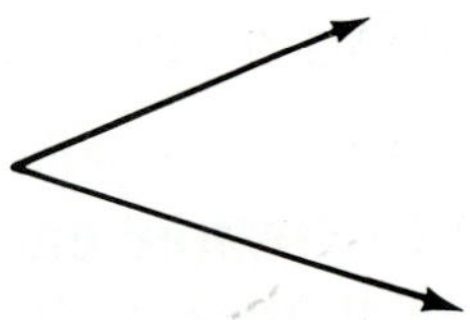

12. 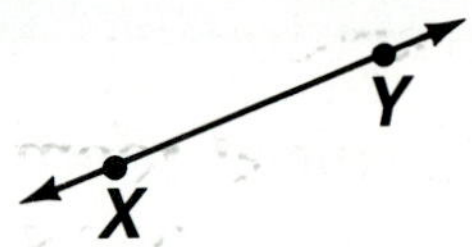

Choose the correct word to complete each sentence.

Words to Know
ray line segment line

13. A ______ is straight and has two endpoints.

14. A ______ is straight and has no endpoints.

15. A ______ has one endpoint and goes on and on in one direction.

Solve.

16. Jane says she can name 6 line segments in this figure. Jack can name 10. How many can you name? Explain.

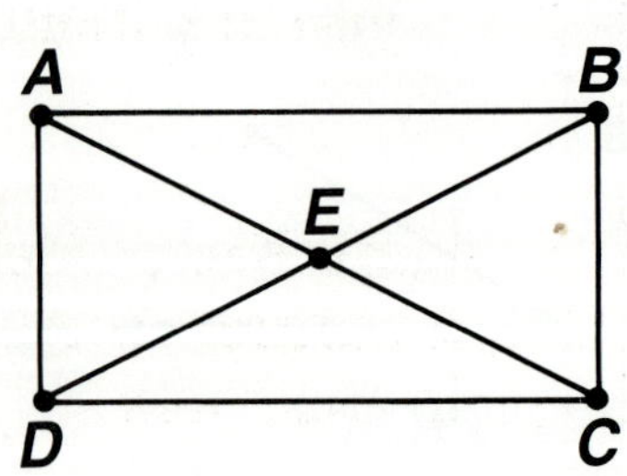

17. Jack is hiding a space figure. He says it has no corners or edges. What figure is it?

Exploring Problem Solving

THINK
EXPLORE
SOLVE
LOOK BACK

Placing the Guards

You must place the guards in the art museum. There must be enough guards so that they can stay in one spot and see all of the walls. They may turn their heads but not turn around.

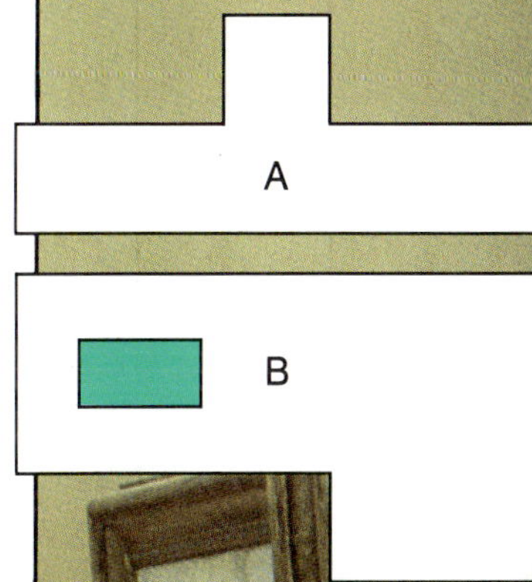

Thinking Critically

Look at the two rooms below. You want to put the fewest guards in each room. Where would you ask each guard to stand? Use Workmat 13 to help you.

Analyzing and Making Decisions

1. Look at room **A** above. Try placing one guard in the room. How can you test to see if the guard can see all the walls?

2. How many guards would you put in room **A**? Where would you ask them to stand?

3. Look at room **B** above. It has a special wall in it. Can one guard see all the walls? Explain.

4. How many guards would you put in room **B**? Where would you ask them to stand?

Look Back Make a drawing of a room. Ask a partner to tell how many guards should be in the room.

Problem Solving Strategies

Guess and Test

Maria wants to make a necklace out of beads. She uses 2 hearts and then 1 flower, then 2 hearts and 1 flower, and so on. She uses 24 beads to make the necklace. How many of the beads are heart-shaped? How many are flower-shaped?

Making a guess and then testing that guess is a good way to start to solve some problems. If you are wrong, you make another guess and test that one. You should keep a record of the guesses you make.

Solving the Problem

Think What are the questions?

Explore What pattern does Maria use? After she has put on 3 flowers, how many hearts are on the necklace? Guess the number of hearts that are needed in all. How many flowers should there be? Will there be 24 beads altogether? If there are 24, your guess worked. If you have too few beads, how should you change your guess?

Solve How many heart-shaped beads are used? How many flower-shaped beads are used?

Look Back How can you check to be sure your answer is correct?

Sharing Your Ideas

1. Why is it important to check your guess?

THINK
EXPLORE
SOLVE
LOOK BACK

Practice

Solve. Use a calculator where appropriate.

CHOICES

2. Jamie made a belt. She used 3 times as many circle links as square links. She used 36 links altogether. How many links were squares? How many were circles?

3. Mrs. Brown sold 23 picture frames. She sold 5 more square frames than circular frames. How many square frames did she sell? How many circular frames did she sell?

4. Larry sells small belts for $5 each, large belts for $9 each, bracelets for $6 each, and necklaces for $7 each. Janet purchased 3 different items for $18. What did she buy?

5. Mr. Frezza sold 19 scarves. He sold 5 more long scarves than square scarves. How many scarves of each kind did he sell?

Mixed Strategy Review

Copy and write the next numbers.

6. 1, 2, 2, 3, 3, 3, ____, ____, ____, ____

Use this information for 7 and 8.

Twenty-one students from Mr. Ryan's class went to the museum. Thirty students from Ms. Lane's class also went to the museum. A total of 14 students from the 2 classes remained in school.

7. How many students in all went to the musuem?

8. How many students are in Mr. Ryan's class?

Create Your Own

Use the information for 9 and 10 to create a problem of your own.

Use this information to solve 9 and 10.

Students were counting off by 3's to make groups. All the 1's were in one group. The 2's were in another, and the 3's in another.

9. Lizette was the 7th student in line. In which group was she?

10. There were 21 students in line. How many students were in each group?

GETTING STARTED

If you made two imprints of your right hand, do you think you could make them the same size and shape? Explain.

Understanding Congruence

Sneakers, invented more than 100 years ago, are more popular than ever. Use a sneaker for an experiment. Make two imprints of your left sneaker. What do you notice about the size and shape of the prints?

▸ Figures with the same size and shape are **congruent.**

Use Workmat 14 or dot paper.

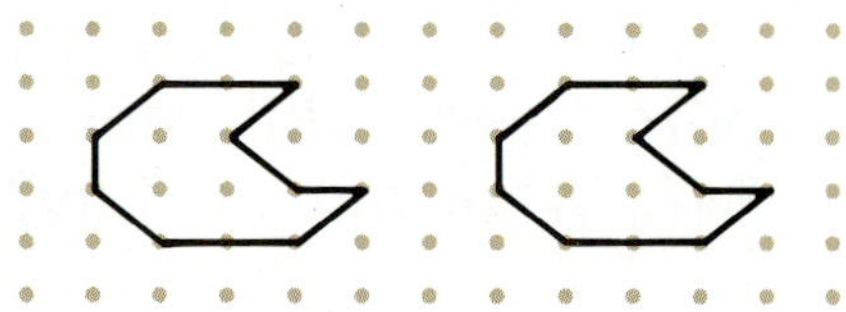

- Connect dots to make a figure.
- Ask a classmate to draw a figure congruent to it.
- Then change roles. Have the classmate make a figure and you draw a figure congruent to it.
- Do this 4 times.

Would each pair of figures match if they were cut out and placed one on the other?

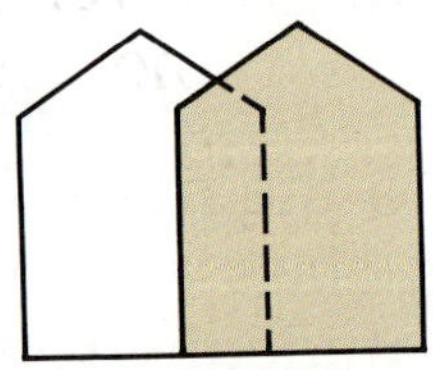

Congruent figures match exactly when placed on one another.

Check Your Understanding

Are the figures in each pair congruent?

1. 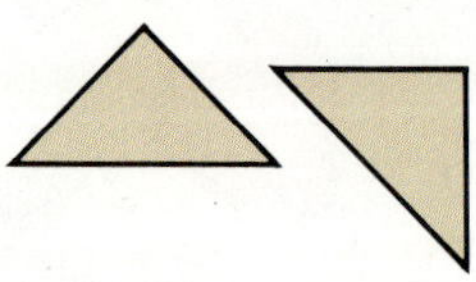

2. 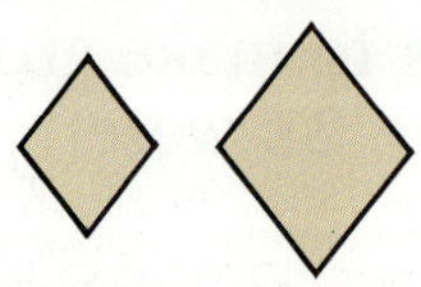

3. 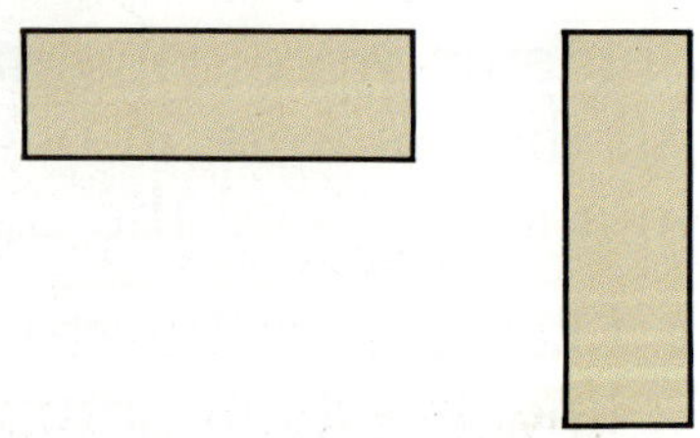

Share Your Ideas Look back at **1**. Are the figures congruent? Why or why not?

Practice

Are the figures in each pair congruent?

4. 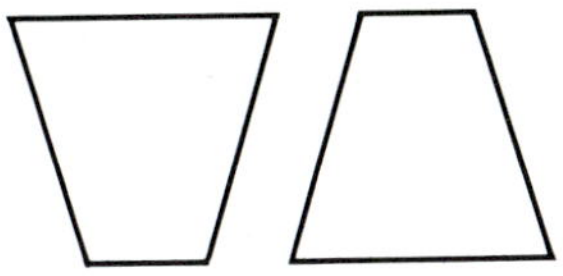

5. 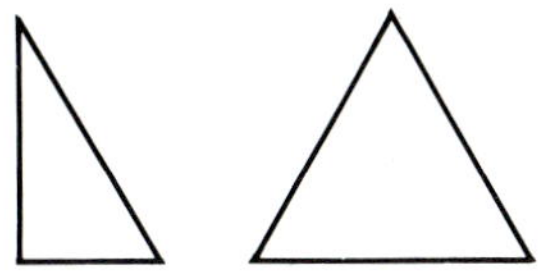

6. 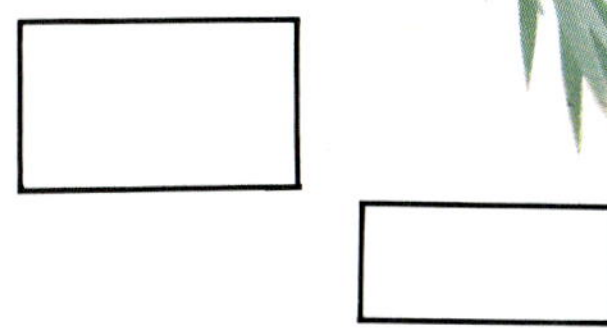

7. 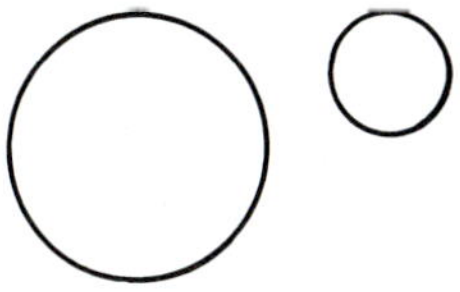

8. 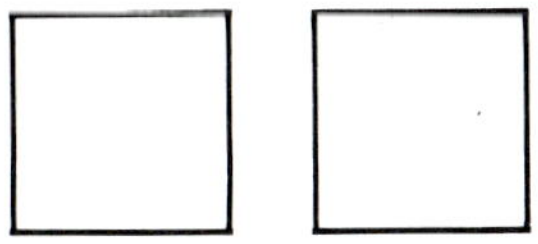

9. 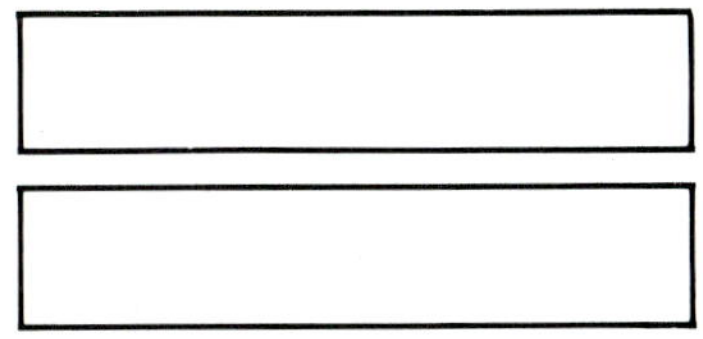

Trace the first figure. Use the tracing to find which figures in 10–13 are congruent to it.

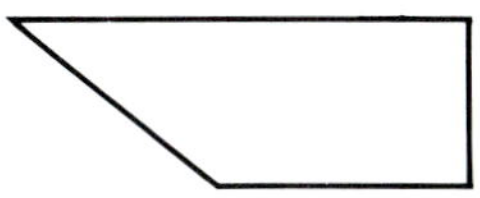

10. 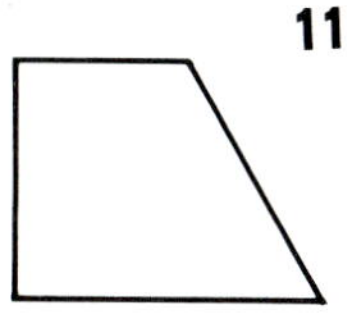

11. 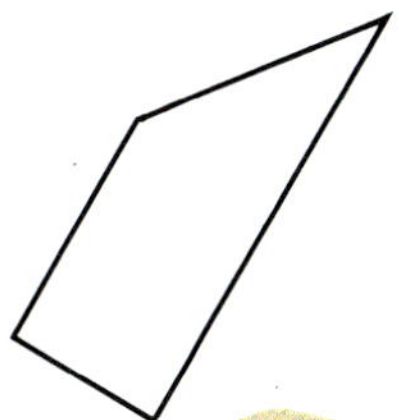

12. 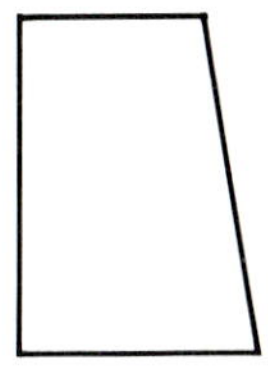

13. 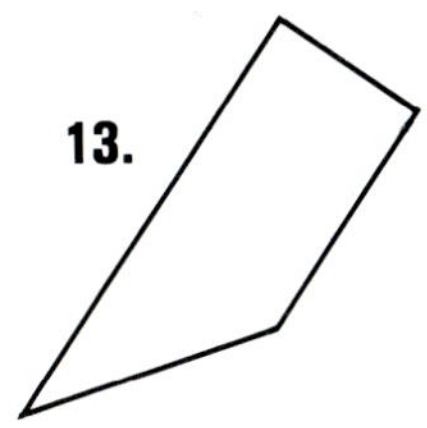

Think and Apply

14. Draw a figure. Ask a classmate to draw another figure congruent to your figure, but in a different position. Check your classmate's work.

15. Congruent figures form the design on some fabrics. Sketch a design for a tee shirt that uses congruent figures.

Logical Thinking

16. If A is congruent to B, and B is congruent to C, then A is congruent to ______.

17. If X is congruent to Z, and Y is congruent to Z, then X is congruent to ______.

a. 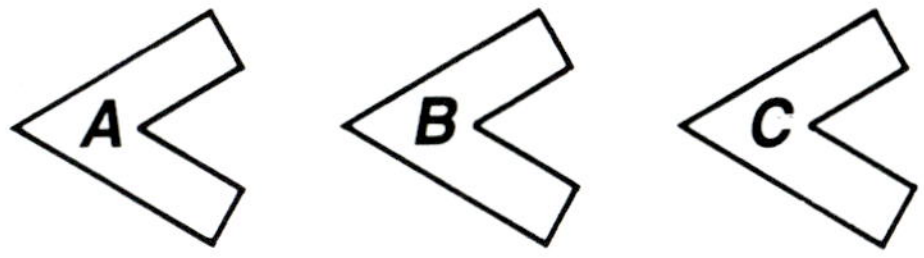

b. 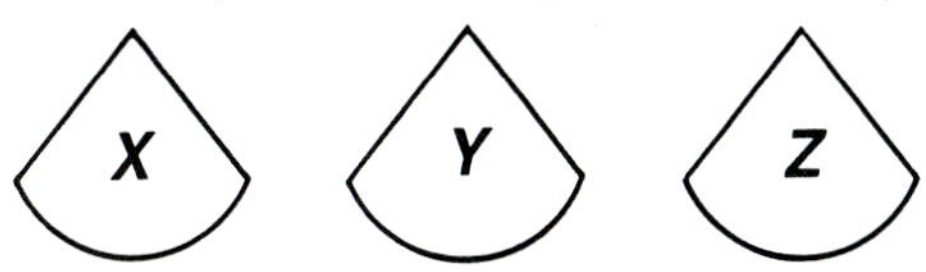

Describe how you can tell whether two figures are congruent to each other.

SUMMING UP

GETTING STARTED

Match the halves to make a complete picture.

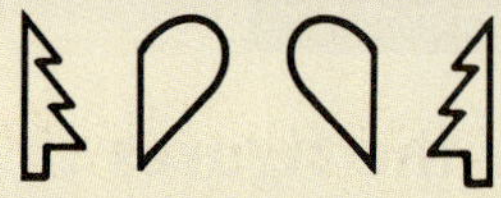

Understanding Symmetry

This kite is the biggest in the world. It is almost as big as a tennis court! One reason the kite can fly is that it is symmetrical.

- A figure has **symmetry** if it can be folded along a line so that the two parts match exactly.
- The line of fold is a **line of symmetry.**

Try this.

- Fold a piece of paper.
- Draw a shape along the fold.
- Cut out the shape and unfold it.
- The cutout is symmetrical.

Make several more cutouts that are symmetrical. Mark the line of symmetry in each.

Check Your Understanding

Is the dashed line a line of symmetry?

1. 

2. 

3. 

Draw the design you will have when you unfold each of these.

4. 

5. 

Share Your Ideas Which objects in your classroom seem to have symmetry?

Practice

Is the dashed line a line of symmetry?

6. 

7. 

8. 

9. 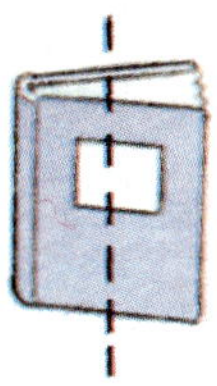

10. 

11. 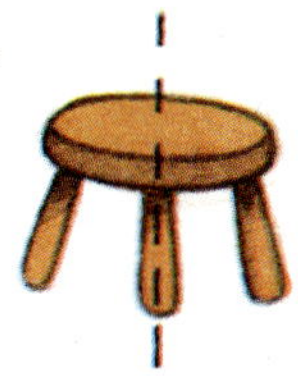

Which letters and digits have a line of symmetry?

12. MATH IS FUN

13. 0 1 2 3 4 5 6 7 8 9

These figures have more than one line of symmetry. Trace each figure. Then draw the lines of symmetry.

14. 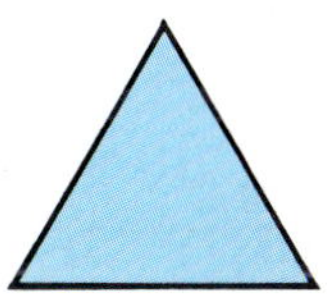

15. 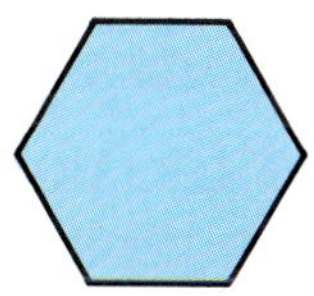

16. 

Think and Apply

17. Draw half of a picture on grid paper. Exchange papers with a classmate. Each of you draw the other half of the picture so the complete picture has symmetry. Return the pictures. Do they have a line of symmetry?

Mathematics and History

18. The ancient Egyptians used symmetry in designing their temples. How do you think the Egyptians made this entranceway symmetrical?

Draw several figures that have symmetry. Write a few sentences to explain why they are symmetrical.

SUMMING UP

Activity

Exploring Perimeter

What if you are making a moccasin and need cord to go around it? How can you find out the length of cord you will need?

Working together

Materials: paper clips, various objects
Record your work.

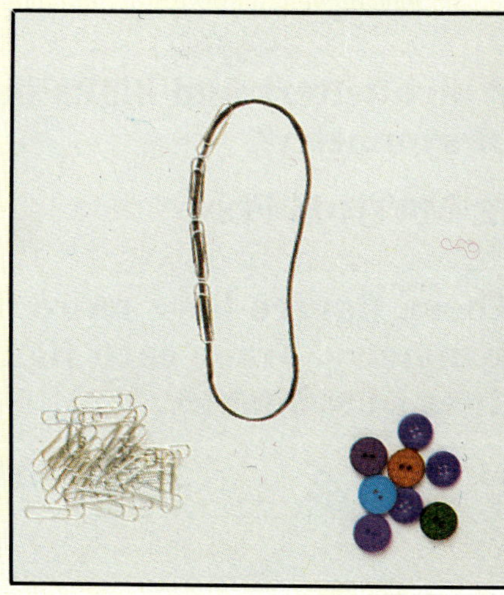

A. Trace your foot. Place paper clips end to end to measure the distance around your footprint.

B. Choose at least 2 other objects as units of measure. Use each to measure around your footprint.

C. Now measure the distance around the cover of your math book. First use paper clips. Then measure with the materials you used in **B**.

Sharing Your Results

DATA **Compare your results with those of your classmates.**

1. Can you tell from your results in **A** which footprint has the greatest distance around? Explain.
2. Can you tell from your results in **B** which footprint has the greatest distance around? Explain.
3. Were some objects more useful for comparing the measurements around your footprint? your math book? Explain.
4. Why were the measurements about the same when you measured with paper clips?
5. Describe how to find the length of cord you will need for your moccasin.

Extending the Activity

▸ The distance around a figure is called the **perimeter.** We use standard units to measure the perimeter.

Work in your group.
Materials: centimeter ruler

6. Use the ruler. Measure each side of the cover of your math book. Add all of the measurements to find the perimeter.

7. Estimate the perimeter of other rectangular objects. Then measure them. How close were you?

▸ To find the perimeter, add the lengths of the sides.

Find each perimeter.

8.
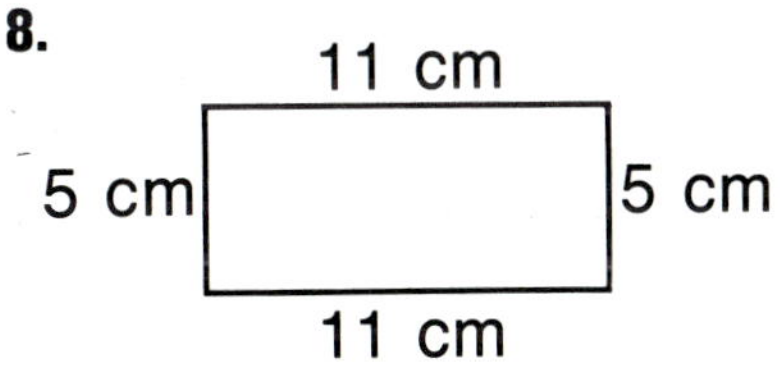

9.
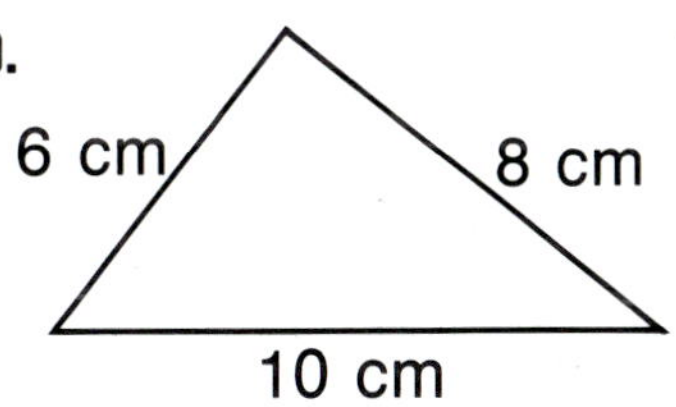

10.
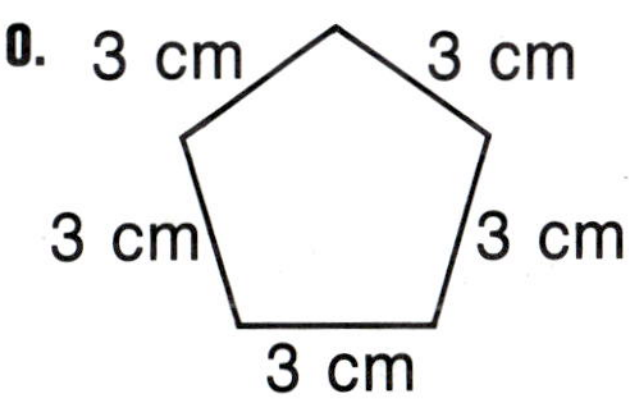

11.
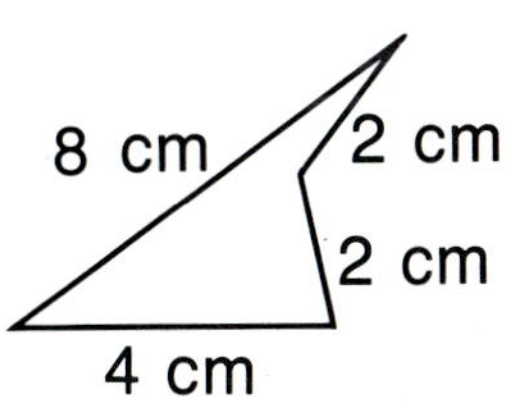

12.
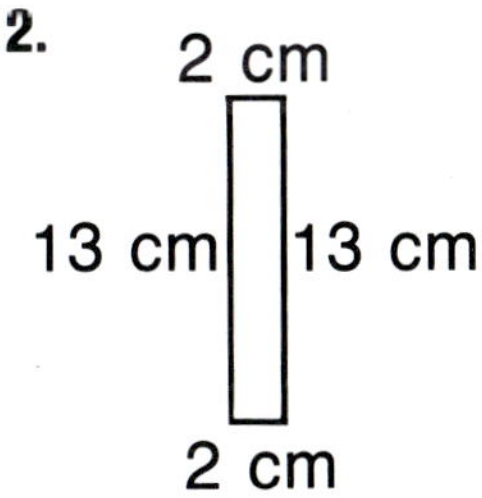

13.
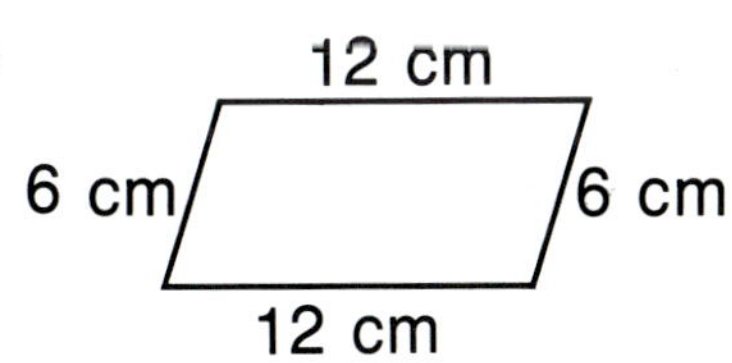

Summing Up

14. **Look back** at 8–13. How does using the same unit of measure help you to discuss and compare measurements?

15. Did you use any shortcuts to find the perimeter of any figures? Explain.

Activity

Exploring Area

How can you decide how much leather you will need for the sole of your moccasin?

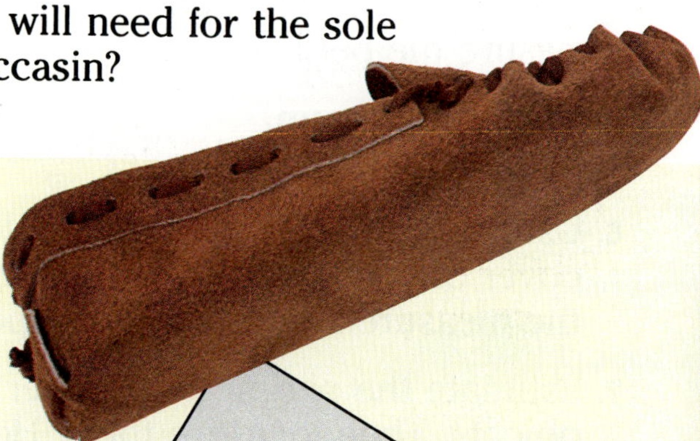

Working together

Materials: counters, square tiles, your footprint on paper

Make a drawing to show what happens each time.

A. Try to completely cover your footprint with a layer of counters.

B. Use the tiles to cover your footprint. Count the number of tiles.

C. Now cover your closed math book with counters and then with square tiles. Count the number of each.

Sharing Your Results

1. Were you able to completely cover your footprint with counters? with square tiles?
2. Which shapes fit together to completely cover the math book with no gaps?
3. Which shape is a better unit of measure for plane figures?
4. Describe how to find the amount of leather you will need for your moccasin.

Extending the Activity

Each square tile that you used is a **square unit.**

1 square unit

▸ The **area** of a figure is the number of square units that cover it.

To find area, count the number of square units.

The area of this figure is 6 square units.

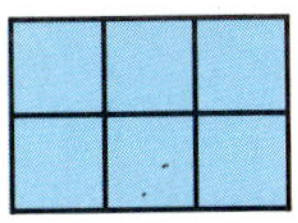

The area of this figure is 9 square units.

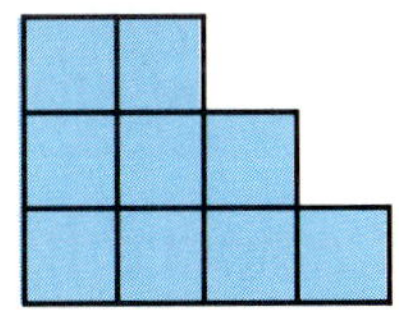

Work in your group.

5. Find the area of each figure. Is there a quick way to do this? Explain.

a. 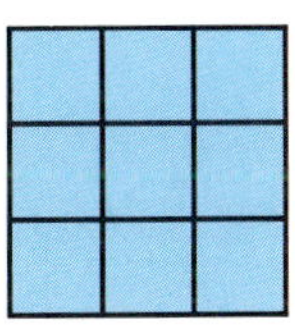

b. 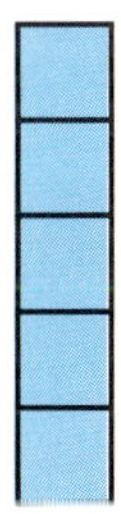

6. Discuss ways to find the area of each figure.

a. 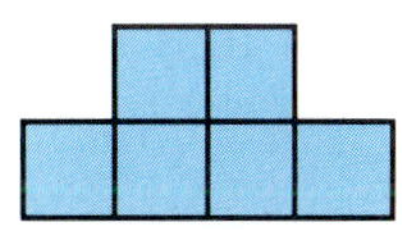

b. 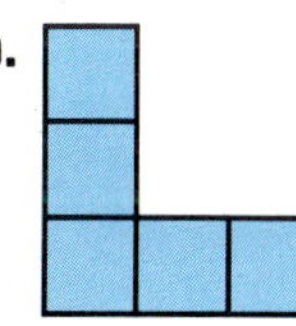

Summing Up

Find the area of each figure. Explain how you did it.

7. 

8. 

9. 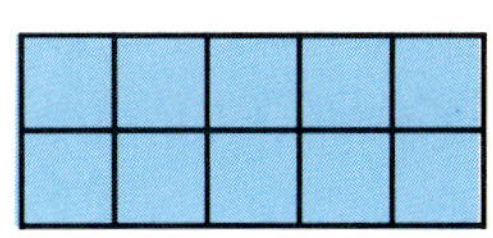

10. 

11. Trace your hand on grid paper. Estimate the area of your hand. Then count the squares. How close were you?

GETTING STARTED

Which box holds more?

Why do you think so?

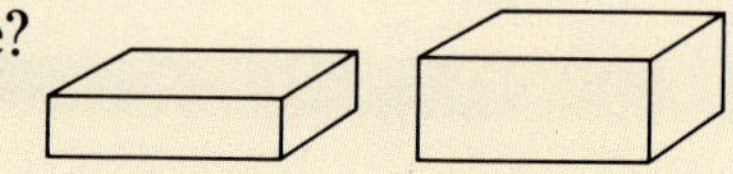

Understanding Volume

Would you need more cubes to make box **a** or box **b**?

a. 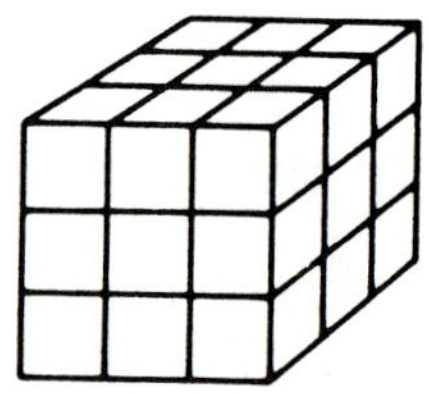

Try it!
Use cubes to build each figure.

How many cubes did you use for each figure? Did you use more cubes for box **a** or box **b**?

b. 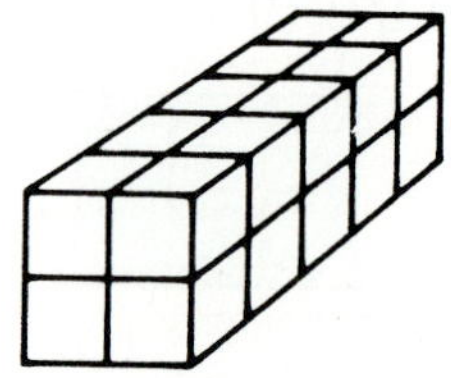

▸ The **volume** of a space figure is the number of cubic units that fit inside it.

The volume of box **a** is greater than the volume of box **b**.

How many cubic units are there in box **c**?

c. 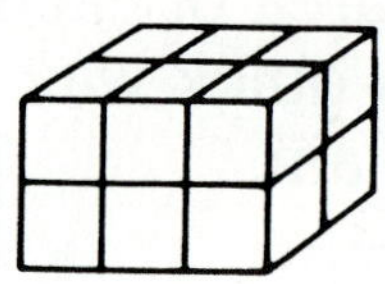

Check Your Understanding

Find the volume of each figure.

1. 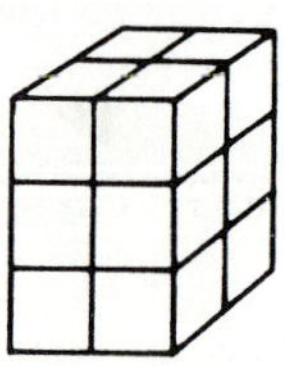

2. 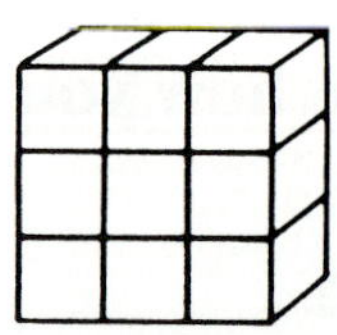

3. 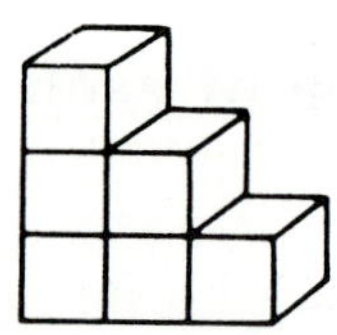

Share Your Ideas Look back at **3**. Could you make a different-shaped figure that has the same volume? Explain.

Practice

Find the volume of each figure.

4. 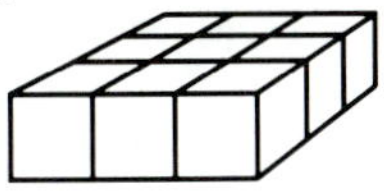

5. 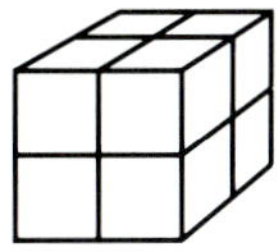

6. 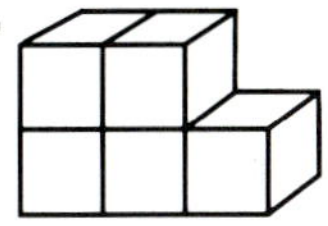

7. 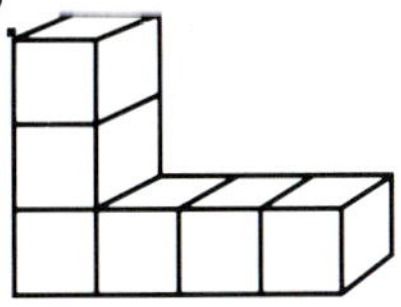

8. 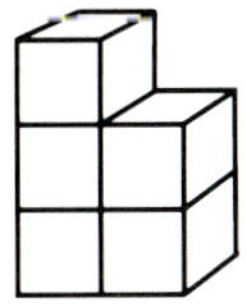

9. 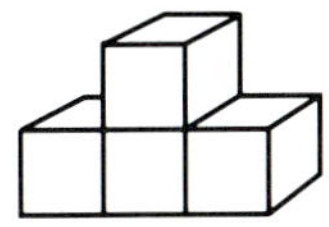

Each figure has two layers. Find the volume.

10. 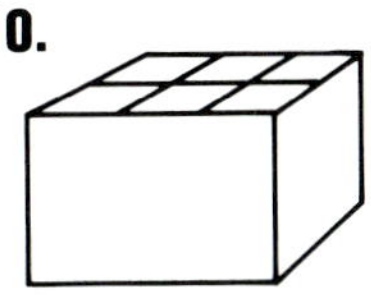

11. 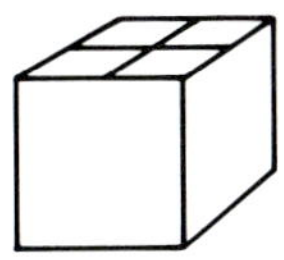

12. 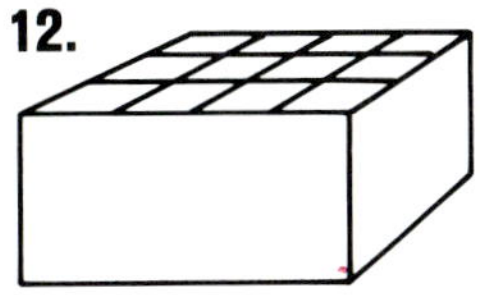

Which figure has the greater volume? Estimate. Then count.

13. a. 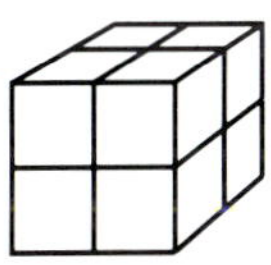
b. 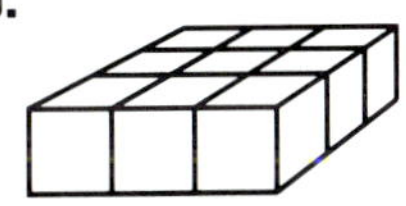

14. a. 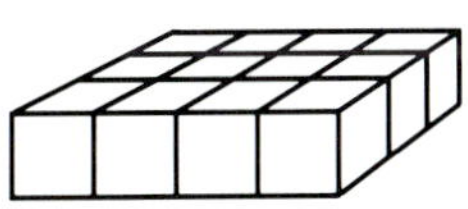
b. 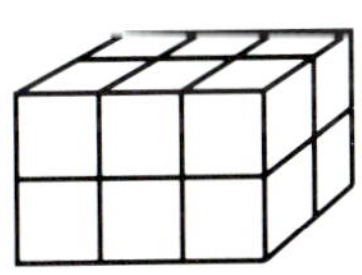

Think and Apply

15. **What if** there is a rectangular prism with a volume of 16 cubic units? Make drawings or use cubes to show possible rectangular prisms with this volume. How many drawings did you make in all?

Look back at **10–12**. Explain how you found the volume of each.

Mixed Review

1. 8×6
2. 1×2
3. 7×9
4. 9×2
5. 7×5
6. 8×8

7. $8 \div 2 = \square$
8. $27 \div 3 = \square$
9. $24 \div 6 = \square$
10. $28 \div 7 = \square$

11. $6\overline{)42}$
12. $7\overline{)56}$
13. $8\overline{)24}$
14. $9\overline{)81}$
15. $3\overline{)15}$
16. $8\overline{)40}$

Write the value.

17. 16 pennies
18. 4 quarters
19. 6 nickels
20. 8 dimes
21. 1 dollar, 1 quarter
22. 4 dollars, 11 nickels

SUMMING UP

Chapter Review

Name each figure. pages 338–339, 342–345

1. 

2. 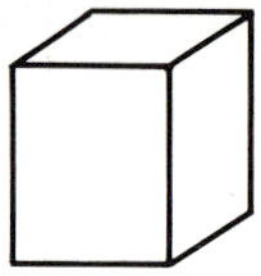

3. 

4. 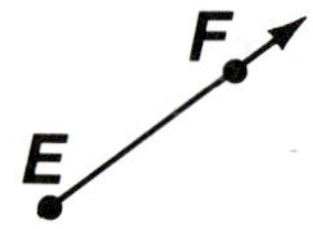

Tell if the figures are congruent. pages 352–353

5. 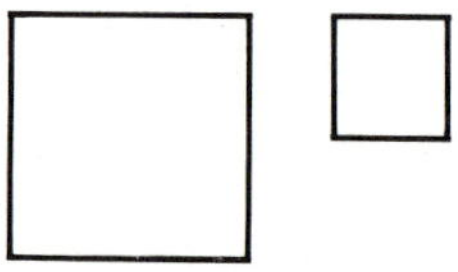

6. 

7. 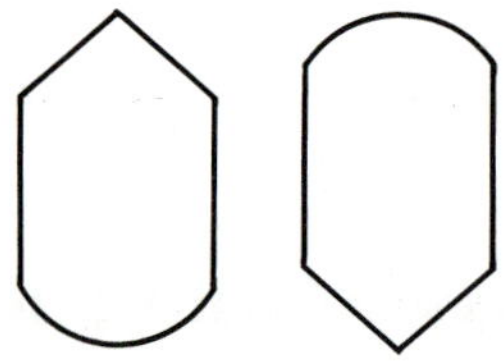

Tell if the dashed line is a line of symmetry. pages 354–355

8. 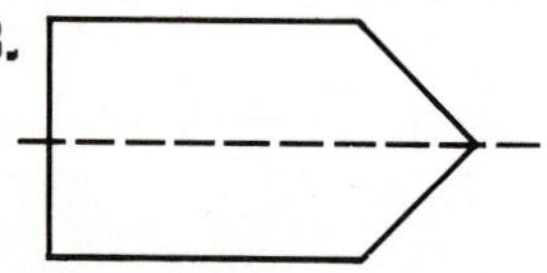

9. 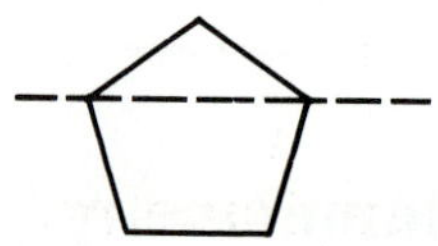

10. 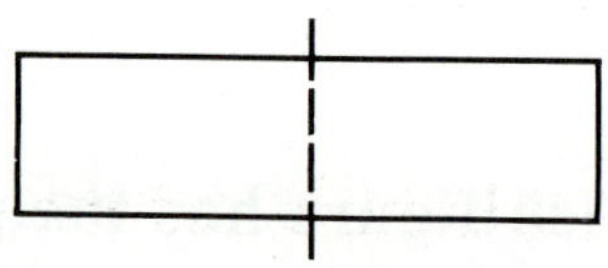

Find the perimeter. pages 356–357

11. 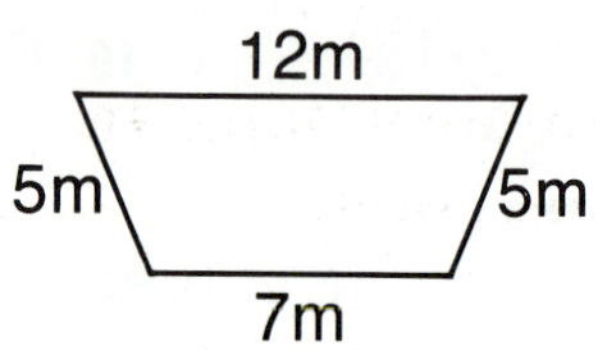

Find the area in square units. pages 358–359

12. 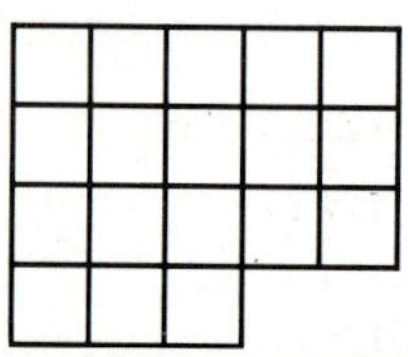

Find the volume in cubic units. pages 360–361

13. 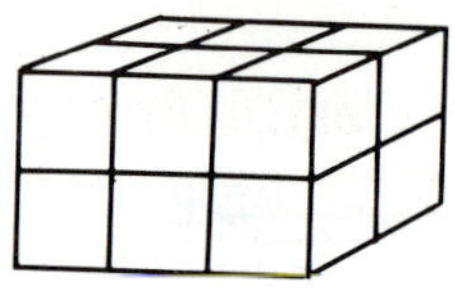

Choose the correct word to complete each sentence.

14. Figures with the same size and shape are _____.

15. The distance around a figure is its _____.

Words to Know
perimeter congruent

Solve. pages 349–351

16. Guess and test. A rectangle has a perimeter of 10 inches. What are the possible lengths of the sides?

17. The perimeter of a rectangular room is 36 feet. What lengths are possible for the sides? Which lengths are sensible?

Chapter Test

Name each figure.

1. 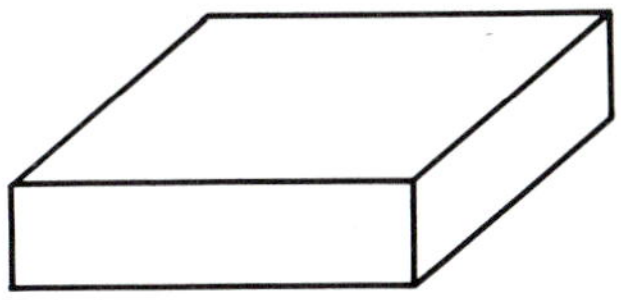

2. 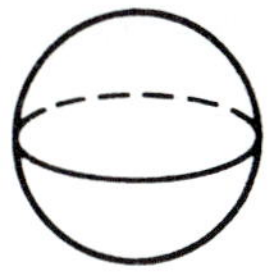

3. 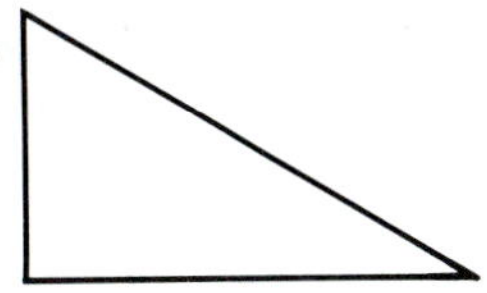

4. 

5. 

6. 

Tell which figures are congruent.

7. 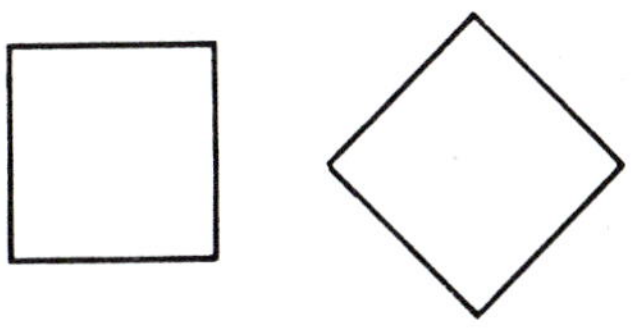

8. 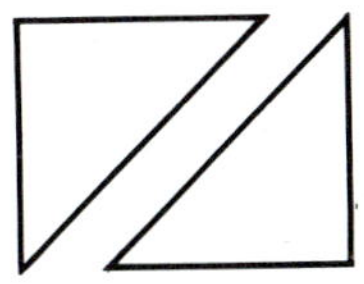

9. 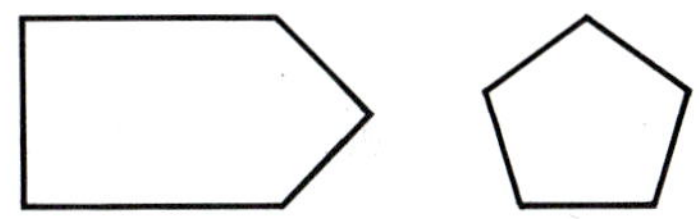

Tell if the dashed line is a line of symmetry.

10. 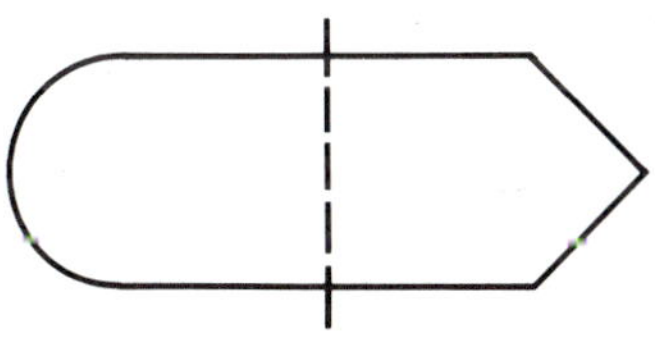

11. 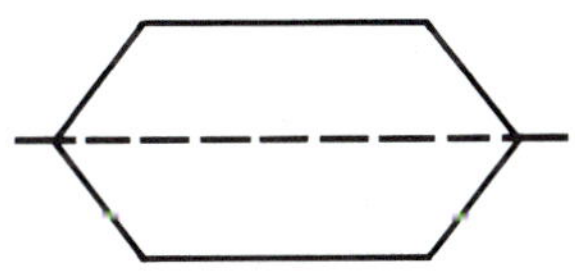

12. 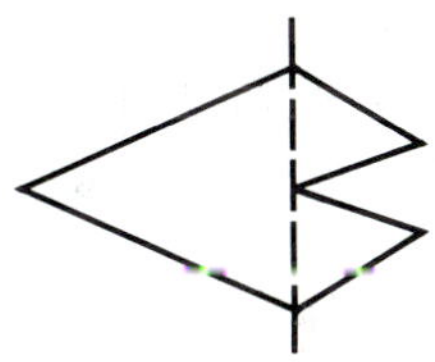

Find the perimeter.

13. 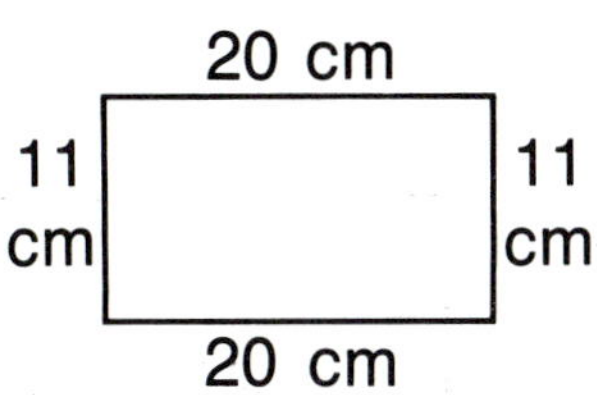

Find the area in square units.

14. 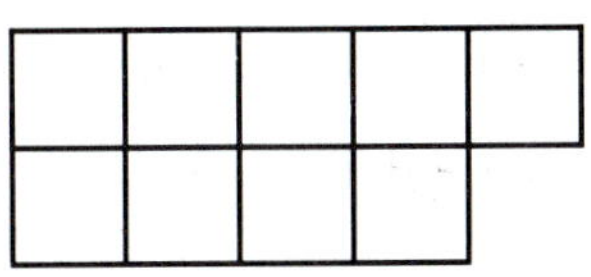

Find the volume in cubic units.

15. 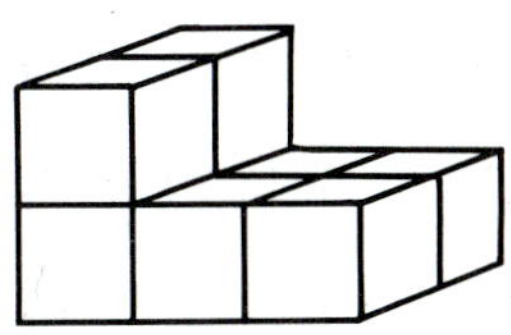

Solve.

16. A triangular sign is 16 inches on each side. What is the perimeter of the sign?

17. Guess and test. A rectangular garden has a perimeter of 12 yards. What are the possible lengths of the sides?

THINK Are all gardens with the same perimeter congruent to each other? Explain.

Computer Link

Angles

The Logo turtle can make a right angle when it turns `RT 90`, or `LT 90`. Use Logo to explore angles.

Materials: Logo

A. Use these commands to draw a right angle as shown. Then draw a right angle in a different position.

```
HOME
FD 50
BK 50
RT 90
FD 50
```

B. Draw a right angle in this position. What number can you use in the `RT` command to turn the turtle?

```
HOME
LT 20
FD 40
BK 40
RT ___
FD 40
```

C. If you first turned `LT 45` to draw one side of a right angle, does that change how much you turn to draw the other side? Explain.

D. Start at `HOME` position. Use the commands to display both angles at once. Are both angles the same size? Why?

To draw angle 1

```
HOME
FD 75
LT 90
FD 75
```

To draw angle 2

```
PU HOME PD
FD 30
LT 90
FD 30
```

Moves the turtle `HOME` without drawing.

Sharing Your Results

1. Does the size of an angle change if you change its position? the length of its sides?
2. Are all right angles the same size? Explain how the size of an angle is determined.

Extending the Activity

Not all angles are right angles. **What if** you want to draw an angle that is less than a right angle? greater than a right angle?

3. What number can you use in the RT command to draw an angle less than a right angle? Check your guess at the computer.

```
FD 30
BK 30        Turn
RT __
FD 30
```

4. Compare your angle with those of your classmates. Name some numbers you can use in a RT command to draw angles that are less than a right angle. What numbers could you use in a LT command?

5. **What if** you want to draw an angle greater than a right angle? Name some possible numbers you could use in a RT or LT command.

6. Draw an angle greater than a right angle. Describe your method.

Summing Up

7. Which of the following determines the size of an angle?
   - Position
   - Length of sides
   - Opening between the sides

8. Explain how you can tell whether each angle is a right angle, less than, or greater than a right angle.

a. 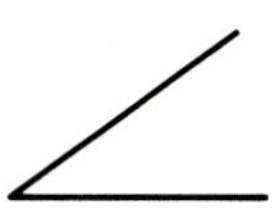
b. 
c. 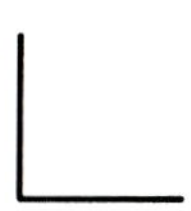
d. 
e. 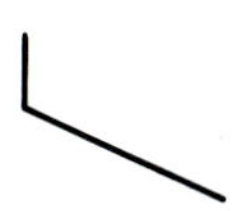
f. 

EXTENSION

Making Shapes

Here's a way to create some familiar and some not-so-familiar geometric shapes!

Materials: 8 yards of yarn or cord tied in a loop

A. Use your loop to make each shape on the list.

B. Make sure each person in the group uses both hands to help make the shape. (Feet, heads, and elbows help, too!)

C. As you go along, compare your group's shapes with those made by other groups.

D. Challenge other groups with questions like this: How can you show that the sides of your square really are equal?

The Shape List

triangle	square
triangle with equal sides	5-sided figure
rectangle	3 congruent squares

Try these.

1. Which shapes were the most difficult to make? Why?
2. Could you make a cube or rectangular prism with your loop? How would these shapes be different from the ones you were making?

Create your own new, unusual shape. Make up a name for your shape. Have fun sharing your shapes with each other!

Maintaining Skills

Choose the correct answer. Write A, B, C, or D.

1. What fraction of the group is red?

A $\frac{5}{8}$ **C** $\frac{3}{5}$

B $\frac{3}{8}$ **D** not given

2. $\frac{1}{4}$ of 20 = □

A 5 **C** 10

B 4 **D** not given

3. Choose a unit to measure the distance from Texas to Florida.

A meter **C** kilometer

B centimeter **D** not given

4. How long is the bar to the nearest $\frac{1}{2}$ inch?

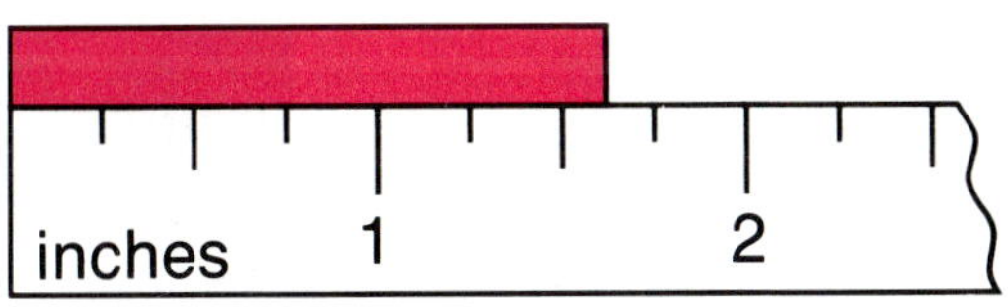

A $\frac{1}{2}$ in. **C** 1 in.

B $1\frac{1}{2}$ in **D** not given

5. What unit would you use to measure the liquid in a glass?

A quart **C** gallon

B pint **D** not given

6. Name the shape.

A sphere **C** cone

B cylinder **D** not given

7. Name the shape.

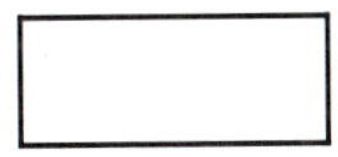

A square **C** triangle

B rectangle **D** not given

8. Find the perimeter.

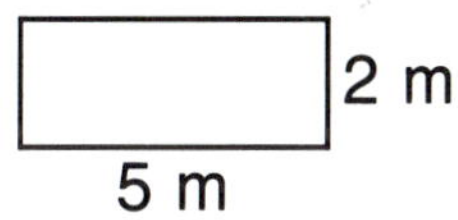

A 14 m **C** 10 m

B 7 m **D** not given

Solve.

9. Pete saw 15 birds on the fence. There were 2 times as many robins as sparrows. How many of the birds were sparrows?

A 7 **C** 10

B 5 **D** not given

10. Amy bought 8 plants for her garden. They cost $2.00 each. How much did she spend for the 8 plants?

A $8.00 **C** $16.00

B $2.00 **D** not given

12 Introducing Decimals

THEME Going Places

Sharing What You Know

Do you enjoy traveling by car? Some people say getting there is half the fun! Yet getting there can also be tricky. Road signs can help you get where you're going. Which road signs have you seen? What do those road signs tell you? How do they help travelers?

Using Language

You have been riding in the family car for hours. You are getting hungry. You see a sign that says, "Food: 3.1 miles." Lucky for you—that's just a little more than 3 miles! The number 3.1 has a **decimal point**. Where else have you seen numbers with a **decimal point**? Talk about what you think these numbers mean.

Words to Know: decimal, tenths, hundredths

Be a Problem Solver

Max is riding his bicycle. The odometer says he has gone 2.8 miles. What number will show next on the odometer?

Write directions telling how to walk from your house to your school. Draw a map to show the path. Include numbers on your map.

Activity
Exploring Decimals

On a trip in New Mexico, Dave saw many interesting southwestern designs. He saw the designs on rugs, blankets, and pottery. In this activity you can create your own designs.

Working together

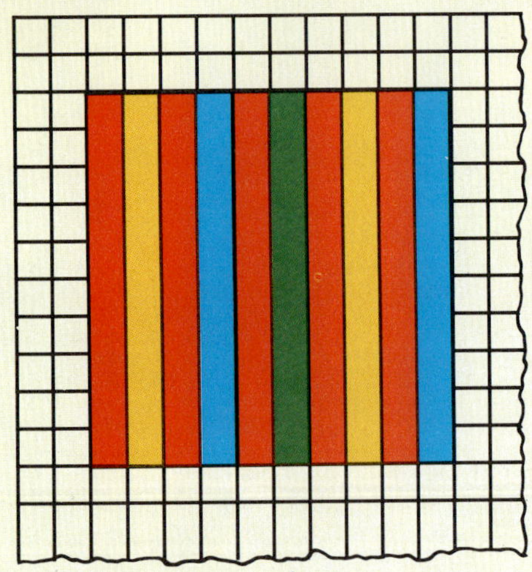

Each column is $\frac{1}{10}$ of the whole.

$\frac{5}{10}$ is red. $\frac{2}{10}$ is yellow.

$\frac{1}{10}$ is green. $\frac{2}{10}$ is blue.

Materials: Workmat 6 or grid paper; red, blue, green, and yellow crayons

A. Outline a 10-by-10 square on the workmat as shown. How many columns are there?

B. Shade each column to create a striped design of your own.

C. Use a fraction. Tell what part of your design is red. blue. green. yellow.

Sharing Your Results

DATA **1.** Compare your design with those of your classmates. Which design has the most red? blue? green? yellow?

2. What if you shaded half of your square red and half blue? How many tenths would be red? blue?

3. What if you shaded your whole square green? How many tenths would be green?

Extending the Activity

You can write the fraction $\frac{7}{10}$ as a decimal.

write 0.7
read seven tenths

4. Make a table like this. **Look back** at your design. Record the fraction and the decimal for each color.

Fraction	Decimal
$\frac{7}{10}$	0.7
$\frac{1}{10}$	0.1

Work in your group.

Materials: Workmat 6 or grid paper with 10-by-10 squares outlined; crayons

5. Take turns shading tenths of squares. Have group members write a decimal for each shaded part.

6. Take turns writing a decimal. Have group members shade tenths of a square for each decimal.

Summing Up

7. **What if** you had one dollar and you wanted to share it equally among 10 classmates? Could you do this with ten coins? Explain.

8. How does thinking about dimes and dollars help you to understand tenths?

Tell how many tenths of a dollar each is. Then write each as a fraction and as a decimal.

9. 4 dimes
10. 6 dimes
11. 8 dimes
12. 1 dime
13. 10 dimes
14. 3 dimes

Activity

Understanding Decimals

Dave noticed that many Navajo designs have geometric patterns.

Now you can create designs by shading small squares.

Working together

Materials: Workmat 6 or grid paper, crayons

A. Outline a 10-by-10 square on the workmat. How many small squares are there?

B. Create a design by shading small squares.

C. Write a fraction to tell which part of the whole design is shaded.

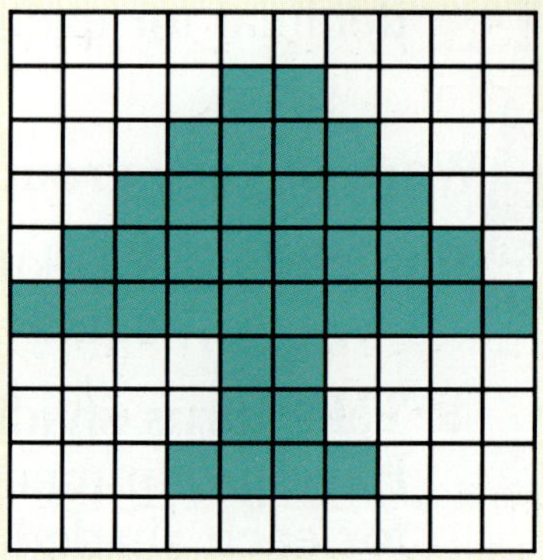

Each small square is $\frac{1}{100}$ of the whole.

$\frac{38}{100}$ of the grid is shaded.

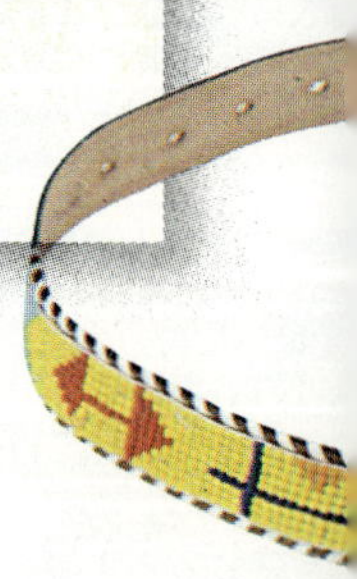

Sharing Your Results

DATA **1.** Compare your design with those of others. Which design used the greatest part of the 10-by-10 square? the least?

2. What if you shaded half of your square green and half yellow? How many hundredths would be green? yellow?

3. What if you shaded your whole square red? How many hundredths would be red?

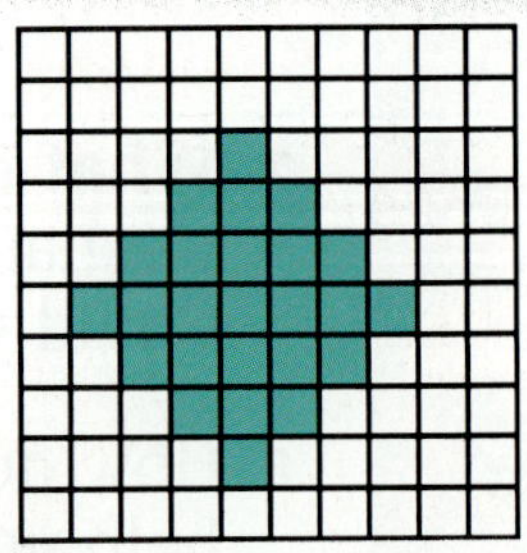

$\frac{25}{100}$ is shaded.

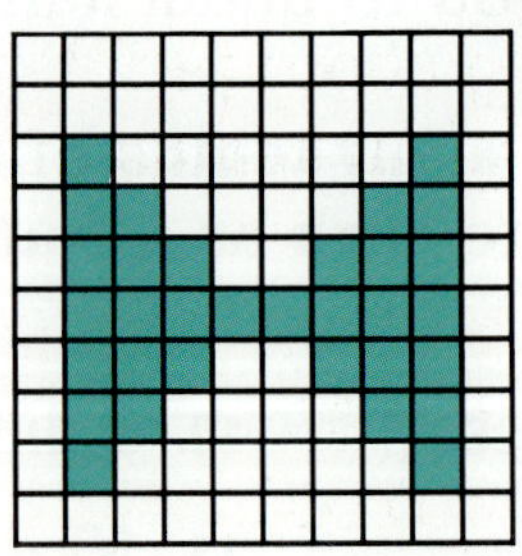

Extending the Activity

You can write the fraction $\frac{16}{100}$ as a decimal.

write 0.16
read sixteen hundredths

4. **Look back** at your design. Write the fraction and the decimal for the shaded part.

Working with money can help you to understand hundredths.

Remember 100 pennies = 1 dollar

Trading Post

Work in your group.

Materials: Workmat 6 and crayons, or Workmat 15 and play pennies

5. Outline a 10-by-10 square to have one small square for each penny in a dollar.
6. Look at the items in the Trading Post. Secretly, choose an item to buy.
7. Shade squares or place play pennies on squares to show how much your item costs.
8. Ask group members which item you chose and what part of a dollar you spent.

Summing Up

9. How does thinking about pennies and dollars help you to understand hundredths?

Explain what each of these means.

10. $.89 0.89	11. $.60 0.60	12. $.35 0.35

GETTING STARTED

You have 10 balloons. What fraction can you use to show that 2 of the balloons are red? that 3 of the balloons are blue?

On summer vacation, Chuck went to Yellowstone National Park. Ten people were waiting to see Old Faithful, a geyser, erupt. Eight of the ten people, or eight tenths, had cameras.

You can write eight tenths as a fraction or as a decimal.

fraction $\frac{8}{10}$

decimal **0.8**

read eight tenths

ones	tenths
0.	8

↑ decimal point

Two tenths of the people did not have cameras.

fraction $\frac{2}{10}$

decimal **0.2**

ones	tenths
0.	2

What do you think the 0 before the decimal point means?

Check Your Understanding

Write a decimal for the blue part.

1. 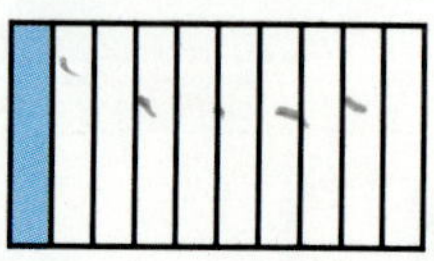

2. 

3. 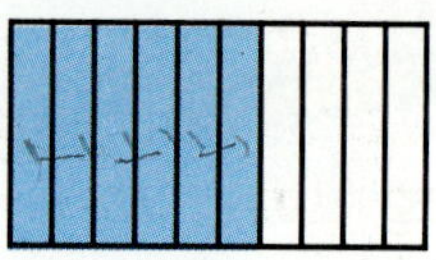

Write each as a decimal.

4. five tenths **5.** 9 out of 10 **6.** $\frac{3}{10}$ **7.** $\frac{4}{10}$ **8.** $\frac{7}{10}$

Share Your Ideas Look back at your answers for **1** and **2**. Which decimal stands for the greater amount? How do you know?

Practice

Write a decimal for the blue part.

9. 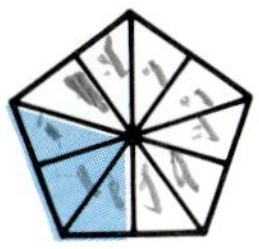

10. 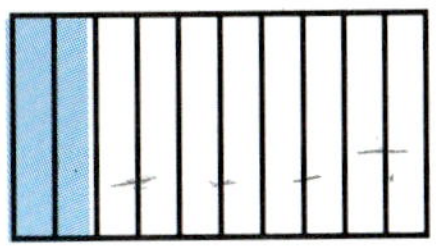

11. 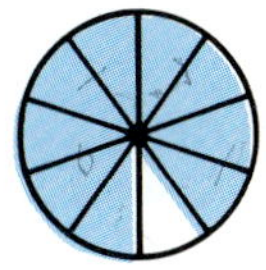

Write each as a decimal.

12. eight tenths **13.** four tenths **14.** one tenth **15.** six tenths

16. 7 out of ten **17.** 3 out of ten **18.** 10 out of ten **19.** 1 out of ten

20. $\frac{9}{10}$ **21.** $\frac{2}{10}$ **22.** $\frac{7}{10}$ **23.** $\frac{8}{10}$ **24.** $\frac{5}{10}$ **25.** $\frac{3}{10}$

Choose the correct decimal for each fraction.

26. $\frac{3}{10}$
a. 0.1 **b.** 0.3
c. 0.10 **d.** 3.0

27. $\frac{6}{10}$
a. 0.6 **b.** 0.06
c. 6.0 **d.** 1.6

28. $\frac{1}{10}$
a. 10.0 **b.** 1.0
c. 0.01 **d.** 0.1

Think and Apply

29. Chuck made a list of the 10 exhibits he wants to see. By noon, he had checked off 6 exhibits. Write a decimal to show what part of the list is checked off.

30. The word *decade* means 10 years. The word *decathlon* means 10 events in an athletic contest. How are these words related to the word *decimal?*

Logical Thinking

Continue each pattern.

31. 0.0, 0.1, 0.2, ___, ___, ___

32. 0.4, 0.5, 0.6, ___, ___, ___

33. 0.9, 0.8, 0.7, ___, ___, ___

34. 0.2, 0.3, 0.4, ___, ___, ___

35. 0.6, 0.5, 0.4, ___, ___, ___

Make a drawing to show that $\frac{8}{10}$ and 0.8 name the same amount.

SUMMING UP

GETTING STARTED

Discuss how to write each of these as a fraction. 16 out of 100 35 out of 100

Decimals: Hundredths

Marge saw a star show at the Adler Planetarium in Chicago. At the show 85 out of 100 seats were filled.

Eighty-five hundredths seats were filled. You can write this amount as a fraction or as a decimal.

fraction $\frac{85}{100}$

decimal 0.85

ones	tenths	hundredths
0.	8	5

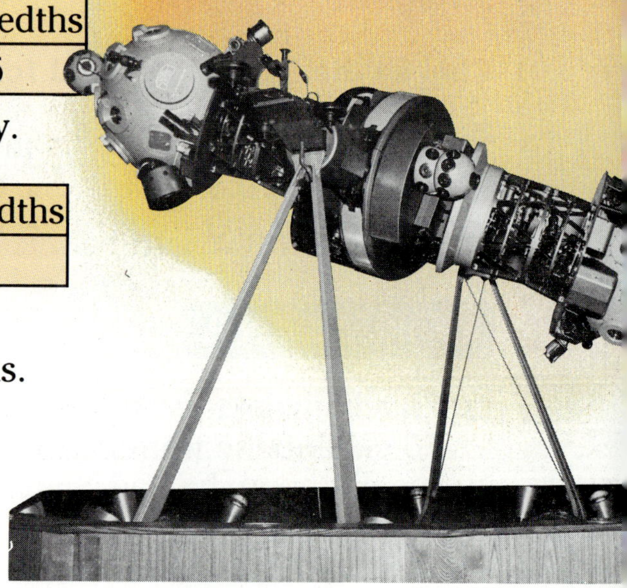

Fifteen hundredths of the seats were empty.

fraction $\frac{15}{100}$

decimal 0.15

ones	tenths	hundredths
0.	1	5

Look at the chart for 0.15. Explain why 15 hundredths = 1 tenth and 5 hundredths.

Check Your Understanding

Write a decimal for the blue part.

1. 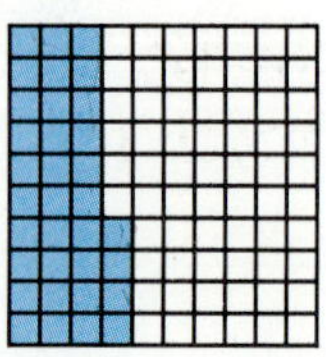

2. 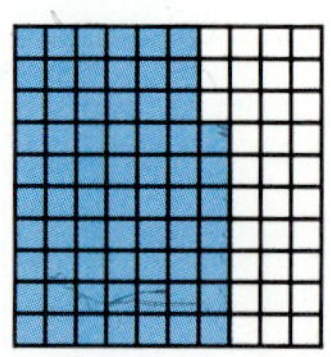

3. 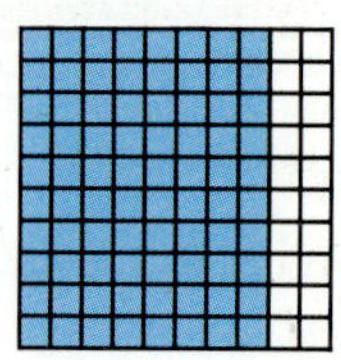

Write each as a decimal.

4. fourteen hundredths
5. 18 out of 100
6. $\frac{29}{100}$
7. $\frac{33}{100}$
8. $\frac{71}{100}$

Share Your Ideas Look back at your answers for **1** and **2**. Which decimal shows the greater amount? How do you know?

Practice

Write a decimal for the blue part.

9. 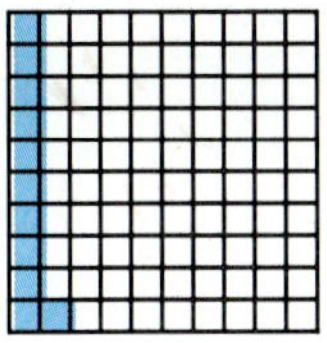

10. 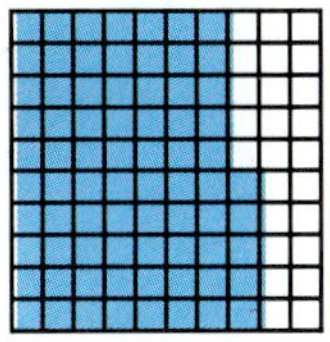

11. 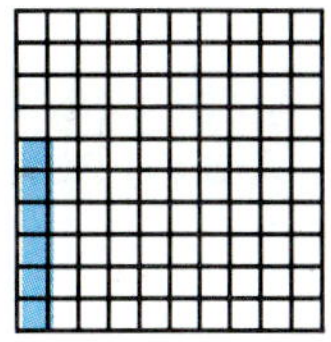

Write each as a decimal.

12. thirty-one hundredths
13. sixty-three hundredths
14. ninety hundredths
15. 24 out of 100
16. $\frac{12}{100}$
17. $\frac{68}{100}$
18. $\frac{3}{100}$
19. $\frac{80}{100}$

Write in words. Name the value of the digit 4.

20. 0.43
21. 0.04
22. 0.84
23. 0.41
24. 0.40

Write each part of a dollar as a decimal.

25. 23 pennies
26. 70 pennies
27. 6 dimes
28. 9 nickels

Think and Apply

29. Use the Logo turtle to draw the largest rectangle you can. Then draw the smallest rectangle you can. Is there a smaller rectangle? Explain.

30. You can use a calculator to find a decimal for a fraction. To find $\frac{3}{10}$, press

What decimal is shown in the display? This time try $\frac{5}{10}$. Now try five other fractions.

31. Shade drawings to show the decimal 0.8 and the decimal 0.80. What do you notice about the shaded parts?

Common Error

32. Why is what Judy wrote incorrect? How could you write the decimal correctly?

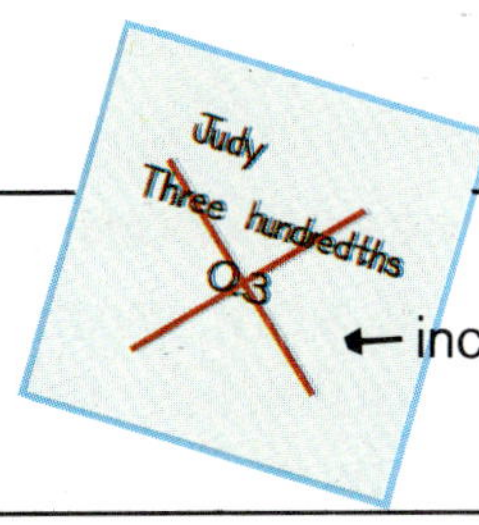

← incorrect

Explain why one tenth is greater than one hundredth. Use a drawing to help you.

SUMMING UP

Using Problem Solving

Writing Decimals in Order

The goal of this game is to use digits to make a decimal that is greater than or less than other decimals.

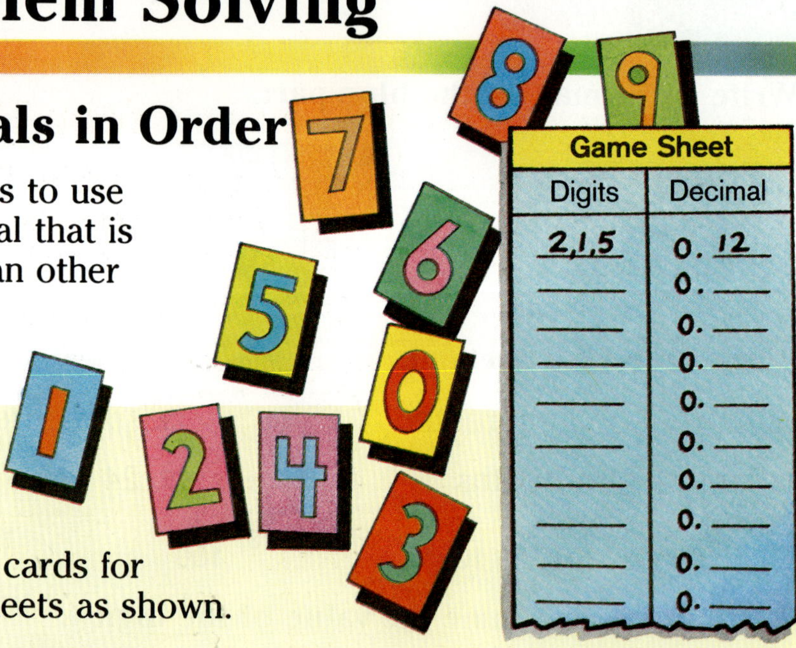

Working together

Materials: a bag with cards for the digits 0–9, game sheets as shown.

Play this game with a partner.

A. Player 1 draws one digit card, writes the digit under *Digits* on the Game Sheet and returns the card to the bag.

B. Player 1 repeats A two more times.

C. Player 1 now chooses the two digits that make the greatest decimal and writes the decimal under *Decimal* on the Game Sheet.

D. Player 2 follows the steps in **A–C** using another Game Sheet.

E. This time Player 1 repeats **A–C**, but tries to write a decimal that is greater than the first decimal. If a greater decimal cannot be written, the player must pass.

F. Player 2 repeats **E**.

G. Play continues until a player writes 9 decimals in order, from least to greatest. The player who does this wins the game.

Sharing Your Ideas

1. **What if** your game sheet looks like this? You draw a 7, 8, and 9. What decimal would you do? Why?

Game Sheet	
Digits	Decimal
3,1,0	0. 13
4,3,2	0. 24
____	0. ____

2. What hints would you give to someone playing this game?

THINK
EXPLORE
SOLVE
LOOK BACK

Extending Your Thinking

3. Play the game again. This time each decimal must be less than the previous decimal.

4. Now play the Up-Down Game. Make a Game Sheet like this. Player 1 draws 3 digit cards and uses two of the digits to write decimals greater than 0.50. Player 2 repeats this, but writes decimals less than 0.50. Each number going down the Game Sheet must be greater than the previous one. Each number going up the Game Sheet must be less than the previous one. The player to fill one half of the Game Sheet first wins.

Up-Down Game Sheet
0.____
0.____
0.____
0.____
0.50
0.____
0.____
0.____
0.____

5. Can you think of another way to play this game? Explain it.

Summing Up

6. Which game did you enjoy the most?

7. **What if** you were playing the Up-Down Game and you and your partner have written the numbers shown at the right? You pick a 5, 6, and 4 and must write the next greatest decimal. What decimal would you write? Explain.

Up-Down Game Sheet
0.____
0.____
0.____
0.37
0.50
0.62
0.____
0.____
0.____

GETTING STARTED

Discuss how each of these numbers is read.
$7\frac{1}{10}$ $1\frac{5}{10}$ $30\frac{98}{100}$
Then write the words for each number.

Decimals Greater Than 1

Doug and his family visited Colonial Williamsburg in Virginia. They walked $2\frac{3}{10}$ miles to see the exhibits.

You can write $2\frac{3}{10}$ as a decimal.

write 2.3

read two and three tenths

ones	tenths
2.	3

Now write $4\frac{25}{100}$ as a decimal.

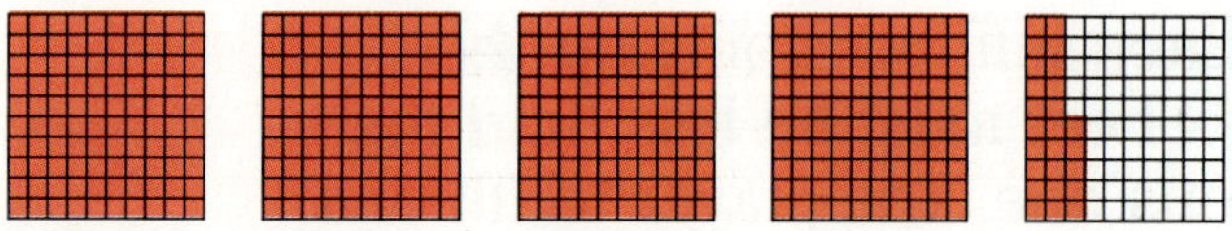

write 4.25

read four and twenty-five hundredths

ones	tenths	hundredths
4.	2	5

How is reading 4.25 like reading $4\frac{25}{100}$?
What does the word *and* separate in each?

Check Your Understanding

Write a decimal for the red part.

1. 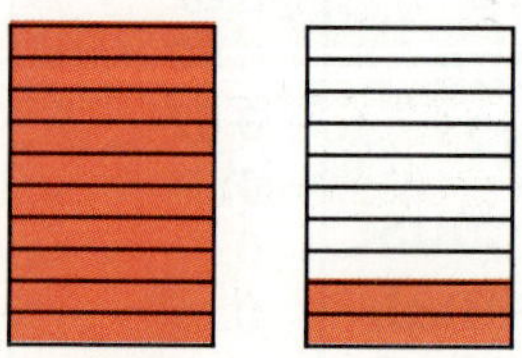

2. 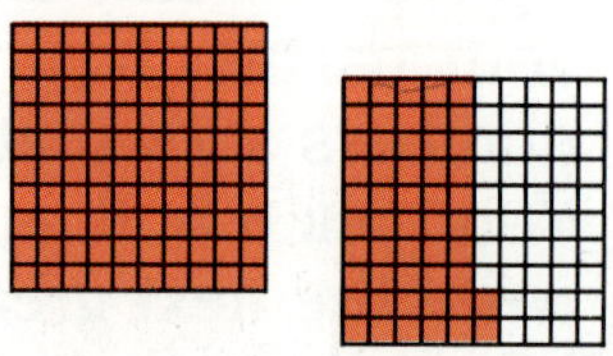

3. 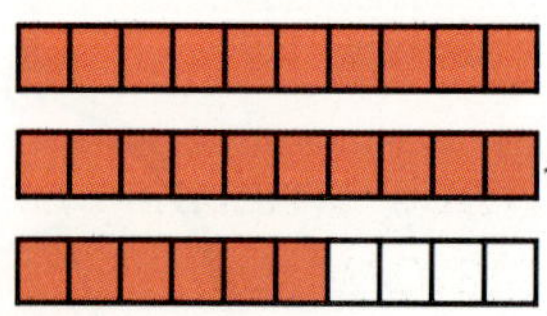

Write each as a decimal.

4. six and thirty-five hundredths
5. one and nineteen hundredths
6. $3\frac{4}{10}$
7. $9\frac{12}{100}$
8. $5\frac{78}{100}$
9. $2\frac{1}{10}$
10. $4\frac{36}{100}$
11. $7\frac{9}{100}$

Share Your Ideas Look back at **1**. Explain how this decimal is different from 1.02.

Practice

Write a decimal for the red part.

12. 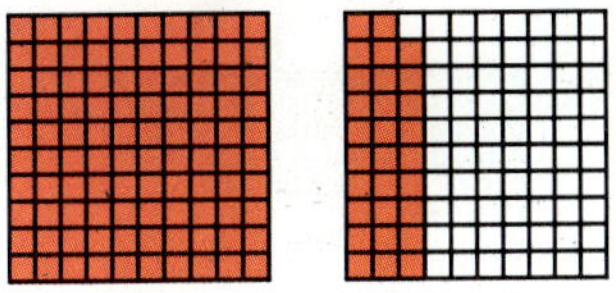

13. 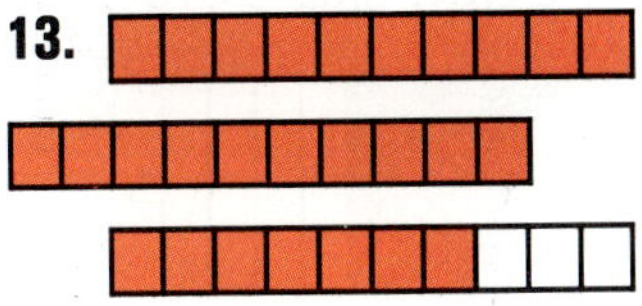

Write each as a decimal.

14. six and twelve hundredths

15. two and eight tenths

16. seven and twenty-one hundredths

17. one and seventy hundredths

18. $5\frac{3}{10}$ **19.** $1\frac{99}{100}$ **20.** $3\frac{62}{100}$

21. $4\frac{80}{100}$ **22.** $\frac{6}{10}$ **23.** $2\frac{7}{100}$

Write in words. Name the value of the digit 5.

24. 2.15 **25.** 5.39 **26.** 0.51

27. 7.5 **28.** 3.05 **29.** 1.59

Choose the greater number.

30. 7.32 or 7.3 **31.** 8.27 or 8.92

32. 2.7 or 2.8 **33.** 3.5 or 3.50

34. 4.68 or 4.61 **35.** 6.13 or 6.03

Think and Apply

36. Doug walked about 1.5 miles around the historic exhibits. He walked 1.25 miles back to the hotel. Which walk was longer?

37. Guess the number. Use these clues: The number has four digits. It is greater than 10 but less than 11. The sum of the digits is 15. A 5 is in the tenths place.

Make a drawing to show that the decimal 1.90 is different from the decimal 1.09.

SUMMING UP

Mixed Review

1. 6,471 − 1,171

2. 9,752 + 2,468

3. 1,000 + 3,999

4. 7,506 − 219

5 2,381 − 1,061

6. 7,492 − 4,687

7. 2,549 − 672

8. 3,000 − 1,042

9. 8,674 + 599

10. 6,429 + 3,298 = ☐

11. 2,832 − 1,946 = ☐

12. 5,000 − 879 = ☐

13. 4,367 + 309 = ☐

Order these from least to greatest.

14. 97, 47, 87, 57

15. 349, 399, 379

16. 9,352; 9,395; 926

17. 3,457; 498; 5,342

18. 5,020; 520; 5,002

Midchapter Review

Write a decimal for the green part. pages 370–371, 374–375

1. 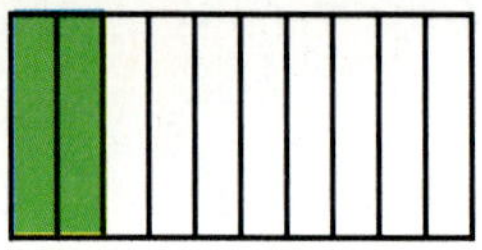

2. 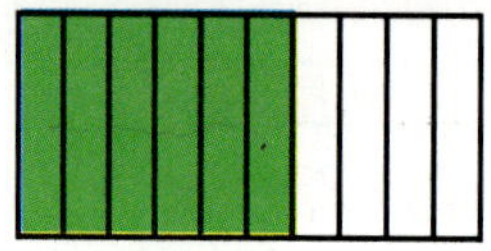

3. 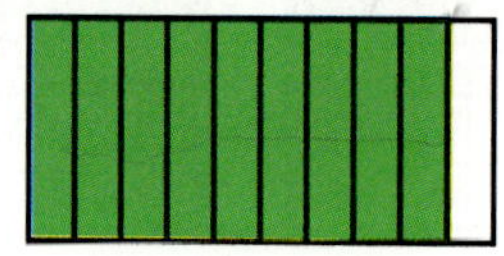

Write each as a decimal. pages 370–371, 374–375

4. 8 out of 10
5. $\frac{1}{10}$
6. $\frac{3}{10}$
7. two tenths
8. $\frac{4}{10}$
9. $\frac{7}{10}$

Write a decimal for the green part. pages 372–373, 376–377, 380–381

10. 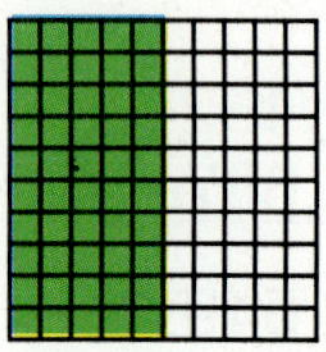

11. 

12. 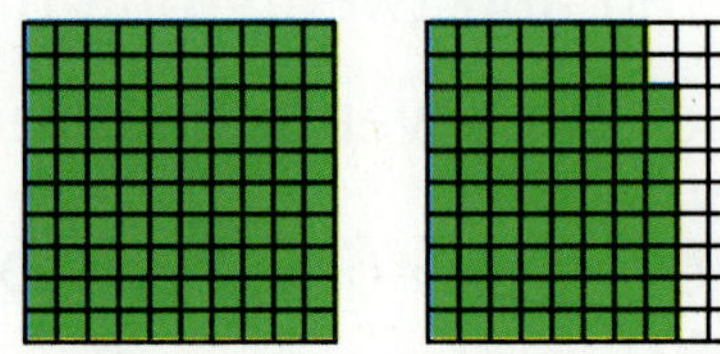

Write each as a decimal. pages 376–377, 380–381

13. nineteen hundredths
14. two and one hundredth
15. 27 out of 100
16. $\frac{13}{100}$
17. $\frac{54}{100}$
18. $8\frac{3}{10}$
19. $9\frac{17}{100}$

Choose the correct word to complete each sentence.

20. The number 0.2 is read two _____.
21. The number 0.65 is read sixty-five _____.
22. $\frac{37}{100}$ can be written as the _____ 0.37.

Words to Know
hundredths tenths decimal

Solve.

23. The Alamo's parking lot has cars in 8 out of 10 spots. Write a decimal to show what part of the lot has cars.

24. Dee surveyed 100 people. Thirty one said they had been to the Alamo. Write a decimal to show what part of the 100 had been to the Alamo.

Exploring Problem Solving

THINK
EXPLORE
SOLVE
LOOK BACK

Will It Fit?

The Kellys have a darkroom so that they can develop their own pictures. They printed their vacation pictures in different sizes. They have envelopes of different sizes. They want to mail their photographs.

Photograph Sizes (inches)	Envelope Sizes (inches)
A. 3.9 by 5	V. 4.2 by 9
B. 4.1 by 6	W. 3.7 by 6.5
C. 3.5 by 4.9	X. 5.5 by 7.5
D. 5 by 7	Y. 3.2 by 4
	Z. 4 by 6.1

Thinking Critically

Which photographs will fit in each of the envelopes?

Analyzing and Making Decisions

1. What does it mean when we say a picture is 3.9 in. by 5 in.? What does it mean when we say an envelope is 4.2 inches by 9 inches?

2. In which envelopes will the 3.9 in. by 5 in. picture fit? Which envelopes are too small for the 3.9 in. by 5 in. picture? Explain.

3. In which envelope would you rather put the 4.1 in. by 6 in. picture: the 3.7 in. by 6.5 in. envelope or the 5.5 in. by 7.5 in. envelope? Explain.

4. Arrange the pictures in order so you can decide which envelopes to put them in.

5. Which picture would you put into each envelope? Are there any envelopes that cannot be used?

Look Back Look at the envelope sizes. The 3.7 in. by 6.5 in. envelope will not fit into the 4 in. by 6.1 in. envelope. What other pairs of envelopes are like this?

Problem Solving Strategies

Making and Using Drawings

There were ten scouts hiking through the woods. Two mosquitoes saw them. The first mosquito said, "I will bite the first hiker. Then I'll skip two hikers and bite the next one, and so on." The second mosquito said, "O.K. I will bite the second hiker. Then I'll skip three hikers and bite the next one, and so on. Let's go." In the next minute, lots of scouts said "Ouch!" Did any hikers say "Ouch!" two times? Did any hikers escape being bitten?

Sometimes a picture can help you solve a problem that is hard to solve in any other way.

Solving the Problem

Think What are the questions?

Explore How many hikers were there? What was the first mosquito going to do? What was the second mosquito going to do? Draw pictures of the hikers. Mark the hikers that the first mosquito bites. Do the same for those the second mosquito bit.

Solve Did any hikers say "Ouch!" two times? If so, which one or ones? Did any hikers escape being bitten?

Look Back How did your drawing help you solve the problem?

Sharing Your Ideas

1. **What if** there were 15 hikers? Who would be bitten two times? Who would not be bitten?

THINK
EXPLORE
SOLVE
LOOK BACK

CHOICES

Practice

Solve. Use a calculator where appropriate.

Use this information to solve 2 and 3.

Beth and Barbara were walking on Main Street from their home to the library. After they had walked 9 blocks, they noticed that they had dropped one book. They walked back 3 blocks and found it. Then they walked 12 blocks to the library.

2. How far from home did they drop the book?

3. How far from home was the library?

Use this information to solve 4 and 5.

A straight road from Pikestown to Mountainville is 50 miles long. Amble and Coatstown are on the road. It is 22 miles from Pikestown to Amble. Between Pikestown and Coatstown it is 15 miles.

4. In what order are the towns from Pikestown to Mountainville?

5. How far is it from Amble to Mountainville?

Create Your Own

Write a problem for which someone needs to draw a picture to solve it.

Mixed Strategy Review

Use the information in the table to solve 6 and 7.

The Martinez family has returned from a 3-day trip. Mrs. Martinez kept track of the family's expenses.

	Food	Lodging	Gas
Day 1	$38	$37	$12
Day 2	$41	$37	$ 6
Day 3	$40	$37	$12

6. How much did they spend in all on food, lodging, and gas?

7. Did they spend more on food or lodging? Explain.

Activity

Exploring Addition with Decimals

Play the game, Sum Cover-Up.

Working together

Materials: Workmat 16; cards numbered 0.1, 0.2, 0.3, 0.4, 0.5, 0.6, 0.7, 0.8, 0.9, 1.0, 1.1; base-ten rods and flats

A. Take turns. Draw two cards each.

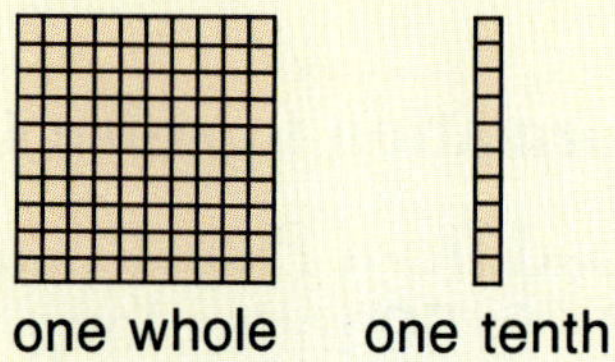

B. Use base-ten flats as one whole and base-ten rods as tenths. Place rods on the workmat to show the amount on the first card. Now show the amount on the second card. Place these rods next to the others. Use as many wholes as needed. How can you decide how much is covered in all?

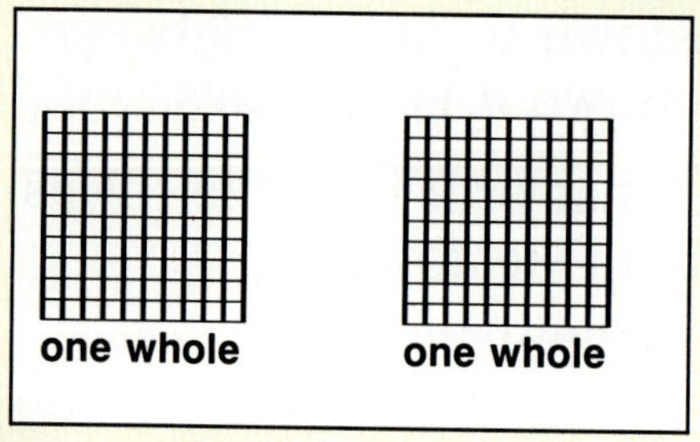

C. Look at your workmat. Compare the total amount of tenths you covered with that of your partner. The player with the greater sum scores one point.

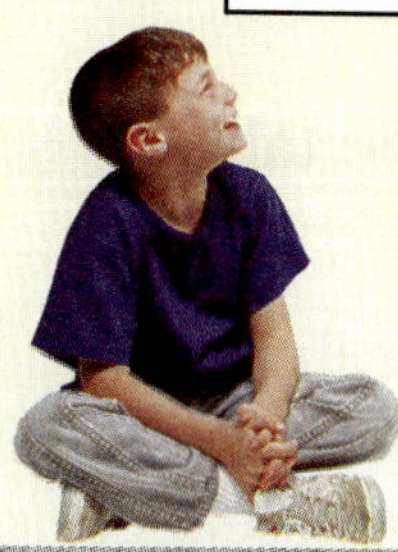

D. Return the cards and play 5 more times each.

Sharing Your Results

1. Think about how you placed the tenths. When did all of the tenths fit on one whole?
2. When did you need to place tenths on another whole?

Extending the Activity

3. Play Sum Cover-Up five times. This time record each addition like this.

ones	tenths
1	
0.	3
+0.	8
1.	1

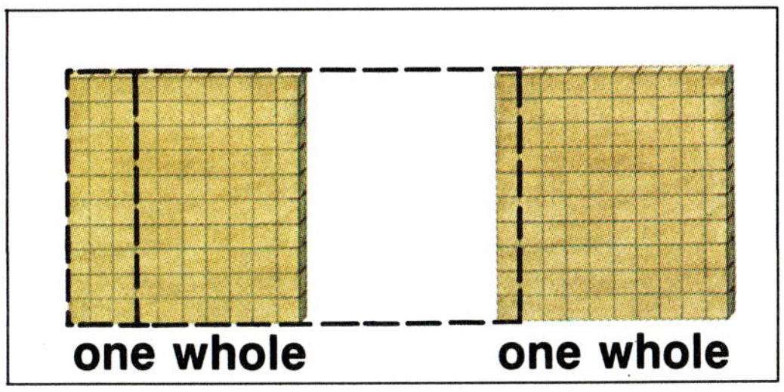

ones	tenths
1	
1.	6
+0.	4
2.	0

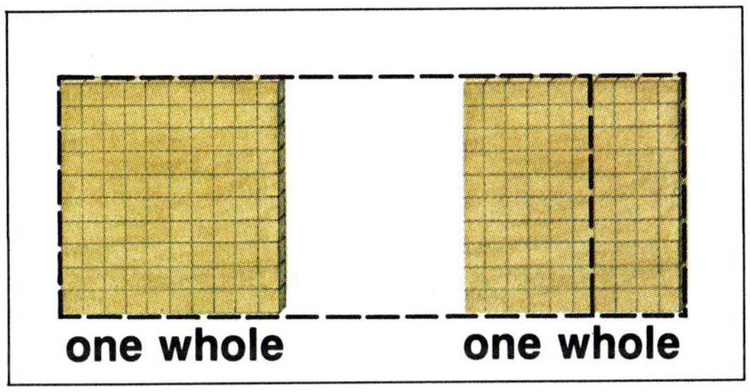

Use what you have learned to find each sum. Use base-ten rods, if you wish.

4. $0.2 + 0.3$

5. $0.5 + 0.4$

6. $0.2 + 0.9$

7. $0.7 + 0.6$

8. $1.0 + 0.8$

9. $1.1 + 0.3$

10. $1.0 + 1.1$

11. $2.3 + 4.5$

12. $3.7 + 1.7$

13. $5.6 + 2.4$

Summing Up

14. How is adding decimals like adding whole numbers? How is it different?

15. Explain how you could add decimals without using base-ten rods.

Activity

Exploring Subtraction with Decimals

Play the game, Subtraction Cover-Up.

Working together

Materials: Workmat 16, cards from Sum Cover-Up, base-ten rods

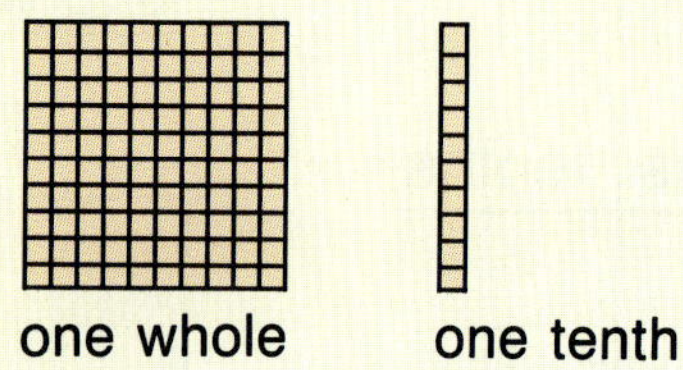

A. Take turns. Draw two cards each. Decide which of your numbers is greater.

B. Use base-ten rods as tenths. Show the greater amount. Place these tenths on your workmat. Cover as many wholes as needed. Now take away tenths that show the lesser amount.

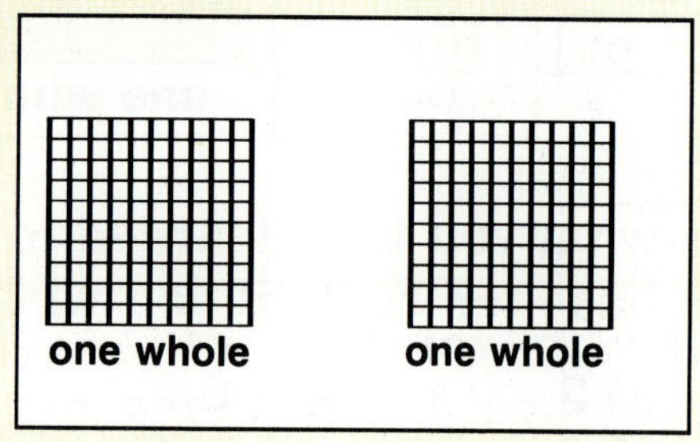

C. Look at your workmat. How many tenths are left? Compare the amount you have left with the amount your partner has left. The player with the greater amount left scores one point.

D. Return the cards and play five times. The player with the highest score wins.

Sharing Your Results

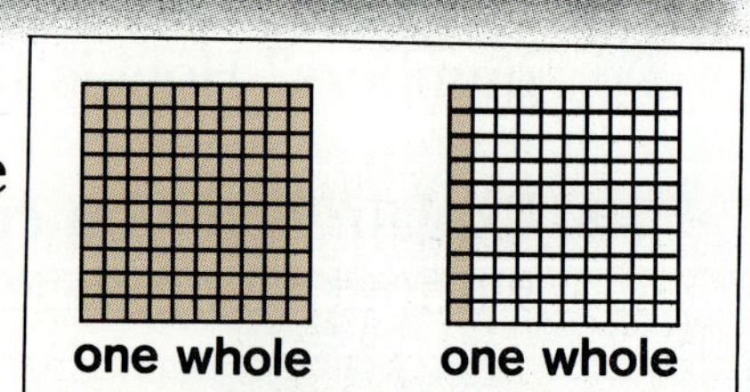

1. **What if** your workmat looked like the one at the right? Draw a picture to show the rods that are left after you take away 0.1 0.2 0.3 .
How did you decide which rods to take away?

Extending the Activity

2. Play Subtraction Cover-Up again. This time record each subtraction like this.

What if you chose 0.9 and 0.5?

Think 0.9 > 0.5

ones	tenths
0.	9
−0.	5
0.	4

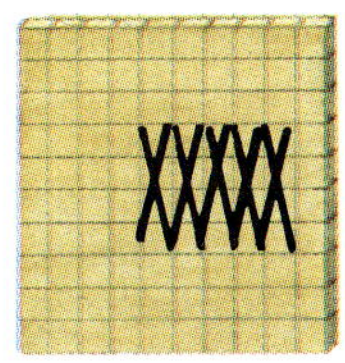

one whole

What if you chose 0.7 and 2.0?

Think 2.0 > 0.7

ones	tenths
1	10
~~2~~.	~~0~~
−0.	7
1.	3

one whole

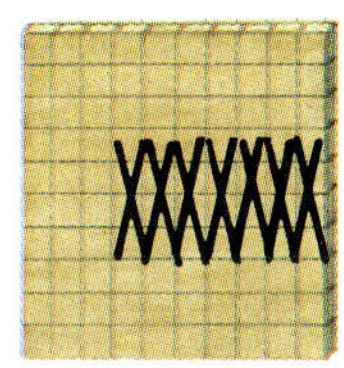

one whole

Use what you have learned to find each difference. Use base-ten rods, if you wish.

3. 0.8 − 0.2	**4.** 0.7 − 0.3	**5.** 0.6 − 0.5	**6.** 0.4 − 0.1	**7.** 1.0 − 0.4
8. 1.1 − 0.6	**9.** 1.1 − 1.0	**10.** 2.5 − 1.3	**11.** 3.2 − 2.4	**12.** 4.6 − 1.9

Summing Up

13. How is subtracting decimals like subtracting whole numbers? How is it different?

14. Explain how you could subtract decimals without using base-ten rods.

Chapter Review

Write a decimal for the green part. pages 370–377, 380–381

1. 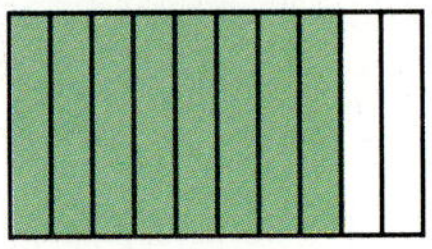

2. 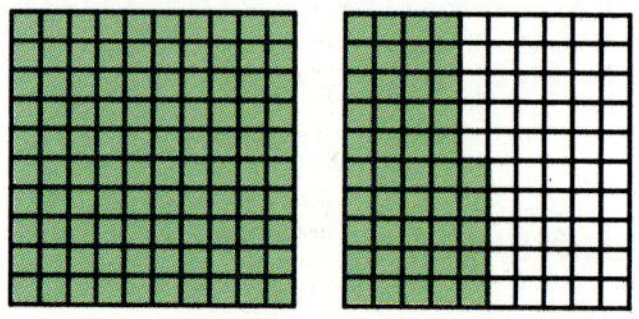

3. 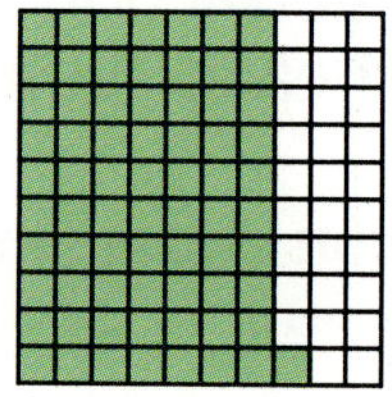

Write each as a decimal. pages 370–377, 380–381

4. six tenths
5. two tenths
6. eight hundredths
7. eighty-two hundredths
8. thirty-six hundredths
9. two and one tenth
10. 6 out of 10
11. 57 out of 100
12. 9 out of 10
13. $\frac{5}{10}$
14. $\frac{8}{100}$
15. $\frac{61}{100}$
16. $3\frac{1}{10}$
17. $4\frac{79}{100}$

Add or subtract. pages 386–389

18. $0.4 + 0.1$
19. $0.5 + 0.2$
20. $1.2 + 0.6$
21. $0.3 + 0.8$
22. $0.6 - 0.1$
23. $0.8 - 0.7$
24. $0.5 - 0.3$
25. $1.4 + 1.3$
26. $1.1 - 0.4$
27. $1.2 - 1.0$

Match the decimal to the words.

28. 6.07 — a. six and seven tenths
29. 6.7 — b. sixty-seven hundredths
30. 0.67 — c. six and seven hundredths

Solve. pages 383–385

31. The sign says that the trail to Mystery Mountain is 2.5 kilometers long. How far is the hike up *and* back?

32. You hike 3.2 kilometers and then 1.8 kilometers. Make a drawing to show the distances. Find the total distance.

Chapter Test

Write a decimal for the green part.

1. 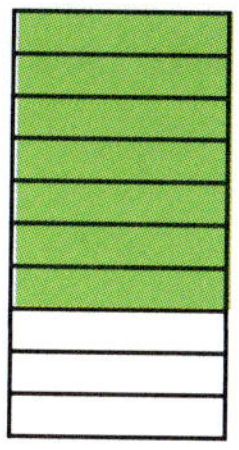

2. 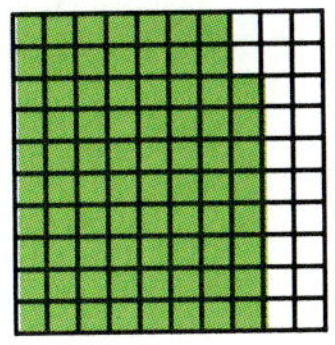

3. 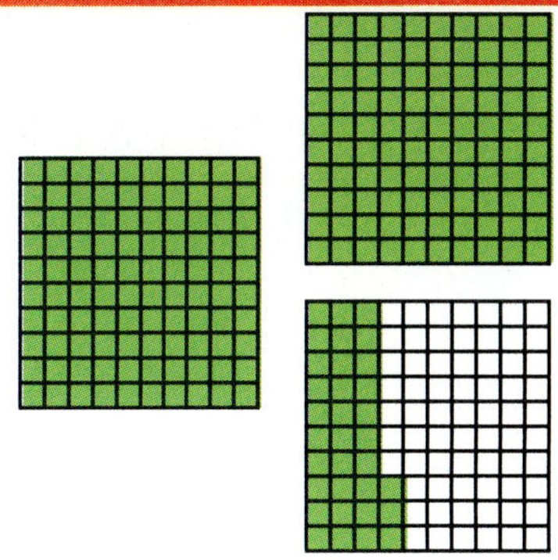

Write each as a decimal.

4. one tenth
5. four tenths
6. two hundredths
7. thirty-two hundredths
8. sixty hundredths
9. thirty-nine hundredths
10. 4 out of 10
11. 31 out of 100
12. 79 out of 100
13. $\frac{2}{10}$
14. $\frac{9}{100}$
15. $2\frac{3}{10}$
16. $8\frac{84}{100}$

Add or subtract.

17. $\begin{array}{r} 0.1 \\ +0.5 \\ \hline \end{array}$
18. $\begin{array}{r} 0.3 \\ +0.3 \\ \hline \end{array}$
19. $\begin{array}{r} 0.4 \\ -0.2 \\ \hline \end{array}$
20. $\begin{array}{r} 0.9 \\ -0.4 \\ \hline \end{array}$
21. $\begin{array}{r} 1.0 \\ +0.6 \\ \hline \end{array}$
22. $\begin{array}{r} 0.7 \\ -0.5 \\ \hline \end{array}$
23. $\begin{array}{r} 1.1 \\ -0.5 \\ \hline \end{array}$
24. $\begin{array}{r} 1.0 \\ +1.2 \\ \hline \end{array}$
25. $\begin{array}{r} 1.0 \\ -0.8 \\ \hline \end{array}$
26. $\begin{array}{r} 1.1 \\ +1.4 \\ \hline \end{array}$

Solve.

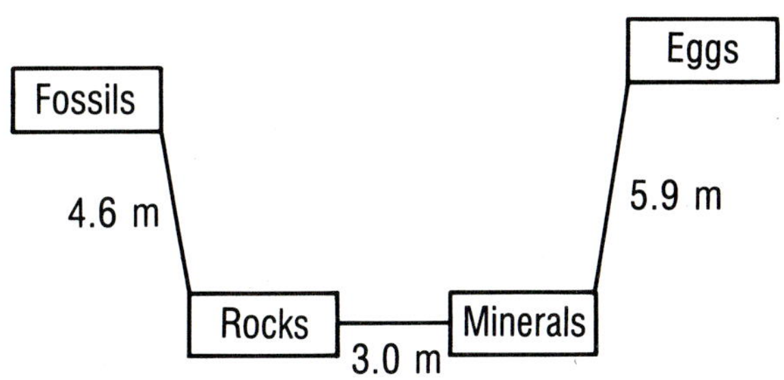

27. The drawing shows the distances between displays at one park. Find the total distance.

28. How much farther is it from the fossils to the rocks than from the rocks to the minerals?

THINK Bill is 2 meters ahead of Tom on Rim Walk at Grand Canyon. Tom is 1 meter behind Al. Al is 1.5 meters behind Grace. Who is first?

EXTENSION

Comparing Decimals: How Close Can You Get?

Which number is closer to 1?

1.35 or 1.53

Explain how you know.
Now play How Close Can You Get?
Build your own numbers close to 1.

Working Together

Materials: number cards 0-9, Workmat 17

1. Shuffle the number cards. Place them face down in a stack.
2. Each player takes 3 cards from the top of the stack.
3. Players use the numbers on the cards to make up a decimal close to 1. Then they record the decimal under Numbers Close to 1 on the workmat.
4. Players compare the numbers to decide whose number is closer to 1. The player whose number is closer to 1 scores a point.
5. Return the cards to the bottom of the stack.
6. Continue playing until all the spaces under Numbers Close to 1 are filled.
7. The player with the most points wins.

NUMBERS CLOSE TO 1

Round	Decimal Number
1.	□.□□
2.	□.□□
3.	□.□□

Create Your Own

Change the rules and make up your own game.

Maintaining Skills

Choose the correct answer. Write A, B, C, or D.

1. How long is the bar to the nearest centimeter?

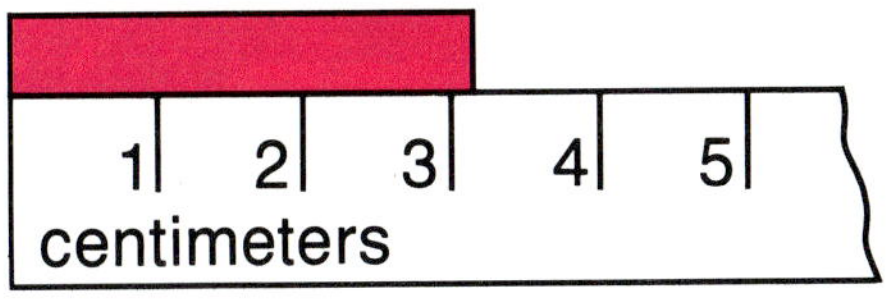

A 5 cm **C** 3 cm

B 1 cm **D** not given

2. Which unit would you use to measure liquid in a bathtub?

A liter **C** meter

B millimeter **D** centimeter

3. Name the figure.

A line segment **C** ray

B line **D** not given

4. Find the perimeter.

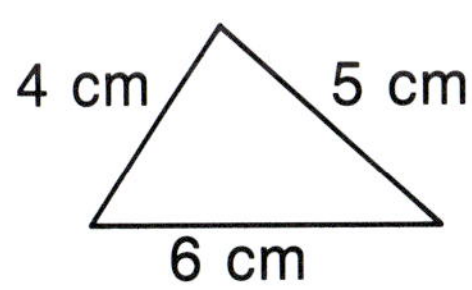

A 120 m **C** 12 cm

B 15 cm **D** not given

5. Find the area.

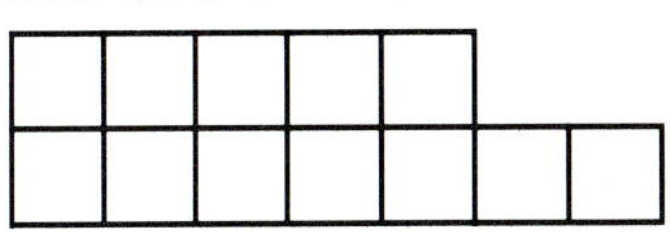

A 9 square units **C** 12 square units

B 10 square units **D** not given

6. Write 3 tenths as a decimal.

A 0.03 **C** 30

B 0.3 **D** not given

7. Write $3\frac{17}{100}$ as a decimal.

A 3.17 **C** 3.017

B 31.7 **D** not given

8.

$$\begin{array}{r} 8.2 \\ +1.4 \\ \hline \end{array}$$

A 9.5 **C** 9.6

B 7.2 **D** not given

9. $9.7 - 2.4 = \square$

A 7.1 **C** 7.3

B 7.5 **D** not given

Solve.

Timmy has pictures of 4 animals. They are birds and cats. Together they have 14 feet.

10. How many birds are there?

A 2 **C** 3

B 1 **D** not given

11. How many cats are there?

A 1 **C** 4

B 2 **D** not given

13 Exploring Multiplication and Division: One-Digit Numbers

THEME Saving Our Planet

Sharing What You Know

There are many kinds of pollution. All pollution harms our planet. Talk about different kinds of pollution. How could your class help in cleaning up the environment? If your class picks up some of the litter in the schoolyard, what would happen to the rest of the litter?

Using Language

One third-grade class spent part of the afternoon raking leaves in the park. The class used the remainder of the afternoon painting park benches. **Remainder** means the part left over. In mathematics, the **remainder** is the number left over after dividing. How are these meanings alike? How are they different?

Words to Know: product, estimate, quotient, remainder

Be a Problem Solver

Hillary had collected fewer than 25 empty cans. When she put the cans in groups of 4, she had 2 cans left over. When she put them in groups of 5, she had 3 cans left over. How many cans did she have?

Write a letter to the President of the United States with your suggestions for protecting the environment.

GETTING STARTED

How is counting by tens from 10 to 100 like counting by hundreds from 100 to 1,000? How is it different?

Multiplication Patterns

Saving our forests is important. Lumber companies plant trees to replace those they cut down. What if pine seedlings come 10 to a box? How many are in 4 boxes?

$4 \times 10 = \square$

▸ Multiplying with 10 is like counting by tens.

10	10	10	10	10	10	10	10	10
× 1	× 2	× 3	× 4	× 5	× 6	× 7	× 8	× 9
10	20	30	40	50	60	70	80	90

There are 40 seedlings in 4 boxes.

▸ Multiplying with 100 is like counting by hundreds.

100	100	100	100	100	100	100	100	100
× 1	× 2	× 3	× 4	× 5	× 6	× 7	× 8	× 9
100	200	300	400	500	600	700	800	900

Explain how using these patterns can help you multiply mentally.

You can use basic multiplication facts to find greater products.

2	20	200	6	60	600
×3	× 3	× 3	×4	× 4	× 4
6	60	600	24	240	2,400

Check Your Understanding

Use patterns to find the products mentally.

1.	4	40	400	**2.**	7	70	700	**3.**	1	10	100
	×2	× 2	× 2		×5	× 5	× 5		×8	× 8	× 8

Share Your Ideas Explain how knowing that $4 \times 8 = 32$ helps you find 4×80. 4×800.

Practice

Use patterns to find the products mentally.

4. 6×7 $\quad 60 \times 7$ $\quad 600 \times 7$
5. 1×5 $\quad 10 \times 5$ $\quad 100 \times 5$
6. 4×4 $\quad 40 \times 4$ $\quad 400 \times 4$
7. 10×9
8. 10×6
9. 100×9
10. 10×0
11. 100×7
12. 100×3
13. 400×6
14. 200×9
15. 300×7
16. 60×9
17. 40×5
18. 700×8
19. $3 \times 30 = \square$
20. $9 \times 400 = \square$
21. $6 \times 600 = \square$

Complete the table.

22.

Number of Dimes	1	2	3	4	5	6	7	8	9
Amount	$.10								

Use the pictograph to answer each question.

23. How many elm trees were planted?
24. How many maple trees were planted?
25. How many trees were planted in all?
26. What would [half tree symbol] stand for?

Each [tree symbol] stands for 50 trees.

Think and Apply

Complete the table to show how many centimeters tall each tree is.
Remember
1 meter = 100 centimeters

	HEIGHTS OF 15-YEAR-OLD TREES		
	Tree	Meters	Centimeters
27.	Juniper	3	
28.	Yew	4	
29.	Oak	8	

Explain how you can use patterns to find $3 \times 4{,}000$.

SUMMING UP

Activity

Exploring Multiplying by One-Digit Numbers

The Topper is the player with the greater amount. **What if** a player draws a card showing 3 groups of 14 and another player draws a card showing 4 groups of 12? Who is the Topper? Try this activity to find out.

Working together

Materials: base-ten blocks, a bag with cards cut out from Workmat 18

A. One player shows 3 groups of 14 with base-ten blocks. The player exchanges ones for tens to use the fewest blocks. The other player shows 4 groups of 12 and exchanges to use the fewest blocks. Players compare blocks. The player with the greater amount takes both cards and is the Topper.

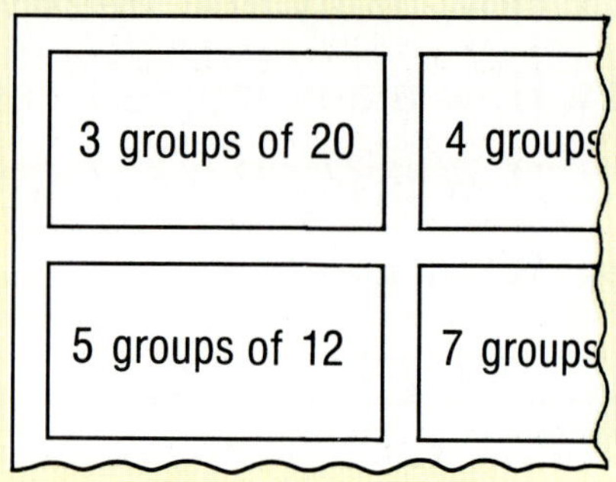

B. To continue playing, each player draws a card from the bag and follows step **A**.

C. Repeat the activity until there are no cards. The player with the most cards wins.

Sharing Your Results

1. What is another way of showing several groups of the same amount?
2. When did you need to exchange ones for tens? tens for hundreds?

Extending the Activity

When you have equal groups, you can multiply.

3 groups of 24 = □

3 × 24 = □

You can record the regrouping this way.

Multiply ones.
Regroup.

$$\begin{array}{r} {\scriptstyle 1} \\ 24 \\ \times\ 3 \\ \hline 2 \end{array}$$

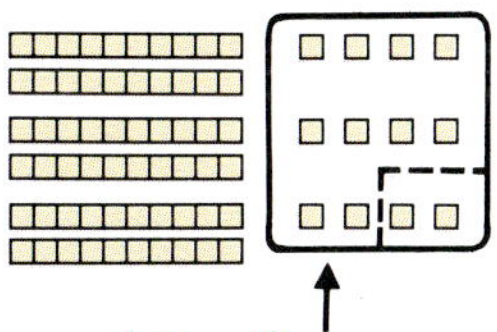

← 12 ones = 1 ten 2 ones

Multiply tens.
Add 1 ten.

$$\begin{array}{r} {\scriptstyle 1} \\ 24 \\ \times\ 3 \\ \hline 72 \end{array}$$

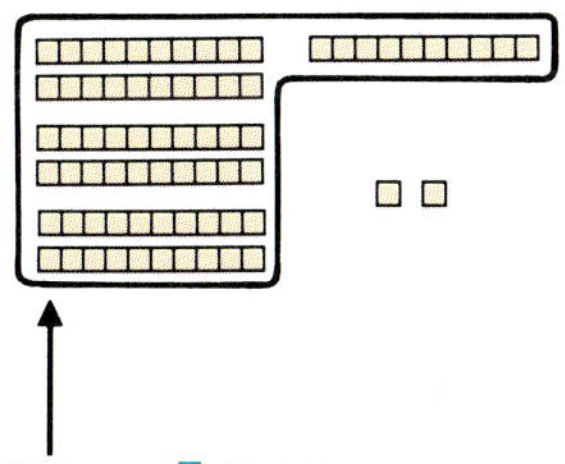

6 tens + 1 ten = 7 tens

Work with your partner. Use base-ten blocks to find each product. Use the example above as a model to record your work.

3. $\begin{array}{r} 42 \\ \times\ 2 \\ \hline \end{array}$ **4.** $\begin{array}{r} 18 \\ \times\ 3 \\ \hline \end{array}$ **5.** $\begin{array}{r} 56 \\ \times\ 2 \\ \hline \end{array}$ **6.** $\begin{array}{r} 39 \\ \times\ 4 \\ \hline \end{array}$

Summing Up

7. How does your record for **3–6** show what happened with the blocks?

8. Explain how you can multiply without using base-ten blocks.

GETTING STARTED

What if there are 2 groups of 14 turtles? What are three ways to find how many there are in all?

Multiplying by One-Digit Numbers

Scientists study how fast green turtles grow. Divers measure the turtles and then tag them.

If each diver makes 12 dives a month for 3 months, how many dives will each diver make?

$3 \times 12 = \square$

Step 1
Multiply ones.

$$\begin{array}{r} 12 \\ \times\ 3 \\ \hline 6 \end{array}$$

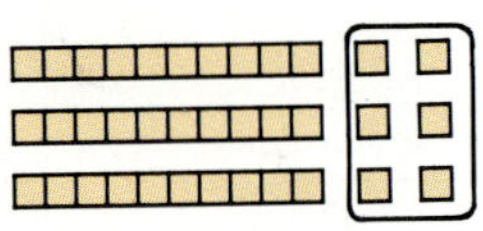

Step 2
Multiply tens.

$$\begin{array}{r} 12 \\ \times\ 3 \\ \hline 36 \end{array}$$

Each diver will make 36 dives.

Find 4 × 18.

Step 1
Multiply ones.
Regroup.

$$\begin{array}{r} {\scriptstyle 3} \\ 18 \\ \times\ 4 \\ \hline 2 \end{array}$$

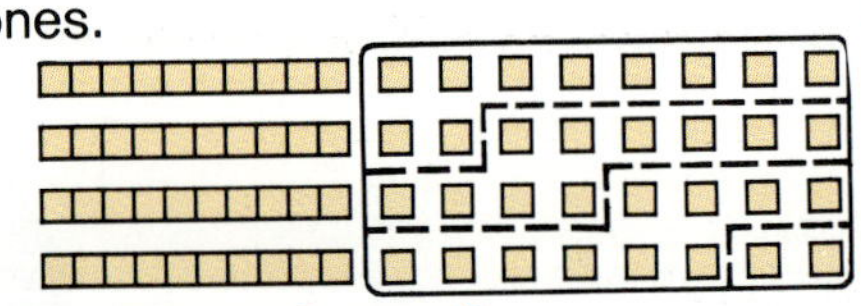

32 ones = 3 tens 2 ones

Step 2
Multiply tens.
Add 3 tens.

$$\begin{array}{r} {\scriptstyle 3} \\ 18 \\ \times\ 4 \\ \hline 72 \end{array}$$

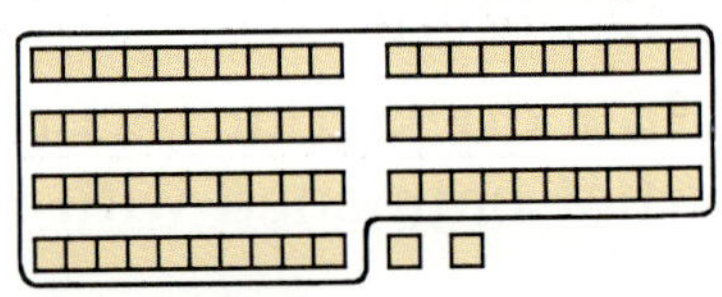

4 tens + 3 tens = 7 tens

Check Your Understanding

Multiply. Use base-ten blocks if you like.

1. $\begin{array}{r} 31 \\ \times\ 3 \\ \hline \end{array}$ **2.** $\begin{array}{r} 11 \\ \times\ 7 \\ \hline \end{array}$ **3.** $\begin{array}{r} 12 \\ \times\ 4 \\ \hline \end{array}$ **4.** $\begin{array}{r} 65 \\ \times\ 1 \\ \hline \end{array}$ **5.** $\begin{array}{r} 21 \\ \times\ 4 \\ \hline \end{array}$ **6.** $\begin{array}{r} 43 \\ \times\ 2 \\ \hline \end{array}$

Share Your Ideas Look back at **1**. How could you use addition to find the answer?

Practice

CHOICES **Multiply. Use base-ten blocks if you like.**

7. $\begin{array}{r} 43 \\ \times\ 2 \\ \hline \end{array}$ 8. $\begin{array}{r} 21 \\ \times\ 4 \\ \hline \end{array}$ 9. $\begin{array}{r} 37 \\ \times\ 2 \\ \hline \end{array}$ 10. $\begin{array}{r} 89 \\ \times\ 1 \\ \hline \end{array}$ 11. $\begin{array}{r} 26 \\ \times\ 2 \\ \hline \end{array}$ 12. $\begin{array}{r} 32 \\ \times\ 3 \\ \hline \end{array}$

13. $\begin{array}{r} 42 \\ \times\ 2 \\ \hline \end{array}$ 14. $\begin{array}{r} 20 \\ \times\ 4 \\ \hline \end{array}$ 15. $\begin{array}{r} 21 \\ \times\ 3 \\ \hline \end{array}$ 16. $\begin{array}{r} 68 \\ \times\ 1 \\ \hline \end{array}$ 17. $\begin{array}{r} 96 \\ \times\ 0 \\ \hline \end{array}$ 18. $\begin{array}{r} 10 \\ \times\ 6 \\ \hline \end{array}$

19. $2 \times 21 = \square$ 20. $2 \times 33 = \square$ 21. $3 \times 20 = \square$

22. $2 \times 24 = \square$ 23. $0 \times 85 = \square$ 24. $1 \times 70 = \square$

CHOICES **Solve each. Choose mental math or paper and pencil. Explain your choices.**

25. Multiply 34 by 2.
26. Multiply 10 by 8.
27. Multiply 8 by 5 and add 6.
28. Multiply 9 by 8 and add 4.

CHOICES **Find each missing digit. Use a calculator if you wish.**

29. $\begin{array}{r} 4\square \\ \times\ 2 \\ \hline 80 \end{array}$ 30. $\begin{array}{r} 2\square \\ \times\ 2 \\ \hline 44 \end{array}$ 31. $\begin{array}{r} 3\square \\ \times\ 3 \\ \hline 96 \end{array}$ 32. $\begin{array}{r} \square\square \\ \times\ 3 \\ \hline 99 \end{array}$ 33. $\begin{array}{r} \square 4 \\ \times\ 1 \\ \hline 6\square \end{array}$ 34. $\begin{array}{r} \square\square \\ \times\ 2 \\ \hline 68 \end{array}$

Think and Apply

35. You have 85 photos of turtles. The case holds 3 rows of 32 photos. Will all the photos fit?

36. Multiply 11 by the numbers 1 through 9. What pattern do you notice in the products?

Logical Thinking

37. **What if** you have the digits [1] [2] [3]? How would you arrange them to make the greatest product? the least product? Explain.

$\begin{array}{r} \square\square \\ \times\ \square \\ \hline \end{array}$

Write a sentence that tells how to multiply by a one-digit number.

SUMMING UP

GETTING STARTED

Use mental math. Multiply 5 × 6 and then add 8. Multiply 6 × 3 and then add 4.

Regrouping Ones or Tens

A forest ranger showed slides of fire damage to 3 classes. There were 26 students in each class. How many students were there in all?

3 × 26 = □

Step 1
Multiply ones.
Regroup.

$$\begin{array}{r} {\scriptstyle 1} \\ 26 \\ \times\ 3 \\ \hline 8 \end{array}$$

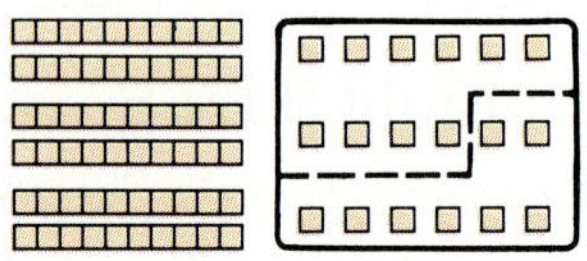

18 ones = 1 ten 8 ones

Step 2
Multiply tens.
Add 1 ten.

$$\begin{array}{r} {\scriptstyle 1} \\ 26 \\ \times\ 3 \\ \hline 78 \end{array}$$

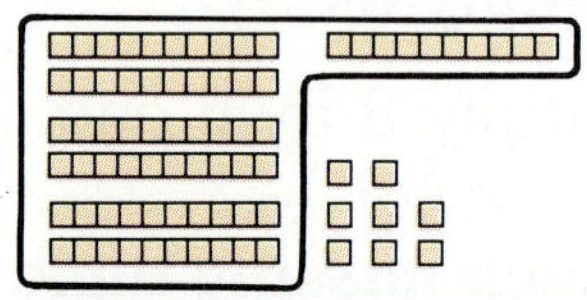

6 tens + 1 ten = 7 tens

There were 78 students in all.

Find 4 × 62.

Step 1
Multiply ones.

$$\begin{array}{r} 62 \\ \times\ 4 \\ \hline 8 \end{array}$$

Step 2
Multiply tens.
Regroup.

$$\begin{array}{r} 62 \\ \times\ 4 \\ \hline 248 \end{array}$$

24 tens = 2 hundreds 4 tens

Explain the steps you would follow to find 4 × 18.

Check Your Understanding

Multiply.

1. $\begin{array}{r} 48 \\ \times\ 2 \\ \hline \end{array}$ **2.** $\begin{array}{r} 12 \\ \times\ 8 \\ \hline \end{array}$ **3** $\begin{array}{r} 15 \\ \times\ 5 \\ \hline \end{array}$ **4.** $\begin{array}{r} 25 \\ \times\ 3 \\ \hline \end{array}$ **5.** $\begin{array}{r} 14 \\ \times\ 7 \\ \hline \end{array}$ **6.** $\begin{array}{r} 56 \\ \times\ 2 \\ \hline \end{array}$

Share Your Ideas Look back at **1**. Is the product greater than or less than 2 × 50? Can you tell without multiplying? Explain.

Practice

Multiply.

7. 16×3	8. 68×4	9. 26×3	10. 12×5	11. 18×2	12. 53×3
13. 36×2	14. 13×7	15. 46×5	16. 12×8	17. 27×3	18. 12×7
19. 21×4	20. 39×5	21. 24×4	22. 12×6	23. 63×2	24. 19×4
25. 31×7	26. 23×4	27. 51×6	28. 71×9	29. 45×3	30. 26×8

31. $2 \times 17 = \square$ 32. $4 \times 33 = \square$ 33. $3 \times 32 = \square$

34. $5 \times 18 = \square$ 35. $3 \times 15 = \square$ 36. $2 \times 75 = \square$

37. $6 \times 16 = \square$ 38. $5 \times 62 = \square$ 39. $3 \times 56 = \square$

Think and Apply

40. Jean and her class are going to Stokes Forest to learn about protecting animals. The one-way trip to Stokes Forest is 46 miles. How many miles is the round trip?

41. A postcard of a condor costs 19 cents. Joe bought 3 cards. Li bought 2 cards. How much did they spend in all? Explain two different ways to solve this problem.

Common Error

42. Joe's answer is incorrect. Explain why. Find the correct product.

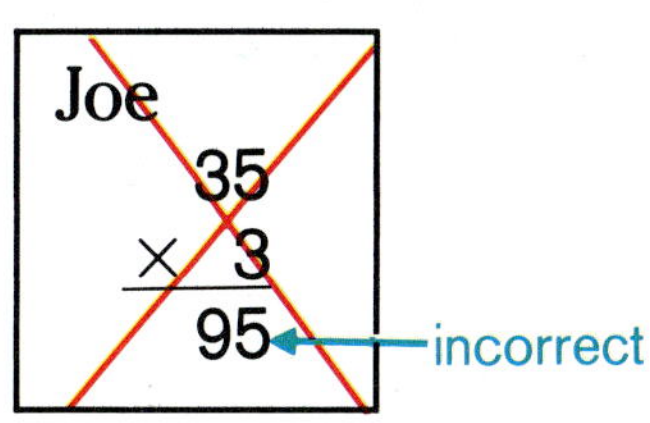

SUMMING UP

Write a multiplication example in which regrouping ones is needed. Give it to a partner to solve.

GETTING STARTED

Use what you know about multiplication to predict where you need to regroup in 3×25 and in 5×25.

Regrouping Ones and Tens

The Florida manatee is the only water mammal that eats just plants. A manatee can eat over 95 pounds of plants each day! How many pounds of plants can it eat in a week? **7 × 95 = □**

Step 1 Multiply ones. Regroup.	**Step 2** Multiply tens. Add 3 tens. Regroup.
$\begin{array}{r} {}^{3} \\ 95 \\ \times\ 7 \\ \hline 5 \end{array}$ 35 ones = 3 tens 5 ones	$\begin{array}{r} {}^{3} \\ 95 \\ \times\ 7 \\ \hline 665 \end{array}$ 63 tens + 3 tens = 66 tens or 6 hundreds 6 tens

A manatee can eat 665 pounds in a week.

Find 6 × 24 = □.

Step 1 Multiply ones. Regroup	**Step 2** Multiply tens. Add 2 tens. Regroup.
$\begin{array}{r} {}^{2} \\ 24 \\ \times\ 6 \\ \hline 4 \end{array}$ 24 ones = 2 tens 4 ones	$\begin{array}{r} {}^{2} \\ 24 \\ \times\ 6 \\ \hline 144 \end{array}$ 12 tens + 2 tens = 14 tens or 1 hundred 4 tens

Check Your Understanding

Find each product.

1. $\begin{array}{r} 52 \\ \times\ 6 \\ \hline \end{array}$ **2.** $\begin{array}{r} 36 \\ \times\ 5 \\ \hline \end{array}$ **3.** $\begin{array}{r} 76 \\ \times\ 2 \\ \hline \end{array}$ **4.** $\begin{array}{r} 63 \\ \times\ 4 \\ \hline \end{array}$ **5.** $\begin{array}{r} 60 \\ \times\ 9 \\ \hline \end{array}$ **6.** $\begin{array}{r} 45 \\ \times\ 3 \\ \hline \end{array}$

Share Your Ideas How can you use base-ten blocks or drawings to show the correct answer in **3**?

Practice

Multiply.

7. 25×8 **8.** 97×3 **9.** 67×4 **10.** 33×8 **11.** 47×6 **12.** 32×5

13. 42×8 **14.** 36×5 **15.** 23×7 **16.** 42×6 **17.** 63×2 **18.** 43×4

19. $6 \times 63 = \square$ **20.** $9 \times 15 = \square$ **21.** $3 \times 39 = \square$

22. $7 \times 46 = \square$ **23.** $6 \times 93 = \square$ **24.** $5 \times 90 = \square$

25. $3 \times 17 = \square$ **26.** $4 \times 45 = \square$ **27.** $0 \times 81 = \square$

Complete. Find each output.

Rule: Multiply by 2.

	Input	Output
28.	25	
29.	35	
30.	45	
31.	55	

Rule: Multiply by 6.

	Input	Output
32.	10	
33.	20	
34.	30	
35.	40	

Rule: Multiply by 4.

	Input	Output
36.	69	
37.	79	
38.	89	
39.	99	

Think and Apply

40. Read the poster. **What if** your class wanted to adopt 8 manatees? How much money would the class need?

ADOPT THE MANATEES!
Adopt a manatee for $15.

Logical Thinking

You can use what you know about doubles to multiply a number by 4.

Think $2 \times 21 = 42$.
4×21 is the double of 42.
The double of 42 is 84.

Try these.

41. 4×32 **42.** 4×12 **43.** 4×14

In multiplication, how is regrouping tens like regrouping ones?

SUMMING UP

GETTING STARTED

There are 5 classes of 22 students. Is the total number of students closer to 100 or to 200? Explain your thinking.

Estimating Products

Jesse uses a cart to take cans to the Recycling Center. The cart holds 72 cans. About how many cans are there in 6 loads?

6 × 72 = □

To estimate the product, round to the nearest ten and then multiply.

$$\begin{array}{r} 72 \\ \times\ 6 \\ \hline \end{array} \xrightarrow{\text{rounds to}} \begin{array}{r} 70 \\ \times\ 6 \\ \hline 420 \end{array} \quad \text{estimated product}$$

There are about 420 cans in 6 loads.

Another Example **6 × 194 = □**

To estimate the product, round to the nearest hundred and then multiply.

$$\begin{array}{r} 194 \\ \times\ \ 6 \\ \hline \end{array} \xrightarrow{\text{rounds to}} \begin{array}{r} 200 \\ \times\ \ 6 \\ \hline 1{,}200 \end{array}$$

The estimated product is 1,200.

Check Your Understanding

Estimate each product.

1. 37×4 **2.** 91×3 **3.** 53×5 **4.** 780×6 **5.** 305×8 **6.** 419×7

Share Your Ideas Look back at **3** and **4**. Is the estimate in each greater or less than the actual product?

Practice

Estimate each product.

7. $\begin{array}{r} 83 \\ \times\ 3 \\ \hline \end{array}$ 8. $\begin{array}{r} 49 \\ \times\ 2 \\ \hline \end{array}$ 9. $\begin{array}{r} 215 \\ \times\ 4 \\ \hline \end{array}$ 10. $\begin{array}{r} 573 \\ \times\ 7 \\ \hline \end{array}$

11. $\begin{array}{r} 75 \\ \times\ 2 \\ \hline \end{array}$ 12. $\begin{array}{r} 94 \\ \times\ 8 \\ \hline \end{array}$ 13. $\begin{array}{r} 188 \\ \times\ 9 \\ \hline \end{array}$ 14. $\begin{array}{r} 720 \\ \times\ 6 \\ \hline \end{array}$

15. $5 \times 63 = \square$ 16. $5 \times 60 = \square$

17. $7 \times 214 = \square$ 18. $4 \times 989 = \square$

Estimate. Write >, <, or = for each ⬬.

19. 6×43 ⬬ 240

20. 5×41 ⬬ 3×57

21. 2×90 ⬬ 6×30

22. 2×580 ⬬ 3×715

Choose the reasonable estimate.

23. The Recycling Center collected 930 cans on Monday. At this rate, about how many cans will be collected in 5 days?

a. 450 b. 10,000 c. 4,500 d. 1,500

Think and Apply

24. Solve this silly problem. Jesse's dog is working on his Pup Scout recycling badge. To get a badge, he must find 22 cans a week for 4 weeks. Estimate how many cans he must find.

Look back at **7–10**. When is your estimate greater than the actual product? less than the actual product?

SUMMING UP

Mixed Review

1. $\begin{array}{r} 22 \\ +83 \\ \hline \end{array}$ 2. $\begin{array}{r} 60 \\ +12 \\ \hline \end{array}$

3. $\begin{array}{r} 81 \\ +34 \\ \hline \end{array}$ 4. $\begin{array}{r} 54 \\ +40 \\ \hline \end{array}$

5. $\begin{array}{r} 98 \\ -21 \\ \hline \end{array}$ 6. $\begin{array}{r} 76 \\ -35 \\ \hline \end{array}$

7. $\begin{array}{r} 80 \\ -68 \\ \hline \end{array}$ 8. $\begin{array}{r} 43 \\ -17 \\ \hline \end{array}$

9. $2\overline{)18}$ 10. $3\overline{)24}$

11. $9\overline{)72}$ 12. $8\overline{)64}$

13. $5\overline{)25}$ 14. $6\overline{)36}$

15. $4\overline{)40}$ 16. $7\overline{)28}$

Write each number in words.

17. 642

18. 407

19. 9,038

20. 43,671

21. 57,002

22. 92,700

23. 105,999

24. 400,004

GETTING STARTED

Predict how many digits will be in each product. 2×31 2×310
Explain your thinking.

Multiplying Three-Digit Numbers

Animals can live in safety on wildlife preserves. If each preserve has 112 acres, how many acres would 6 preserves have?

$6 \times 112 = \square$

Estimate. 6×112 is about 600.
Now multiply.

Step 1	Step 2	Step 3
Multiply ones. Regroup.	Multiply tens. Add 1 ten.	Multiply hundreds.
$\begin{array}{r} ^{1} \\ 112 \\ \times 6 \\ \hline 2 \end{array}$ 12 ones = 1 ten 2 ones	$\begin{array}{r} ^{1} \\ 112 \\ \times 6 \\ \hline 72 \end{array}$ 6 tens + 1 ten = 7 tens	$\begin{array}{r} ^{1} \\ 112 \\ \times 6 \\ \hline 672 \end{array}$

There are 672 acres.
Compare the product with your estimate.
Does it make sense?

More Examples

a. $\begin{array}{r} ^{1} \\ 215 \\ \times 3 \\ \hline 645 \end{array}$

b. $\begin{array}{r} ^{4} \\ 161 \\ \times 7 \\ \hline 1{,}127 \end{array}$

c. $\begin{array}{r} ^{1} \\ 520 \\ \times 8 \\ \hline 4{,}160 \end{array}$

Check Your Understanding

Estimate. Then find each product.

1. $\begin{array}{r} 411 \\ \times 7 \\ \hline \end{array}$
2. $\begin{array}{r} 136 \\ \times 2 \\ \hline \end{array}$
3. $\begin{array}{r} 501 \\ \times 9 \\ \hline \end{array}$
4. $\begin{array}{r} 162 \\ \times 4 \\ \hline \end{array}$
5. $\begin{array}{r} 108 \\ \times 8 \\ \hline \end{array}$

Share Your Ideas Explain how you multiplied in **4**.

Practice

Estimate. Then find each product.

6. $\begin{array}{r} 103 \\ \times \quad 8 \\ \hline \end{array}$
7. $\begin{array}{r} 215 \\ \times \quad 3 \\ \hline \end{array}$
8. $\begin{array}{r} 141 \\ \times \quad 7 \\ \hline \end{array}$
9. $\begin{array}{r} 274 \\ \times \quad 2 \\ \hline \end{array}$
10. $\begin{array}{r} 225 \\ \times \quad 4 \\ \hline \end{array}$
11. $\begin{array}{r} 150 \\ \times \quad 5 \\ \hline \end{array}$
12. $\begin{array}{r} 121 \\ \times \quad 6 \\ \hline \end{array}$
13. $\begin{array}{r} 900 \\ \times \quad 9 \\ \hline \end{array}$
14. $\begin{array}{r} 283 \\ \times \quad 3 \\ \hline \end{array}$
15. $\begin{array}{r} 248 \\ \times \quad 2 \\ \hline \end{array}$
16. $\begin{array}{r} 105 \\ \times \quad 9 \\ \hline \end{array}$
17. $\begin{array}{r} 208 \\ \times \quad 4 \\ \hline \end{array}$
18. $\begin{array}{r} 18 \\ \times \; 5 \\ \hline \end{array}$
19. $\begin{array}{r} 999 \\ \times \quad 1 \\ \hline \end{array}$
20. $\begin{array}{r} 311 \\ \times \quad 8 \\ \hline \end{array}$
21. $7 \times 111 = \square$
22. $9 \times 401 = \square$
23. $4 \times 210 = \square$
24. $6 \times 200 = \square$
25. $7 \times 42 = \square$
26. $3 \times 252 = \square$

Solve. Choose mental math, paper and pencil, or a calculator. Explain your choices.

CHOICES

27. $(2 \times 5) \times 80 = \square$
28. $(6 \times 9) \times 10 = \square$
29. $(3 \times 3) \times 800 = \square$
30. $(8 \times 7) \times 100 = \square$

Think and Apply

You can multiply with a calculator.

$7 \times 853 = \square$

Press 7 × 8 5 3 =.

The product is shown in the display.

5971

Use a calculator to find each product.

31. $8 \times 927 = \square$
32. $9 \times 485 = \square$
33. Count the number of animals you see in a day. **What if** you saw the same number of animals each day? How many animals would you see in a year? Hint: 1 year = 365 days

Make up a multiplication problem in which you will not need to regroup. How did you choose your numbers?

SUMMING UP

Midchapter Review

Multiply. pages 396–397

1. 60×3
2. 80×7
3. 900×4
4. 300×2
5. 500×8
6. $5 \times 700 = \square$
7. $2 \times 40 = \square$
8. $8 \times 600 = \square$

Find each product. pages 398–405

9. 71×8
10. 63×5
11. 25×9
12. 41×7
13. 14×6
14. 24×8
15. 83×2
16. 97×2
17. 42×4
18. 16×6
19. $3 \times 77 = \square$
20. $5 \times 82 = \square$
21. $9 \times 54 = \square$

Estimate each product. pages 406–407

22. 62×3
23. 35×5
24. 112×6
25. 704×8
26. 271×9

Multiply. pages 408–409

27. 121×3
28. 402×4
29. 790×5
30. 891×6
31. 107×9

Choose the correct word to complete each sentence.

Words to Know
product estimate

32. When you do not need an exact answer, you can _____.
33. The answer in multiplication is the _____.

Solve.

34. Each week the ranger drives 230 miles to inspect the wildlife preserve. How many miles does she drive in 2 weeks? in 3 weeks? in 4 weeks?
35. Make drawings to show 3 rows of 25 pine trees and 4 rows of 17 spruce trees. Which drawing has more trees? How many more does it have?

Exploring Problem Solving

THINK
EXPLORE
SOLVE
LOOK BACK

How Many Birds Can Be Saved?

The oil spill did great damage to the shore. Some people are saving the birds by finding them and then using towels to wipe them free of oil. One person can gather one bird in 10 minutes. Two people, working together, can clean one bird in 30 minutes.

Thinking Critically

If 16 people come to the shore, what would you have each person do? How many birds can the people clean in 4 hours?

Analyzing and Making Decisions

Use a calculator where appropriate.

CHOICES

1. What is the first thing that needs to be done at the clean up site? How might jobs change after the beginning?
2. How many birds can one person gather in 1 hour? in 4 hours? How many birds can two people clean in 1 hour? in 4 hours?
3. **What if** you formed teams for the cleanup? How many people would be on a team? What would each person do?
4. If 16 people came to the clean up site, how would you divide the work? About how many birds would be cleaned in 4 hours?

Look Back What if people needed to rest during the working time? How would that change your cleanup plans?

Problem Solving Strategies

Logic

The Nature Club is planting four different groups of trees on a steep hillside. The oak grove is next to the spruce grove and to the maple grove. The pine grove is higher than the spruce grove. The maple grove is at the bottom of the hill. In what order are the groves being planted on the hillside?

To solve some problems you do not have to use any numbers. You must read carefully and think like a detective. You need to put the clues together to solve the problem. Sometimes a drawing or diagram may help.

Solving the Problem

Make a drawing to help you solve this problem.

Think What is the question?

Explore The oak grove is next to what other two groves? Show them in a drawing. Where is the pine grove? Now can you label the top and bottom of the hill? Explain.

Solve In what order, from the bottom of the hill to the top of the hill, are the groves?

Look Back Did you need to change your drawing? Explain.

Share Your Ideas

1. **What if** you only knew that the oak grove is next to the spruce grove? Then what might be the order of the groves?
2. A drawing can make a problem easier to understand and to solve. Do you agree or disagree? Explain.

THINK
EXPLORE
SOLVE
LOOK BACK

Practice

Solve. Use a calculator where appropriate.

CHOICES

3. Three kinds of salmon are being raised on a fish farm. The tank of coho salmon is on one end. The tank of king salmon is between the sockeye salmon and the coho salmon. In what order are the fish tanks?

4. When released, the king salmon are 10 cm longer than the coho salmon. The sockeye salmon are 6 cm smaller than the king salmon. Which of these three kinds of salmon is the smallest when released?

Which pair of statements cannot be true?

5. The elm tree is taller than the spruce tree. The spruce tree is taller than the elm tree.

6. The oak tree is taller than the maple tree. The maple tree is shorter than the pine tree.

Create Your Own

Look back at **9**. Write a problem in which you must compare the total number of cans that Jim and June had.

Mixed Strategy Review

Use the information in the picture to solve 7 and 8.

7. You have a \$5 bill and a \$10 bill. Can you purchase all 3 tools? Explain.

8. You have a \$10 bill. How much change will you get if you buy a hammer and a screwdriver?

9. June and her brother are picking up cans and bottles. Together they have 32 cans. June has 6 more cans than her brother Jim. How many cans do they each have?

Activity

Exploring Division Facts with Remainders

The Save-Our-Planet Club is making posters about endangered animals. Members will put posters in packages of 5 and take them to schools. **What if** they have 23 posters? How many packages of 5 can they make?

Working together

Materials: 45 counters, Workmat 19
Record your work each time.

Total	Number in Each Group	Number of Groups	Number Left Over
23	5		
44	5		

A. Use 23 counters for the 23 posters. Show as many groups of 5 as you can. How many groups of 5 did you make? How many counters were left over?

B. **What if** you have 44 counters? Predict how many groups of 5 you can make. Then try it. Record the number of groups and how many counters are left over.

C. Now explore with 5 different numbers of counters. Predict, try it, and record your work.

Sharing Your Results

1. Talk about how you made your predictions. Did multiplication facts help you?
2. For which numbers could you make groups of 5 with no counters left over?
3. When you had counters left over, how many were left over each time?
4. When you are dividing by 5, can you have 6 counters left over? Explain your answer.

Extending the Activity

We divide to find how many groups. We show our work this way.

44 ÷ 5 = □

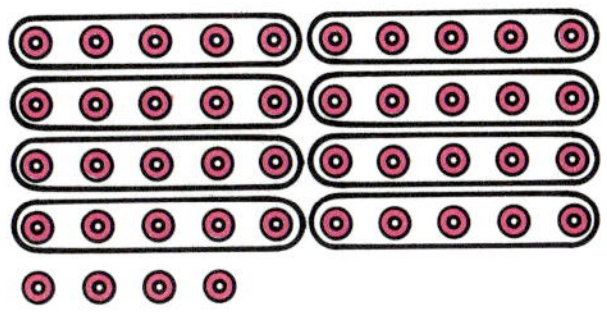

number of groups ↓
8 remainder **4**
number in each group → **5)44**
−40
4 number left over

8 groups of 5, 4 left over

We also divide to find how many in each group.

44 ÷ 5 = □

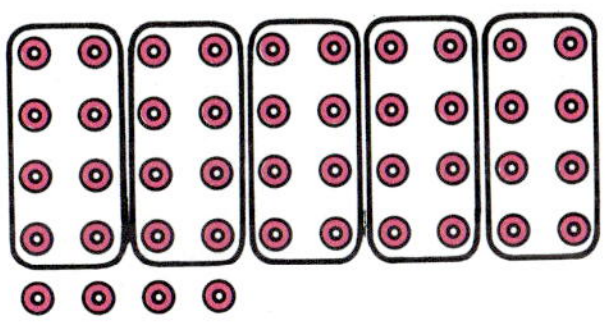

number in each group ↓
8 remainder **4**
number of groups → **5)44**
−40
4 number left over

5 groups of 8, 4 left over

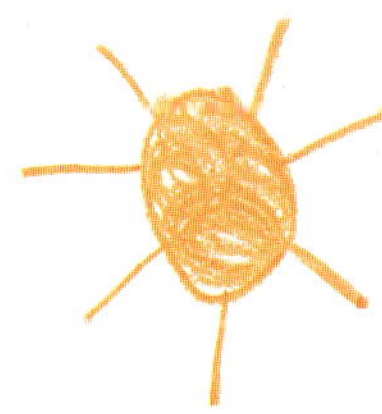

Use counters to find each quotient. Use the examples above as models to record your work.

5. 3)25 **6.** 2)19 **7.** 9)12 **8.** 7)30

9. 4)14 **10.** 6)56 **11.** 4)33 **12.** 8)44

Summing Up

13. How does your record for **5–12** show what happened with the counters?

14. Explain how you can divide without using counters.

GETTING STARTED

How are these products the same? How are they different?

$\begin{array}{r} 8 \\ \times 4 \\ \hline 32 \end{array}$ $\begin{array}{r} 80 \\ \times 4 \\ \hline 320 \end{array}$ $\begin{array}{r} 800 \\ \times 4 \\ \hline 3{,}200 \end{array}$

Division Facts and Patterns

Garbage in our oceans is a threat to sea life. Sixty people joined a rally to stop ocean dumping. They rode to the state capital in 2 buses. Each bus had the same number of people. How many people were on each bus?

60 ÷ 2 = □

Basic facts can help you divide tens mentally.

$\begin{array}{r} 3 \\ 2\overline{)6} \end{array}$ $\begin{array}{r} 30 \\ 2\overline{)60} \end{array}$

There were 30 people on each bus.

Basic facts can help you divide hundreds. Describe the patterns in each.

$\begin{array}{r} 3 \\ 2\overline{)6} \end{array}$ $\begin{array}{r} 30 \\ 2\overline{)60} \end{array}$ $\begin{array}{r} 300 \\ 2\overline{)600} \end{array}$

$\begin{array}{r} 5 \\ 3\overline{)15} \end{array}$ $\begin{array}{r} 50 \\ 3\overline{)150} \end{array}$ $\begin{array}{r} 500 \\ 3\overline{)1{,}500} \end{array}$

$\begin{array}{r} 6 \\ 5\overline{)30} \end{array}$ $\begin{array}{r} 60 \\ 5\overline{)300} \end{array}$ $\begin{array}{r} 600 \\ 5\overline{)3{,}000} \end{array}$

Check Your Understanding

Find each quotient. Use mental math.

1. $4\overline{)8}$ $4\overline{)80}$ $4\overline{)800}$

2. $3\overline{)24}$ $3\overline{)240}$ $3\overline{)2{,}400}$

3. $3\overline{)9}$ $3\overline{)90}$ $3\overline{)900}$

4. $6\overline{)36}$ $6\overline{)360}$ $6\overline{)3{,}600}$

Share Your Ideas Look back at **2**. Explain how a basic fact and patterns helped you.

Practice

Divide. Use mental math.

5. $3\overline{)6}$ $3\overline{)60}$ $3\overline{)600}$
6. $4\overline{)20}$ $4\overline{)200}$ $4\overline{)2,000}$
7. $8\overline{)8}$ $8\overline{)80}$ $8\overline{)800}$
8. $7\overline{)42}$ $7\overline{)420}$ $7\overline{)4,200}$
9. $5\overline{)35}$ $5\overline{)350}$ $5\overline{)3,500}$
10. $3\overline{)27}$ $3\overline{)270}$ $3\overline{)2,700}$
11. $9\overline{)54}$ $9\overline{)540}$ $9\overline{)5,400}$
12. $8\overline{)16}$ $8\overline{)160}$ $8\overline{)1,600}$
13. $280 \div 7 = \square$
14. $1,800 \div 2 = \square$
15. $160 \div 4 = \square$
16. $250 \div 5 = \square$
17. $300 \div 6 = \square$
18. $1,500 \div 5 = \square$

Find each missing number.

19. $4 \div \square = 1$
20. $40 \div \square = 10$
21. $400 \div \square = 100$
22. $\square \div 3 = 20$
23. $\square \div 6 = 80$
24. $\square \div 5 = 200$

Use patterns to solve each.

25. $3\overline{)9}$ $3\overline{)90}$ $3\overline{)900}$ $3\overline{)9,000}$ $3\overline{)90,000}$
26. $4\overline{)28}$ $4\overline{)280}$ $4\overline{)2,800}$ $4\overline{)28,000}$ $4\overline{)280,000}$

Think and Apply

27. There are 3 buses waiting to take 120 people to the environmental center. Each bus will take the same number of people. How many people will go in each bus?

28. Use mental math to find the mystery number.
Start with 400.
Divide by 5 and then add 20.
Divide by 2.
Divide by 5.

SUMMING UP

Make up a division problem that can be solved by using a basic fact and patterns.

Activity

Exploring Dividing by One-Digit Numbers

Dr. Brower needs help banding 52 monarch butterflies. **What if** 4 children share the work? How many butterflies will each child band?

Working together

Record your work.

Materials: base-ten blocks, Workmat 20

A. Show 52, using the fewest possible blocks. Starting with the tens, make 4 equal groups. Exchange tens for ones when necessary. Record the number in each group and the number left over.

Total Number		Number of Groups	Number in Each Group		Number Left Over
tens	ones		tens	ones	
5	2	4			

B. **What if** 6 children share the work? Show 52 again. This time, make 6 equal groups. What must you do right at the start?

C. Share 52 equally among 3 groups. Then try 5 groups.

Sharing Your Results

1. Describe how you distributed the blocks into groups.
2. When did you need to exchange tens for ones? When did you need to exchange right at the start?

Extending the Activity

Now work with your partner to explore other numbers. Record your work on Workmat 20.

3. Take turns choosing a number from 20 to 99.

4. Show the number, using the fewest possible blocks.

5. Show how to share the blocks among 7, 8, and 9 groups.

6. Repeat this activity at least four times.

Summing Up

7. When blocks were left over, why were you unable to divide them into the groups you formed?

8. When were no blocks left over? How can you use multiplication to explain why no blocks were left over?

9. The 87 blocks below are divided into 7 groups. The leftover blocks are hidden by the butterfly. How can you tell how many blocks are hidden by the butterfly? Explain.

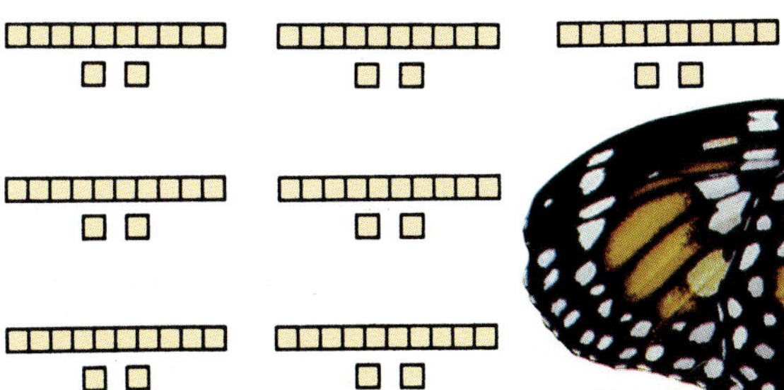

Activity

Understanding Dividing by One-Digit Numbers

Jim and Jan like to play Share and Share Alike. Try their game.

Working together

Record your work.

Materials: base-ten blocks, a cube with numbers 1–6, Workmat 20

A. Toss the number cube two times. Form a two-digit number.

B. Show the number, using the fewest possible blocks.

C. Toss the cube again. Share the blocks equally among the number of groups shown on the cube.

D. Repeat the activity four times.

Total Number		Number of Groups	Number in Each Group		Number Left Over
tens	ones		tens	ones	

Sharing Your Results

1. Use the information on your workmat to explain how the blocks were shared.
2. Without using blocks, could you fill in the workmat for 75 items shared by 9 people? Explain.

Extending the Activity

When you divide, you can record your work as shown here.

$63 \div 4 = \square$

Step 1 Divide the 6 tens into 4 groups. Each group gets 1 ten. That uses 4 tens. How many tens are left?

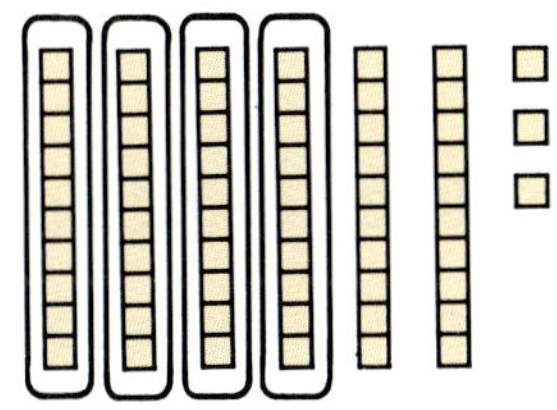

$$\begin{array}{r} 1 \\ 4\overline{)63} \\ -4 \\ \hline 2 \end{array}$$

Step 2 Exchange 2 tens for 20 ones. Add the 3 ones. How many ones are there in all?

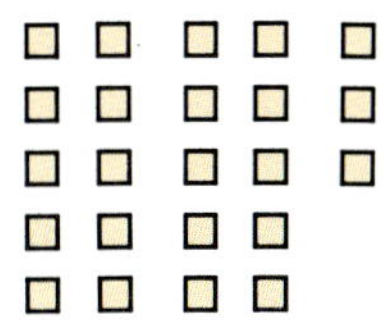

$$\begin{array}{r} 1 \\ 4\overline{)63} \\ -4\downarrow \\ \hline 23 \end{array}$$

Step 3 Divide the 23 ones into 4 groups. Each group gets 5 ones. That uses 20 ones. How many ones are left?

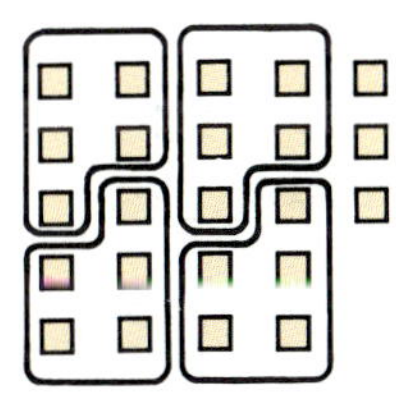

$$\begin{array}{r} 15 \\ 4\overline{)63} \\ -4 \\ \hline 23 \\ -20 \\ \hline 3 \end{array}$$

Each group has 15. There are 3 left.

Work with a partner. Divide. Use base-ten blocks. Use the example above as a model to record your work.

3. $5\overline{)75}$ **4.** $4\overline{)52}$ **5.** $3\overline{)33}$ **6.** $6\overline{)96}$ **7.** $7\overline{)85}$

8. $2\overline{)29}$ **9.** $8\overline{)96}$ **10.** $4\overline{)68}$ **11.** $5\overline{)87}$ **12.** $9\overline{)99}$

Summing Up

13. **Look back** at what you recorded. How does it show what happened with the blocks?

14. Explain how you can divide without using base-ten blocks.

GETTING STARTED

If 6 students share 60 stickers, how many will each student get? How can your answer help you estimate the quotient for 79 ÷ 6?

Dividing by One-Digit Numbers

After an oil spill, 6 workers will clean oil from 79 ducks. How many ducks will each worker clean if they share the work equally? How many ducks will be left to be cleaned?

79 ÷ 6 = □ Estimate. 79 ÷ 6 is about 10.

You can divide by using base-ten blocks.

79 = 7 tens 9 ones = 6 tens 19 ones

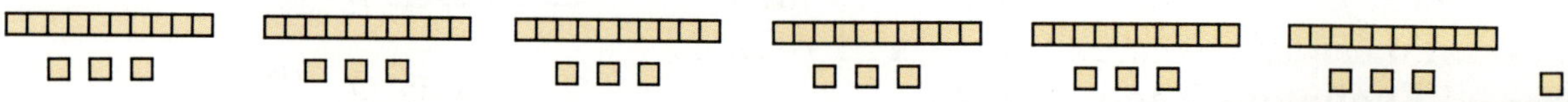

6 groups of 13, 1 left over

You can divide without using blocks.

Step 1 Divide tens. Multiply. Subtract and compare.	**Step 2** Bring down ones. Divide ones. Multiply. Subtract and compare.
$\begin{array}{r} 1 \\ 6\overline{)79} \\ -6 \\ \hline 1 \end{array}$ Think $6\overline{)7}$ $1 < 6$	$\begin{array}{r} 13\text{ R1} \\ 6\overline{)79} \\ -6 \\ \hline 19 \\ -18 \\ \hline 1 \end{array}$ Think $6\overline{)19}$ $1 < 6$ Write the remainder in the quotient.

Each worker will clean 13 ducks and 1 duck will be left.

Check Your Understanding

Find each quotient. Use blocks if you wish.

1. $5\overline{)67}$ **2.** $4\overline{)93}$ **3.** $6\overline{)84}$ **4.** $3\overline{)46}$ **5.** $2\overline{)39}$

Share Your Ideas Look back at **2.** Explain how blocks can be used to show the division.

Practice

Find each quotient. Use blocks if you wish.

6. $4\overline{)61}$
7. $9\overline{)48}$
8. $8\overline{)66}$
9. $2\overline{)45}$
10. $4\overline{)56}$
11. $3\overline{)71}$
12. $5\overline{)62}$
13. $6\overline{)83}$
14. $3\overline{)57}$
15. $7\overline{)79}$
16. $8\overline{)81}$
17. $9\overline{)98}$
18. $65 \div 5 = \square$
19. $97 \div 3 = \square$
20. $72 \div 4 = \square$
21. $89 \div 2 = \square$

Compare. Use >, <, or = for each ●. Choose mental math or paper and pencil. Explain your choices.

CHOICES

22. $80 \div 5$ ● $80 \div 4$
23. $86 \div 4$ ● $26 \div 4$
24. $48 \div 3$ ● 48×3
25. 60×5 ● $60 \div 5$

Think and Apply

26. The wildlife committee reported that 8 workers shared the tagging of 96 ducks. How many ducks did each worker tag?
27. Al had 73 wildlife stamps. He displayed them 4 to a frame. How many frames did he fill? How many stamps were left over?

SUMMING UP

Make up a story problem that can be solved by using $45 \div 2$. Then solve your problem.

Mixed Review

Write a fact family for each set of numbers.

1. 4, 6, 24
2. 3, 5, 15
3. 2, 7, 14
4. 8, 3, 24
5. 5, 9, 45
6. 7, 6, 42
7. 4, 8, 32
8. 6, 2, 12
9. 8, 6, 48
10. 7, 9, 63

Find the next 3 numbers in the pattern.

11. 2, 4, 6, . . .
12. 10, 20, 30, . . .
13. 5, 10, 15, . . .
14. 21, 31, 41, . . .
15. 3, 6, 9, . . .
16. 25, 30, 35, . . .
17. 110, 120, 130, . . .
18. 100, 150, 200, . . .
19. 200, 400, 600, . . .
20. 160, 170, 180, . . .

Using Problem Solving

Interview: Calculators in Recycling

Tom Goodrich is a manager at a recycling center. Tom uses a calculator to find out how much to pay a customer. Here is what he pays for the newspaper and aluminum cans.

Prices			
	Per Pound	Per 100 Pounds	Per Ton
Newspaper	—	$.60	$12
Aluminum	$.25	$25	$500

A. A customer brought in 9 pounds of aluminum. The recycling center bought the aluminum for $.25 a pound. How much did the recycling center pay the customer? You can use a calculator to find the answer.

9 × . 2 5 = 2.25

B. You are buying the paper and aluminum. Use the prices in the chart and a calculator to determine how much you should pay each customer. Keep a record of your work.

Customer	Material	Weight	Money Paid
1	Aluminum	7 lb	
2	Newspaper	300 lb	
3	Aluminum	15 lb	
4	Aluminum	25 lb	
5	Newspaper	500 lb	
6	Newspaper	600 lb	
7	Aluminum	5 lb	
8	Newspaper	2 tons	

Sharing Your Ideas

1. If you could use a calculator for finding either the newspaper or the aluminum prices, which would you find? Explain.
2. How much money in all was paid out to the customers?

Practice

3. It takes Brad one week to save a stack of newspapers 4 inches thick. A 4-inch stack weighs about 10 lb. How many weeks will it take him to save a stack 40 inches tall? How much will the stack weigh? How much money will he make when he sells it?

4. Look back at **3**. How many pounds of stacked newspaper are as tall as you? as tall as your teacher? as tall as your desk? How much money would each of these stacks be worth?

5. Tiffany brought in 72 aluminum cans. They weighed 3 lb. About how many cans are in a pound? How much money did she receive?

6. Tiffany saves about 2 cans a day. About how long does it take her to save a pound of aluminum?

7. How long does it take her to save enough aluminum to make $1?

Summing Up

8. Look back at **4**. How did you find out how much each stack of newspapers would weigh?

9. Why do you think the recycling company pays much more for the aluminum than for the paper?

Chapter Review

Find each product. pages 396–405, 408–409

1. 70×6	2. 54×2	3. 28×6	4. 15×5	5. 17×9
6. 76×5	7. 125×3	8. 237×4	9. 99×2	10. 43×8
11. 400×9	12. 925×3	13. 121×7	14. 309×9	15. 230×8

Estimate each product. pages 406–407

16. 14×4	17. 61×7	18. 23×9	19. 72×8	20. 310×5

Find each quotient. pages 414–423

21. $2\overline{)22}$	22. $9\overline{)81}$	23. $7\overline{)63}$	24. $3\overline{)41}$	25. $4\overline{)49}$
26. $6\overline{)23}$	27. $3\overline{)86}$	28. $8\overline{)95}$	29. $5\overline{)50}$	30. $3\overline{)17}$
31. $8\overline{)80}$	32. $2\overline{)47}$	33. $4\overline{)46}$	34. $6\overline{)66}$	35. $5\overline{)60}$

Choose the correct word to complete each sentence.

36. When you divide, the number left over is called the ____.

37. In the division fact $68 \div 4 = 17$, 17 is called the ____.

Words to Know
quotient remainder

Solve. pages 411–413

38. For 70 days, workers tried to clean up an oil spill that reached the shoreline. How many weeks did they work?

39. The Recycling Center keeps bins along the road. Cans are 5 m from paper. Plastics are 4 m from bottles. Bottles are 3 m from cans. How far is the paper from the plastic?

Chapter Test

Find each product.

1. $\begin{array}{r} 30 \\ \times\ 9 \\ \hline \end{array}$
2. $\begin{array}{r} 21 \\ \times\ 3 \\ \hline \end{array}$
3. $\begin{array}{r} 45 \\ \times\ 6 \\ \hline \end{array}$
4. $\begin{array}{r} 82 \\ \times\ 7 \\ \hline \end{array}$
5. $\begin{array}{r} 24 \\ \times\ 4 \\ \hline \end{array}$
6. $\begin{array}{r} 56 \\ \times\ 9 \\ \hline \end{array}$
7. $\begin{array}{r} 49 \\ \times\ 8 \\ \hline \end{array}$
8. $\begin{array}{r} 68 \\ \times\ 7 \\ \hline \end{array}$
9. $\begin{array}{r} 700 \\ \times\ 4 \\ \hline \end{array}$
10. $\begin{array}{r} 309 \\ \times\ 3 \\ \hline \end{array}$
11. $\begin{array}{r} 127 \\ \times\ 7 \\ \hline \end{array}$
12. $\begin{array}{r} 819 \\ \times\ 5 \\ \hline \end{array}$

Estimate each product.

13. $\begin{array}{r} 74 \\ \times\ 4 \\ \hline \end{array}$
14. $\begin{array}{r} 86 \\ \times\ 7 \\ \hline \end{array}$
15. $\begin{array}{r} 309 \\ \times\ 5 \\ \hline \end{array}$
16. $\begin{array}{r} 780 \\ \times\ 8 \\ \hline \end{array}$

Find each quotient.

17. $3\overline{)27}$
18. $2\overline{)62}$
19. $3\overline{)78}$
20. $5\overline{)55}$
21. $4\overline{)89}$
22. $6\overline{)83}$
23. $7\overline{)51}$
24. $9\overline{)87}$
25. $8\overline{)44}$
26. $9\overline{)90}$
27. $2\overline{)40}$
28. $4\overline{)43}$
29. $6\overline{)75}$
30. $5\overline{)81}$
31. $3\overline{)34}$

Solve.

32. The Scouts collected 78 bundles of newspapers. They can take 6 bundles at a time to the Recycling Center. How many trips will they make?

33. Sal put recycling bins at the curb. The cans were next to the bottles and the plastics. The papers were next to the bottles and closest to the corner. Starting at the corner, in what order were the bins?

THINK At the Wildlife Outdoor Theater, there are 10 rows of seats. The first 6 rows have 12 seats in a row. The rest of the rows have 20 seats in a row. How many seats are there?

EXTENSION

Divisible by 3

When you divide a number by 3 and the remainder is 0, we say that the number is divisible by 3.

Here is a quick way to tell if a number is divisible by 3. Think about the number 96.

- Add the digits in 96.
- Continue to add until there is only one digit.
- 6 is divisible by 3, so 96 is divisible by 3.

$96 \rightarrow 9 + 6 = 15$
$15 \rightarrow 1 + 5 = 6$

A number is divisible by 3 if the sum of its digits is divisible by 3.

Play All's Fair When Three Can Share

Work in a group. Take turns.

Materials: cube with the numbers 1–6

1. Toss the number cube two times. Form a two-digit number.
2. Is your number divisible by 3? If it is, you score a point.
3. The player with the most points wins.

Predict whether each number is divisible by 3. Then use a calculator to find out.

4. 81 5. 112 6. 150 7. 175

8. How might you test to see if a four-digit number is divisible by 3? Try your idea.

Family Math

In Chapters **11–13** your child has been working with geometry, decimals, multiplication, and division. With this project, you can help your child have fun with geometric shapes.

These are masks that are made by the ancient Japanese art of paper folding.

Origami Masks

Materials: one $8\frac{1}{2}$ in.-square piece of paper for each mask, string

Cat Mask

1. Fold forward at *AB*.

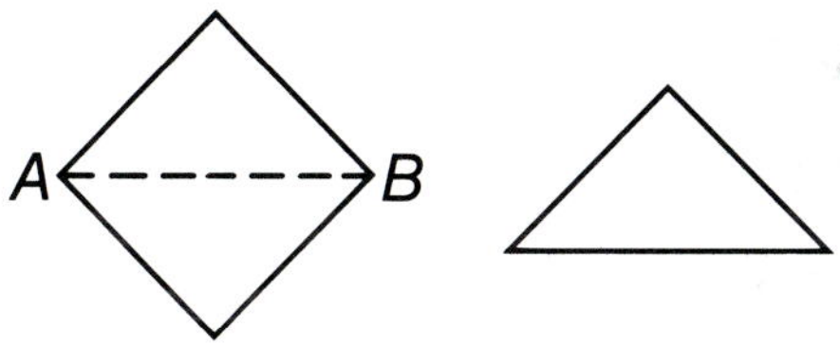

2. Fold forward at *CD*.

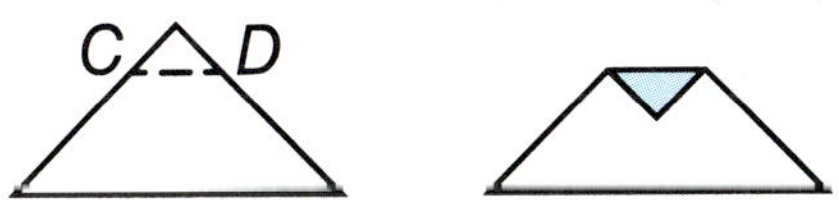

3. Fold forward at *EF* and at *FG*.

4. Turn the mask over and draw a face. Attach strings for ties.

Dog Mask

1. Fold forward at *AB*.

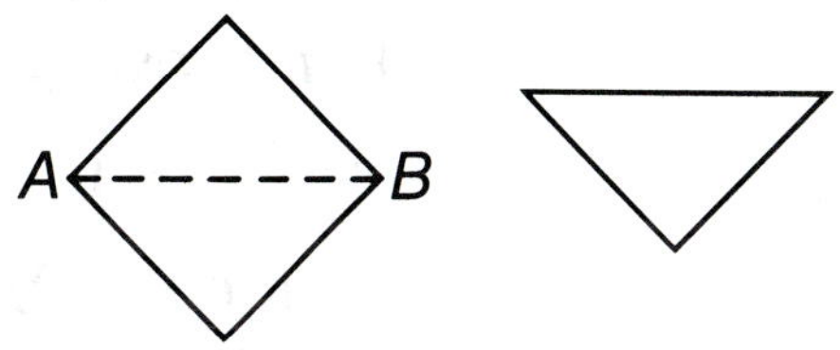

2. Fold forward at *CD* and at *DE*.

3. Fold back at *FG* and at *HI*.

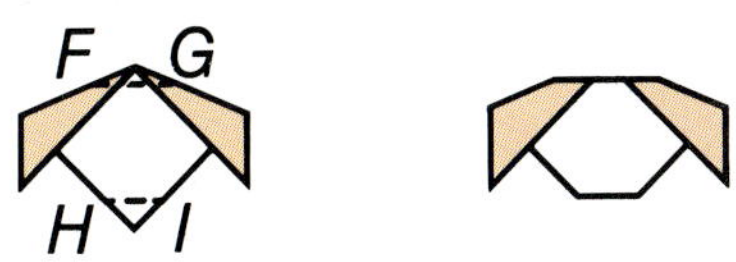

4. Draw a face. Attach strings for ties.

Final Review

Choose the correct answer. Write A, B, C, or D.

1. $16 - 7 = \square$

A 10 **C** 9

B 8 **D** not given

2. Round 38 to the nearest ten.

A 30 **C** 40

B 38 **D** not given

3. Choose the number for $3{,}000 + 400 + 80 + 6$.

A 3,846 **C** 3,406

B 3,086 **D** not given

4. $582 + 206 = \square$

A 788 **C** 784

B 789 **D** not given

5.

$$\begin{array}{r} \$38.54 \\ -\ 21.63 \\ \hline \end{array}$$

A \$17.11 **C** \$17.91

B \$16.91 **D** not given

6. What time is it?

A 9:00 **C** 12:45

B 11:40 **D** not given

7. $8 \times 6 = \square$

A 14 **C** 36

B 48 **D** not given

8. $6 \times 2 \times 3 = \square$

A 24 **C** 15

A 11 **D** not given

9. $64 \div 8 = \square$

A 8 **C** 7

A 6 **D** not given

10. What is the fraction for the shaded part?

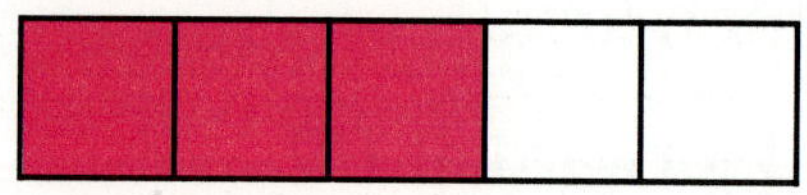

A $\frac{2}{3}$ **C** $\frac{3}{5}$

B $\frac{2}{5}$ **D** not given

11. Compare. $\frac{4}{5}$ ⬬ $\frac{4}{6}$

A $<$ **C** $=$

B $>$ **D** not given

12. Which unit would you use to measure a pencil?

A meter **C** liter

B kilometer **D** not given

Choose the correct answer. Write A, B, C, or D.

13. How long is the bar to the nearest $\frac{1}{2}$ inch?

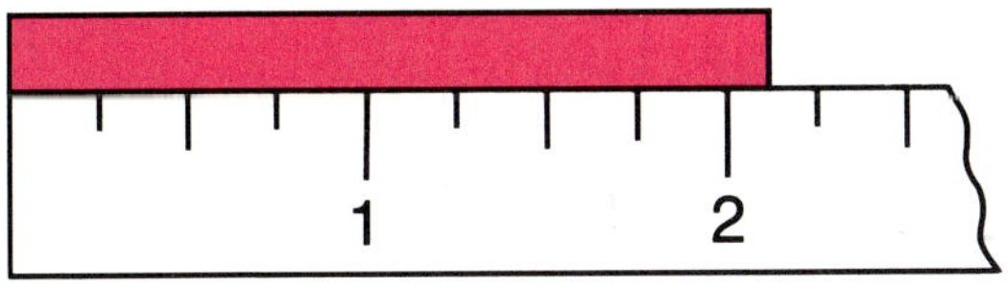

A $1\frac{1}{2}$ in. **C** 1 in.
B 2 in. **D** not given

14. Name the shape.

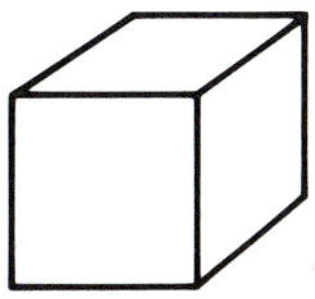

A cube **C** cone
B pyramid **D** not given

15. Find the area.

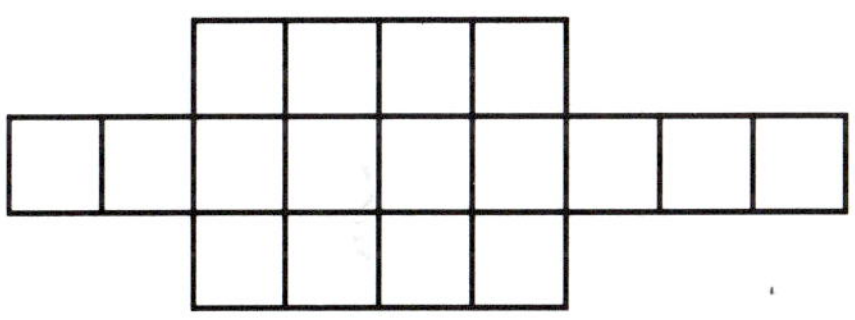

A 15 sq units **C** 19 sq units
B 24 sq units **D** not given

16. Write 3 and 1 hundredth as a decimal.

A 3.1 **C** 3.001
B 3.01 **D** not given

17. $3\overline{)252}$

A 84 **C** 74
B 83 **D** not given

Solve.

18. Bryan solved 10 problems correctly on the test. How many problems were on the test?

A 20 **C** cannot tell
B 15 **D** not given

Make a list to solve.

19. Bicycle license plates can be made with the letters A, B, and C. How many different license plates can be made using 3 letters each time?

A 9 **C** 27
B 30 **D** not given

Use a pattern to solve.

20. Decals cost 65¢ for 1, 60¢ each for 2, and 55¢ each for 3. If the pattern continues how much will each of 4 decals cost?

A 50¢ **C** 35¢
B 40¢ **D** not given

Use a drawing to solve.

21. Jan threw a beanbag forward $1\frac{1}{2}$ yd and then backward $\frac{1}{2}$ yd. How far from where it started did the beanbag finally land?

A $\frac{1}{2}$ yd **C** 1 yd
B 2 yd **D** not given

Extra Practice

Set A

Find each sum. pages 2–5

1. $\begin{array}{r} 7 \\ +6 \\ \hline \end{array}$ 2. $\begin{array}{r} 5 \\ +7 \\ \hline \end{array}$ 3. $\begin{array}{r} 9 \\ +8 \\ \hline \end{array}$

4. $6 + 5 = \square$
$5 + 6 = \square$

5. $8 + 7 = \square$
$7 + 8 = \square$

Set B

Add. pages 6–7

1. $\begin{array}{r} 4 \\ 2 \\ +6 \\ \hline \end{array}$ 2. $\begin{array}{r} 3 \\ 6 \\ +5 \\ \hline \end{array}$ 3. $\begin{array}{r} 3 \\ 1 \\ +7 \\ \hline \end{array}$ 4. $\begin{array}{r} 9 \\ 0 \\ +4 \\ \hline \end{array}$ 5. $\begin{array}{r} 3 \\ 6 \\ +3 \\ \hline \end{array}$ 6. $\begin{array}{r} 1 \\ 4 \\ +2 \\ \hline \end{array}$ 7. $\begin{array}{r} 2 \\ 6 \\ +5 \\ \hline \end{array}$

8. $7 + 2 + 6 = \square$ 9. $2 + 5 + 8 = \square$ 10. $4 + 5 + 8 = \square$

Set C

Find each missing addend. Then find each difference. pages 12–13

1. $\begin{array}{r} 4 \\ +\square \\ \hline 12 \end{array}$ $\begin{array}{r} 12 \\ -\ 4 \\ \hline \end{array}$ 2. $\begin{array}{r} 6 \\ +\square \\ \hline 8 \end{array}$ $\begin{array}{r} 8 \\ -6 \\ \hline \end{array}$ 3. $\begin{array}{r} 3 \\ +\square \\ \hline 10 \end{array}$ $\begin{array}{r} 10 \\ -\ 3 \\ \hline \end{array}$ 4. $\begin{array}{r} 8 \\ +\square \\ \hline 16 \end{array}$ $\begin{array}{r} 16 \\ -\ 8 \\ \hline \end{array}$

5. $7 + \square = 13$
$13 - 7 = \square$

6. $0 + \square = 8$
$8 - 0 = \square$

7. $6 + \square = 15$
$15 - 6 = \square$

8. $5 + \square = 11$
$11 - 5 = \square$

Set D

Subtract. pages 14–17

1. $\begin{array}{r} 12 \\ -\ 4 \\ \hline \end{array}$ 2. $\begin{array}{r} 8 \\ -5 \\ \hline \end{array}$ 3. $\begin{array}{r} 7 \\ -0 \\ \hline \end{array}$ 4. $\begin{array}{r} 16 \\ -\ 8 \\ \hline \end{array}$ 5. $\begin{array}{r} 9 \\ -9 \\ \hline \end{array}$ 6. $\begin{array}{r} 11 \\ -\ 7 \\ \hline \end{array}$ 7. $\begin{array}{r} 9 \\ -8 \\ \hline \end{array}$

8. $14 - 8 = \square$ 9. $\square = 6 - 2$ 10. $\square = 9 - 5$ 11. $17 - 8 = \square$

Complete each fact family. pages 18–19

12. $\begin{array}{r} 8 \\ +3 \\ \hline \end{array}$ $\begin{array}{r} 3 \\ +8 \\ \hline \end{array}$ $\begin{array}{r} 11 \\ -\ 3 \\ \hline \end{array}$ $\begin{array}{r} 11 \\ -\ 8 \\ \hline \end{array}$ 13. $\begin{array}{r} 5 \\ +9 \\ \hline \end{array}$ $\begin{array}{r} 9 \\ +5 \\ \hline \end{array}$ $\begin{array}{r} 14 \\ -\ 5 \\ \hline \end{array}$ $\begin{array}{r} 14 \\ -\ 9 \\ \hline \end{array}$

Set E

Tell whether each answer is odd or even. pages 20–21

1. $3 + 9 = \square$ 2. $7 + 4 = \square$ 3. $8 - 5 = \square$ 4. $11 - 3 = \square$

5. $6 - 2 = \square$ 6. $4 + 4 = \square$ 7. $13 - 6 = \square$ 8. $2 + 9 = \square$

Set A

Write the number. pages 28–29, 40–41, 48–49

1. 6 hundreds 8 tens 3 ones
2. two hundreds seven tens
3. 4 thousands 1 hundred 5 ones
4. 8 thousands 9 tens 9 ones
5. 73 thousands 4 hundreds
6. 277 thousands 1 hundred 1 one
7. 2,000 + 900 + 5
8. 83,000 + 700 + 40 + 2

Write a word name for each. pages 28–29, 40–41, 48–49

9. 308 10. 7,420 11. 6,111 12. 85,200 13. 416,384

Set B

Write the number that comes before and after. pages 30–31

1. 89 2. 207 3. 386 4. 600 5. 840 6. 777

Set C

Write the ordinal name for each. pages 32–33

1. 8th 2. 12th 3. 90th 4. 25th 5. 63rd

Set D

Round to the nearest ten. pages 34–37

1. 23 2. 86 3. 14 4. 74 5. 35 6. 48 7. 94

Round to the nearest hundred.

8. 240 9. 602 10. 759 11. 237 12. 861 13. 349 14. 777

Set E

Compare. Use $<$ or $>$ for ●. pages 46–47

1. 34 ● 43
2. 285 ● 279
3. 630 ● 6,300
4. 8,100 ● 8,010
5. 28,431 ● 29,341
6. 47,800 ● 48,799

Extra Practice

Set A

Write each value. pages 54–55

1. 2 quarters, 3 nickels
2. 6 dimes, 1 nickel, 6 pennies
3. 4 one-dollar bills, 2 quarters
4. 2 five-dollar bills, 7 dimes
5. 1 five-dollar bill, 3 one-dollar bills, 2 quarters, 1 dime
6. 3 ten-dollar bills, 3 five-dollar bills, 3 one-dollar bills
7. 4 nickels, 2 dimes, 1 quarter, 2 one-dollar bills

Set B

Add. pages 66–67

1. 34 + 51
2. 72 + 16
3. 26 + 47
4. 13 + 64
5. 84 + 7
6. 53 + 35
7. 12 + 17
8. 43 + 49
9. 28 + 58
10. 71 + 20
11. 37 + 23
12. 63 + 24
13. 25 + 42 = □
14. 67 + 15 = □
15. 85 + 5 = □

Add. Check by adding up. pages 68–69

16. 72 + 49
17. 57 + 14
18. 97 + 8
19. 47 + 74
20. 88 + 88
21. 35 + 64
22. 66 + 19
23. 84 + 92
24. 48 + 96
25. 78 + 72
26. 79 + 81
27. 4 + 99
28. 79 + 98 = □
29. 47 + 57 = □
30. 85 + 76 = □

Set C

Add. Check by adding up. pages 72–73

1. 14 + 21 + 33
2. 34 + 75 + 29
3. 82 + 14 + 7
4. 34 + 35 + 36
5. 26 + 70 + 84
6. 45 + 52 + 79
7. 28 + 42 + 31
8. 43 + 19 + 50
9. 72 + 7 + 43
10. 38 + 47 + 65
11. 25 + 14 + 3
12. 80 + 42 + 73
13. 47 + 18 + 93 + 52 = □
14. □ = 58 + 35 + 9 + 76

Extra Practice

Set A

Subtract. Check by adding. pages 118–119

1. $203 - 142$
2. $705 - 437$
3. $108 - 99$
4. $900 - 472$
5. $603 - 507$
6. $200 - 154$
7. $820 - 574$
8. $300 - 198$
9. $808 - 709$
10. $700 - 287$
11. $200 - 106$
12. $600 - 222$
13. $405 - 98 = \square$
14. $900 - 456 = \square$
15. $800 - 795 = \square$

Set B

Subtract. Check by adding. pages 120–123

1. $1,857 - 423$
2. $7,648 - 1,315$
3. $9,205 - 3,102$
4. $8,164 - 5,034$
5. $5,729 - 3,625$
6. $6,172 - 4,044$
7. $3,216 - 1,721$
8. $4,031 - 2,974$
9. $6,000 - 5,550$
10. $8,185 - 3,992$
11. $8,467 - 5,109 = \square$
12. $7,109 - 2,316 = \square$
13. $6,341 - 5,730 = \square$

Set C

Subtract. Check by adding. pages 124–125

1. $\$1.74 - .82$
2. $\$5.63 - 1.45$
3. $\$17.39 - 9.67$
4. $\$9.14 - 3.43$
5. $\$50.68 - 17.91$
6. $\$21.00 - 15.80$
7. $\$2.03 - 1.98$
8. $\$10.00 - 1.37$
9. $\$20.50 - 17.43$
10. $\$95.83 - 78.94$
11. $\$42.50 - \$21.75 = \square$
12. $\$16.50 - \$15.89 = \square$

Set D

Write each time in two ways. pages 140–143

1. 
2. 
3. 
4. 

Tell how much time has passed. pages 144–145

5. start 8:15
   end 11:15
6. start 1:00
   end 5:30
7. start 7:05
   end 7:45
8. start 2:30
   end 4:00

Extra Practice

Set A

Use the calendar to answer these questions. pages 148–149

1. Name the day of the week for June 17.
2. Name the date for the third Tuesday.
3. Name the date for one week after June 2.
4. Name the day of the week for June 23.

June

S	M	T	W	T	F	S
						1
2	3	4	5	6	7	8
9	10	11	12	13	14	15
16	17	18	19	20	21	22
23	24	25	26	27	28	29
30						

Set B

Use the pictograph to answer each question. pages 154–157

1. Which color was liked the most? the least?
2. What does ● stand for?
3. How many named red?
4. How many named yellow?

Each ● stands for 2 students.

Set C

Use the grid to answer each question. pages 160–161

1. What ordered pair names the location of the blue dot?
2. What ordered pair names the location of the yellow dot?
3. What is located at (5, 6)?
4. What is located at (2, 1)?
5. What is located at (4, 3)?

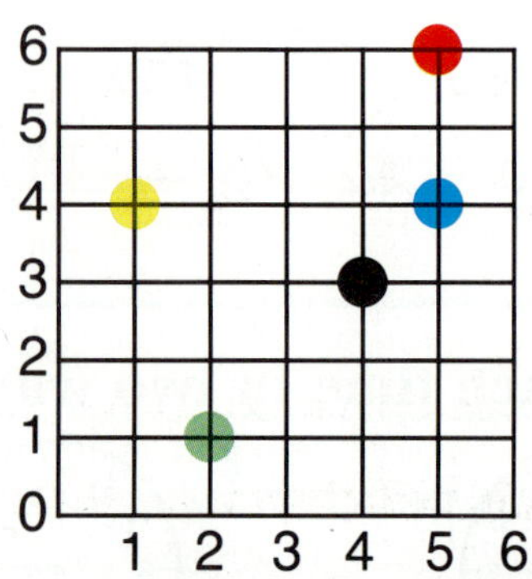

Set D

Complete. pages 176–177

1. If $3 \times 8 = 24$, then $8 \times 3 = \square$.
2. If $6 \times 5 = 30$, then $5 \times 6 = \square$.
3. If $7 \times 4 = 28$, then $4 \times 7 = \square$.
4. If $8 \times 9 = 72$, then $9 \times 8 = \square$.

Extra Practice

Set A

Find each product. pages 178–181

1. 3×2 2. 7×2 3. 2×6 4. 4×2 5. 9×2 6. 5×2 7. 2×8

8. 4×3 9. 3×9 10. 7×3 11. 7×2 12. 6×3 13. 3×3 14. 3×8

15. $2 \times 2 = \square$ 16. $5 \times 3 = \square$ 17. $\square = 3 \times 2$ 18. $\square = 1 \times 3$

Set B

Find each missing number. pages 182–183

1. $4 \times \square = 4$ 2. $\square \times 12 = 0$ 3. $1 \times 8 = \square$ 4. $99 \times 0 = \square$

5. $\square \times 33 = 33$ 6. $7 \times \square = 0$ 7. $\square \times 1 = 1$ 8. $76 \times \square = 0$

9. $0 \times 0 = \square$ 10. $\square \times 12 = 0$ 11. $43 \times \square = 43$ 12. $61 \times 1 = \square$

Set C

Find each product. pages 188–191

1. 3×4 2. 8×4 3. 4×5 4. 4×4 5. 7×3 6. 1×4 7. 4×9

8. 6×5 9. 5×2 10. 5×5 11. 2×9 12. 9×5 13. 5×7 14. 8×5

15. $8 \times 4 = \square$ 16. $\square = 5 \times 3$ 17. $5 = \square \times 5$ 18. $\square = 4 \times 0$

Set D

Find each product. pages 192–193

1. 2×3 2. 5×0 3. 9×4 4. 1×6 5. 3×7 6. 5×9 7. 8×2

8. 4×4 9. 2×2 10. 3×3 11. 5×5 12. 4×2 13. 6×5 14. 0×3

15. $5 \times 8 = \square$ 16. $4 \times \square = 4$ 17. $\square = 4 \times 8$ 18. $\square = 9 \times 3$

Extra Practice

Set A

Find each product. pages 204–209

1. 6×6	2. 8×6	3. 0×6	4. 6×4	5. 8×3	6. 9×6	7. 6×7
8. 4×7	9. 7×9	10. 5×4	11. 6×7	12. 8×7	13. 7×7	14. 3×7
15. 6×8	16. 9×8	17. 8×7	18. 8×0	19. 8×8	20. 5×5	21. 2×8
22. 4×9	23. 9×1	24. 9×9	25. 6×7	26. 9×4	27. 7×9	28. 8×9

29. $8 \times 5 = \square$ 30. $7 \times 6 = \square$ 31. $5 \times 9 = \square$ 32. $6 \times 5 = \square$

33. $\square = 9 \times 0$ 34. $\square = 1 \times 8$ 35. $\square = 4 \times 6$ 36. $\square = 4 \times 8$

Set B

Use the multiplication table on page 216. pages 216–217

Find all the facts that

1. have a product of 12.
2. have a product of 9.
3. have a product of 8.
4. have a product of 35.
5. have a product greater than 26 and less than 31.

Set C

Find each product. pages 218–219

1. $3 \times 8 \times 1 = \square$
2. $4 \times 2 \times 3 = \square$
3. $3 \times 2 \times 5 = \square$
4. $7 \times 1 \times 6 = \square$
5. $2 \times 3 \times 4 = \square$
6. $3 \times 3 \times 2 = \square$
7. $4 \times 2 \times 3 = \square$
8. $9 \times 8 \times 1 = \square$
9. $2 \times 2 \times 2 = \square$

Extra Practice

Set A

Divide. pages 234–241

1. $16 \div 2 = \square$ 2. $10 \div 2 = \square$ 3. $2 \div 2 = \square$ 4. $14 \div 2 = \square$

5. $9 \div 3 = \square$ 6. $27 \div 3 = \square$ 7. $18 \div 3 = \square$ 8. $12 \div 3 = \square$

9. $36 \div 4 = \square$ 10. $20 \div 4 = \square$ 11. $8 \div 4 = \square$ 12. $28 \div 4 = \square$

13. $40 \div 5 = \square$ 14. $10 \div 5 = \square$ 15. $25 \div 5 = \square$ 16. $5 \div 5 = \square$

17. $2\overline{)8}$ 18. $2\overline{)12}$ 19. $2\overline{)6}$ 20. $3\overline{)21}$ 21. $3\overline{)24}$

22. $3\overline{)3}$ 23. $4\overline{)16}$ 24. $4\overline{)12}$ 25. $4\overline{)32}$ 26. $5\overline{)45}$

27. $5\overline{)15}$ 28. $5\overline{)35}$ 29. $2\overline{)2}$ 30. $4\overline{)24}$ 31. $3\overline{)6}$

32. $5\overline{)20}$ 33. $4\overline{)4}$ 34. $3\overline{)15}$ 35. $5\overline{)30}$ 36. $2\overline{)18}$

Set B

Find each quotient. pages 246–247

1. $1\overline{)7}$ 2. $4\overline{)0}$ 3. $1\overline{)9}$ 4. $6\overline{)0}$ 5. $1\overline{)2}$

6. $0 \div 5 = \square$ 7. $3 \div 1 = \square$ 8. $0 \div 1 = \square$ 9. $4 \div 4 = \square$

Divide. pages 248–255

10. $6\overline{)18}$ 11. $6\overline{)30}$ 12. $6\overline{)6}$ 13. $6\overline{)42}$ 14. $6\overline{)12}$

15. $7\overline{)28}$ 16. $7\overline{)42}$ 17. $7\overline{)21}$ 18. $7\overline{)49}$ 19. $7\overline{)14}$

20. $8\overline{)16}$ 21. $8\overline{)48}$ 22. $8\overline{)72}$ 23. $8\overline{)24}$ 24. $8\overline{)40}$

25. $9\overline{)54}$ 26. $9\overline{)27}$ 27. $9\overline{)81}$ 28. $9\overline{)18}$ 29. $9\overline{)45}$

30. $54 \div 6 = \square$ 31. $36 \div 6 = \square$ 32. $56 \div 7 = \square$ 33. $35 \div 7 = \square$

34. $64 \div 8 = \square$ 35. $32 \div 8 = \square$ 36. $0 \div 9 = \square$ 37. $63 \div 9 = \square$

38. $48 \div 6 = \square$ 39. $72 \div 9 = \square$ 40. $63 \div 7 = \square$ 41. $8 \div 8 = \square$

42. $0 \div 7 = \square$ 43. $56 \div 8 = \square$ 44. $24 \div 6 = \square$ 45. $36 \div 9 = \square$

Extra Practice

Set A

Write the other facts in each family. pages 256–257

1. $3 \times 7 = 21$
2. $9 \times 4 = 36$
3. $15 \div 5 = 3$
4. $7 + 9 = 16$
5. $72 \div 8 = 9$
6. $4 \times 4 = 16$
7. $42 \div 7 = 6$
8. $2 \times 1 = 2$

Set B

Write a fraction for the part that is red. pages 266–269

1. 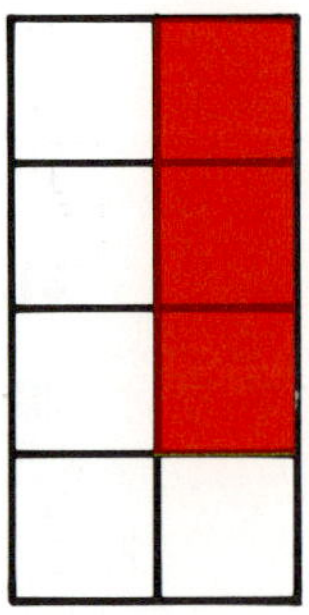
2. 
3. 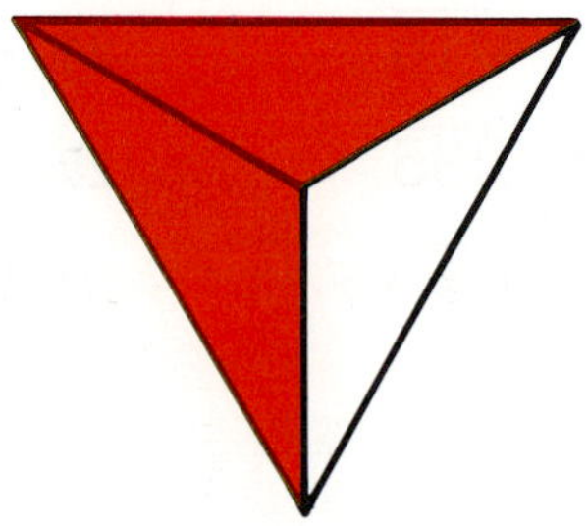
4. 
5. 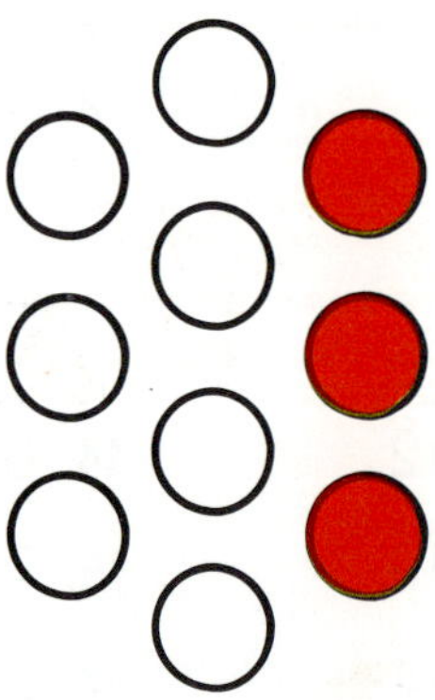
6. 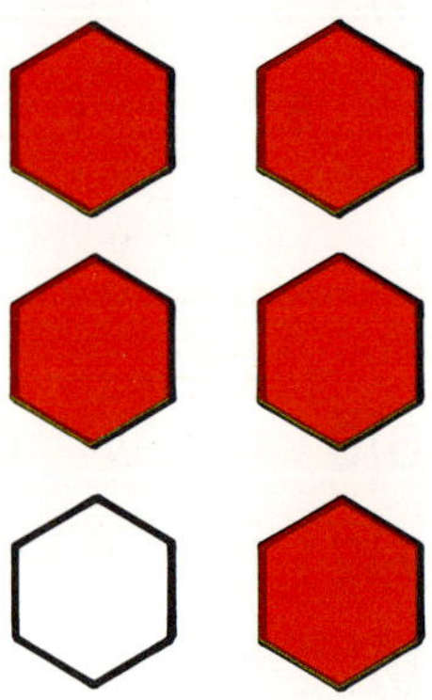
7. 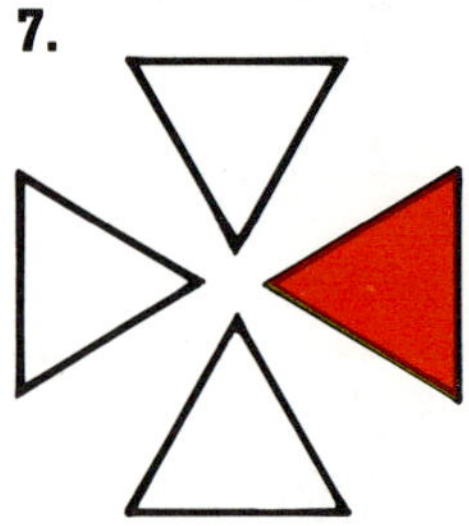
8. 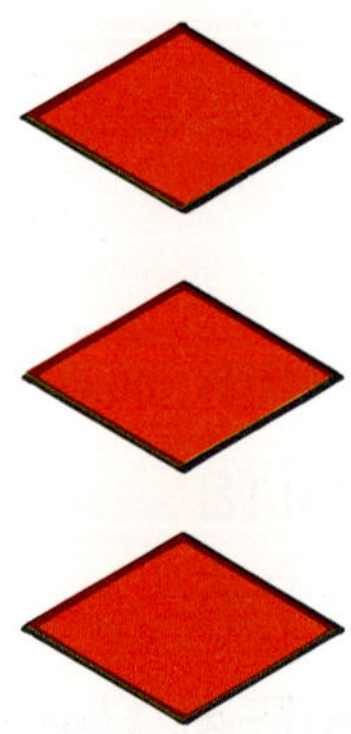

Complete. pages 270–271

9. $\frac{1}{2}$ of $4 = \square$
10. $\frac{1}{3}$ of $3 = \square$
11. $\frac{1}{5}$ of $15 = \square$
12. $\frac{1}{4}$ of $12 = \square$
13. $\frac{1}{7}$ of $14 = \square$
14. $\frac{1}{6}$ of $12 = \square$
15. $\frac{1}{2}$ of $18 = \square$
16. $\frac{1}{8}$ of $40 = \square$

Extra Practice

Set A

Complete. pages 272–273

1.

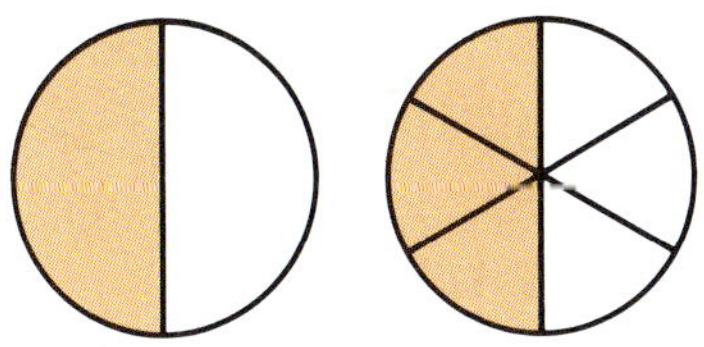

$\frac{1}{2} = \frac{\square}{6}$

2.

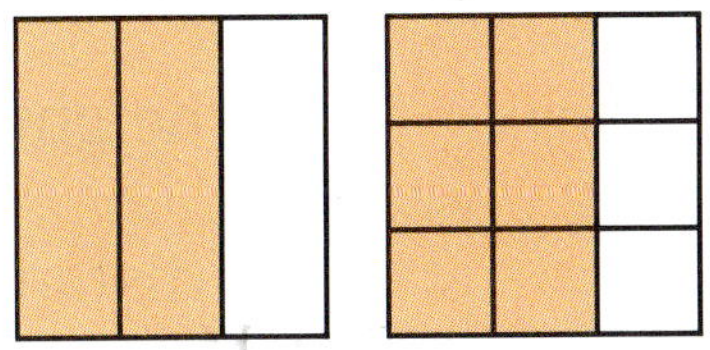

$\frac{2}{3} = \frac{\square}{\square}$

3.

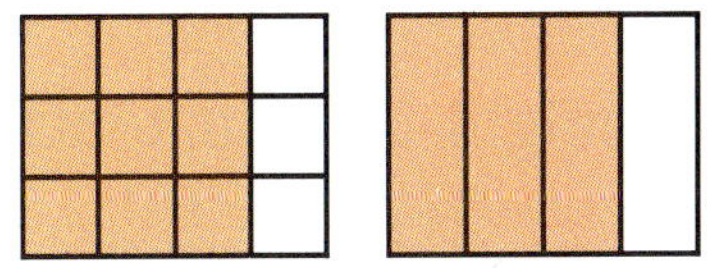

$\frac{9}{12} = \frac{\square}{\square}$

Use the chart on page 278 to write >, <, or =. pages 276–279

4. $\frac{7}{8}$ ⬬ $\frac{5}{8}$
5. $\frac{1}{6}$ ⬬ $\frac{3}{6}$
6. $\frac{7}{10}$ ⬬ $\frac{8}{10}$
7. $\frac{2}{2}$ ⬬ $\frac{1}{2}$
8. $\frac{3}{8}$ ⬬ $\frac{1}{4}$
9. $\frac{2}{4}$ ⬬ $\frac{3}{6}$
10. $\frac{1}{2}$ ⬬ $\frac{4}{10}$
11. $\frac{1}{3}$ ⬬ $\frac{3}{8}$

Write a mixed number for the part that is red. pages 280–281

12.

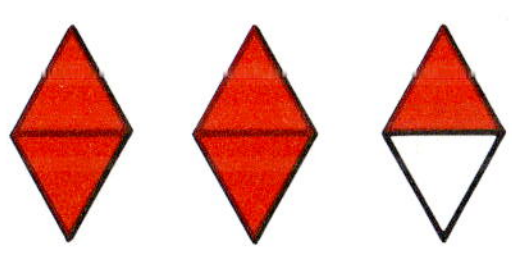

13.

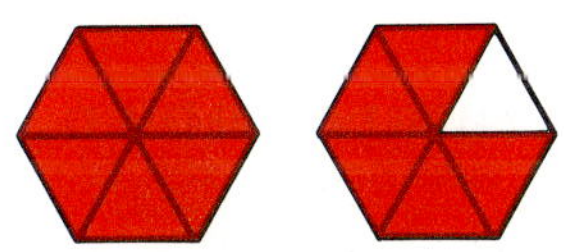

14.

Look at the spinner. pages 288–289

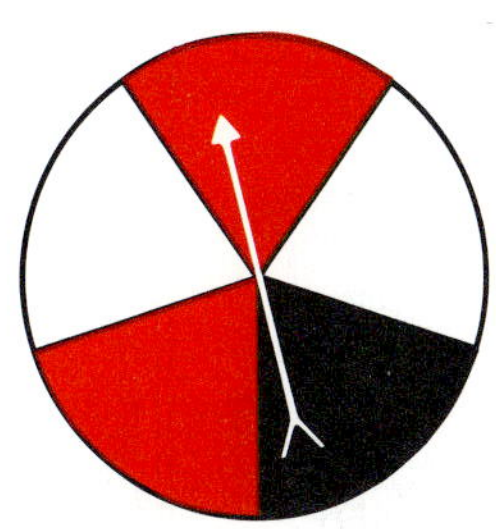

15. Name the possible outcomes of a spin.
16. What is the chance of the pointer stopping on each color?
17. If you spin *many* times, will you get more reds or blacks? Explain.

Extra Practice

Set A

Choose cm, dm, m, or km to measure each. pages 300–303

1. length of a piece of chalk
2. length of a car
3. height of your teacher's desk
4. height of a flagpole
5. width of a crayon
6. distance a bus travels
7. length of a video cassette tape
8. length of a river

Choose liters or milliliters to measure each. pages 306–307

9. a drop of water
10. cider in a jug
11. sand in a sandbox
12. milk container with school lunch

Choose grams or kilograms to weigh each. pages 308–311

13. a nail
14. a bag of oranges
15. ten dimes
16. three mathematics books

Write each Celsius temperature shown. pages 312–313

17.
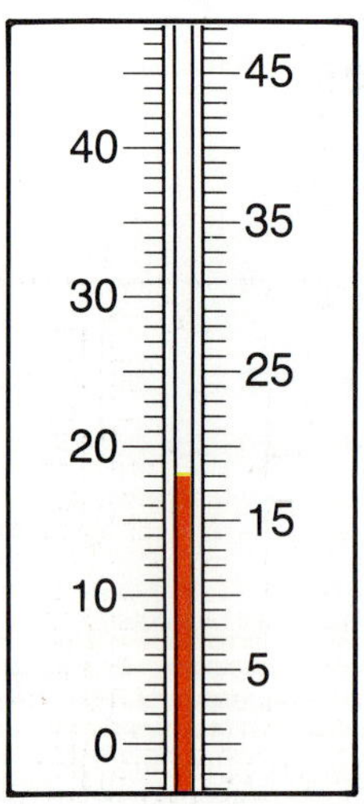

18.
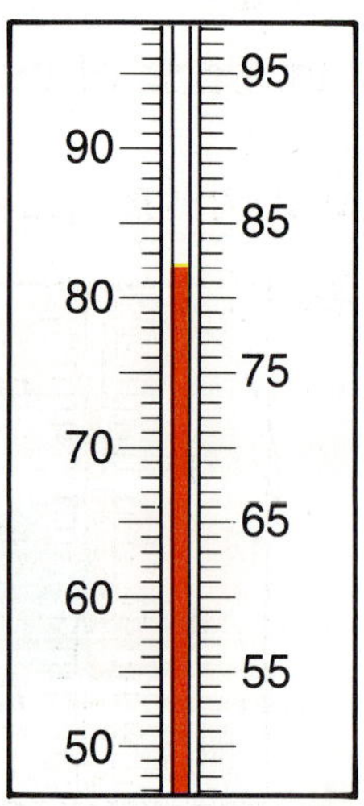

19.
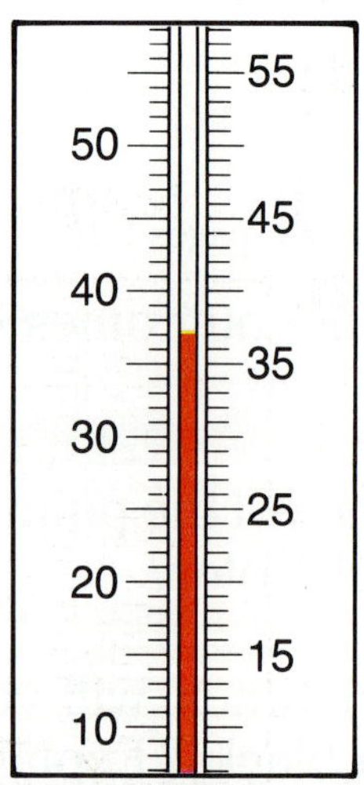

20.
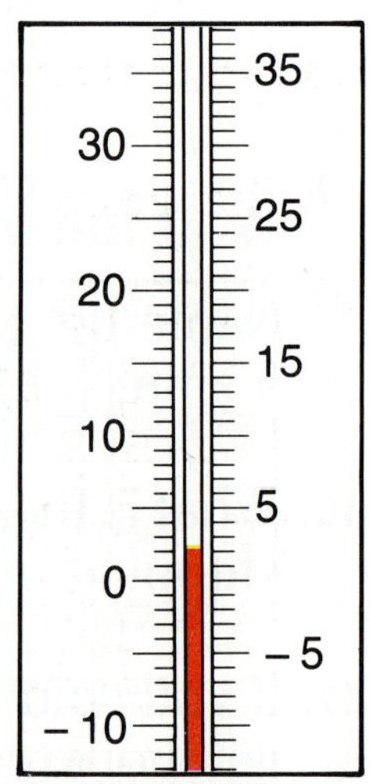

Extra Practice

Set A

Estimate each length. Then measure to the nearest inch and nearest $\frac{1}{4}$ inch. pages 318–319

1. ______________________

2. ______________________

3. ______________________

Choose inch, foot, yard, or mile to measure each. pages 318–321

4. width of a post card
5. distance across a playground
6. height of your teacher
7. width of a door
8. distance across North Dakota
9. height of a satellite above earth

Choose cup, pint, quart, or gallon to measure each. pages 322–323

10. lemonade in a pitcher
11. milk in a baby bottle
12. gasoline in a tank truck
13. a can of house paint
14. juice in a bottle
15. water in a mug

Choose ounce or pound to measure the weight of each.
pages 324–327

16. an orange
17. three pencils
18. box of crayons
19. pair of eyeglasses
20. your teacher's desk
21. sack of potatoes

Write each Fahrenheit temperature shown. pages 328–329

22. 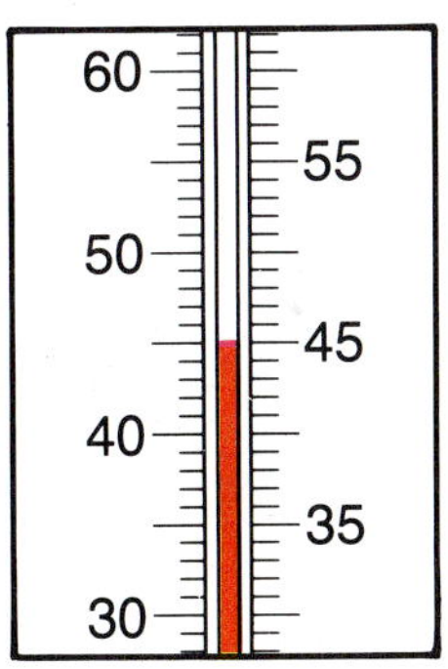

23. 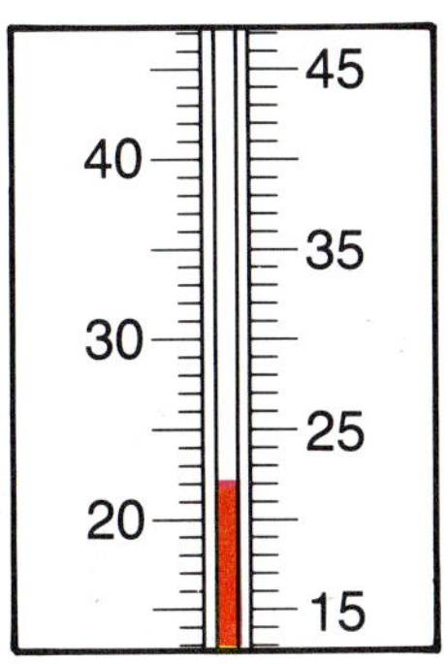

24. 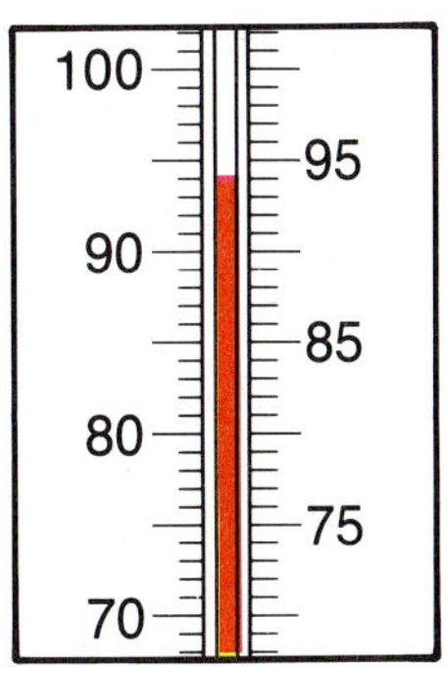

25. 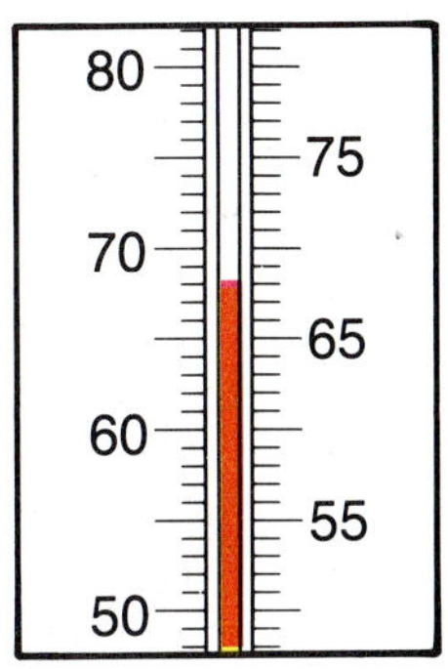

Extra Practice

Set A

Name each shape. pages 338–339

1. 

2. 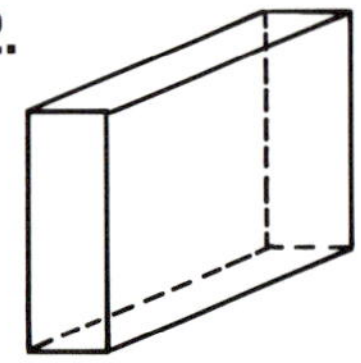

3. 

4. 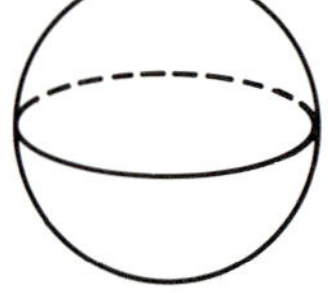

5. 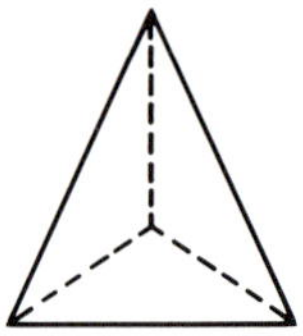

Name each plane figure. pages 342–347

6. 

7. 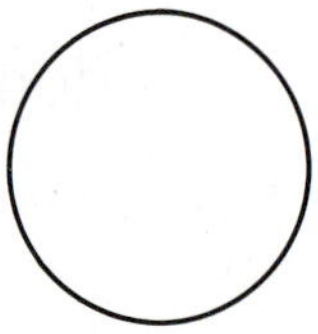

8. 

9. 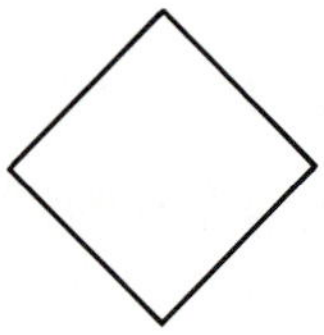

10. 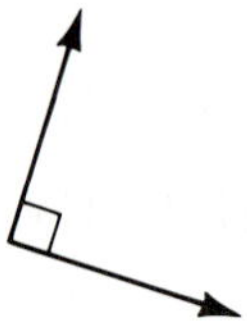

11. 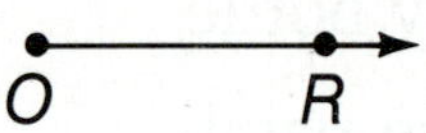

12. 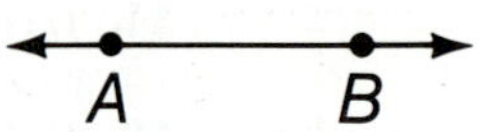

13. 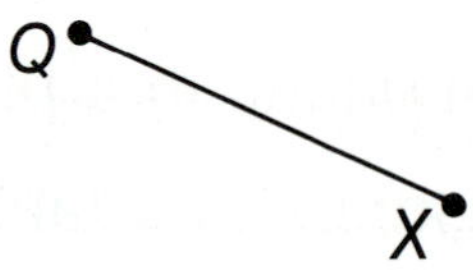

Set B

Are the figures congruent? pages 352–353

1. 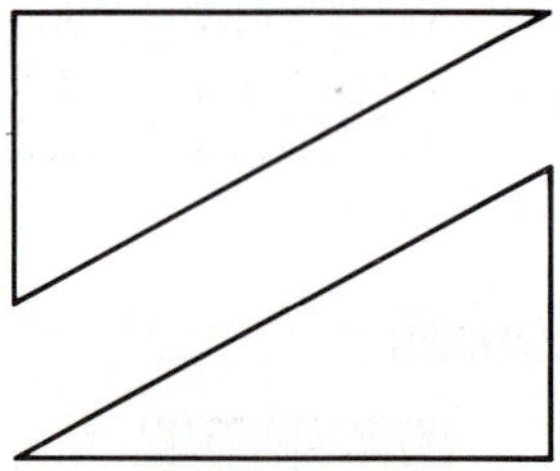

2. 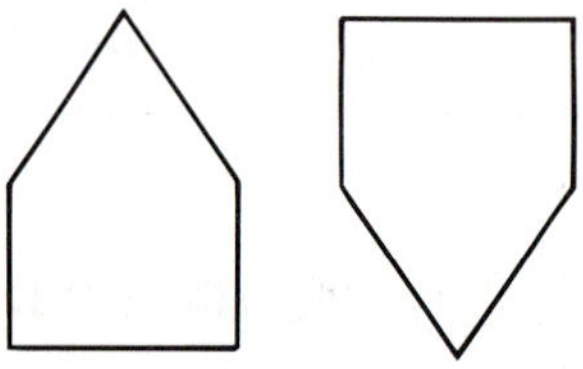

3. 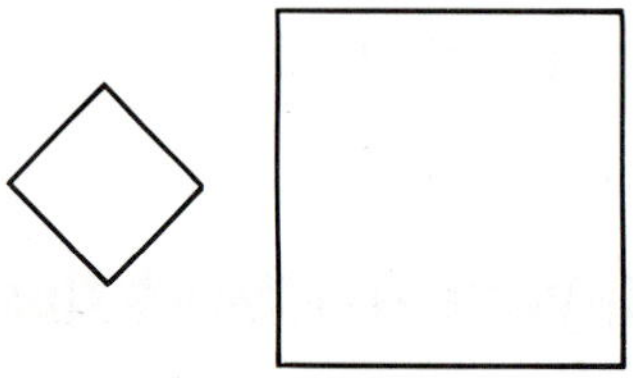

Is the dashed line a line of symmetry? pages 354–355

4. 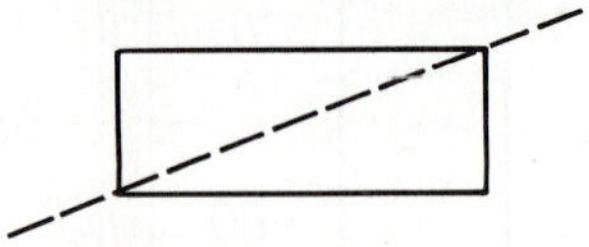

5. 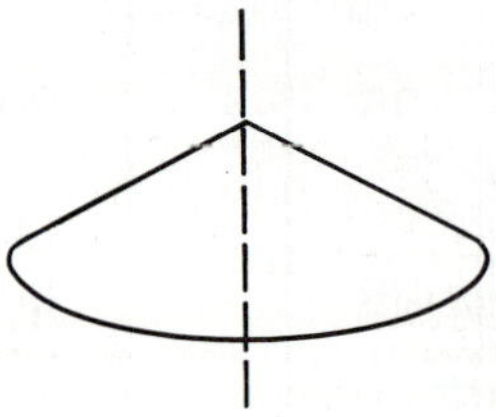

6. 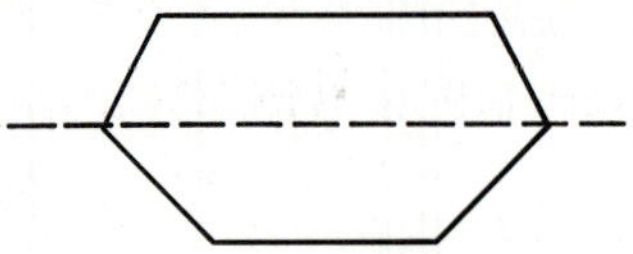

Extra Practice

Set A

Find the perimeter of each figure. pages 356–357

1. 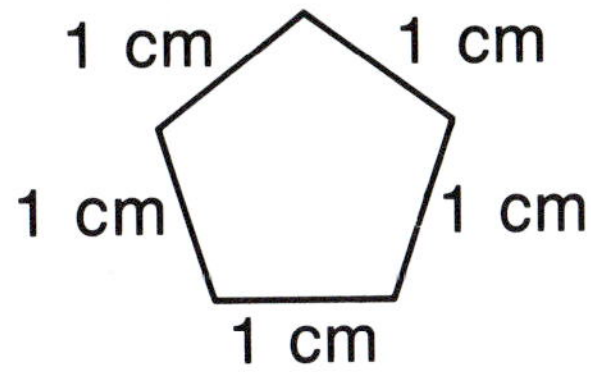

2. 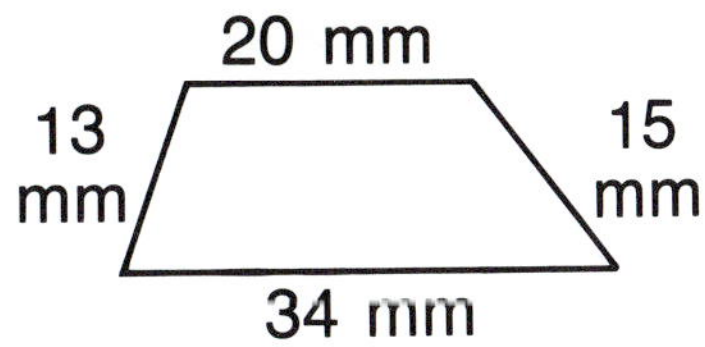

3. 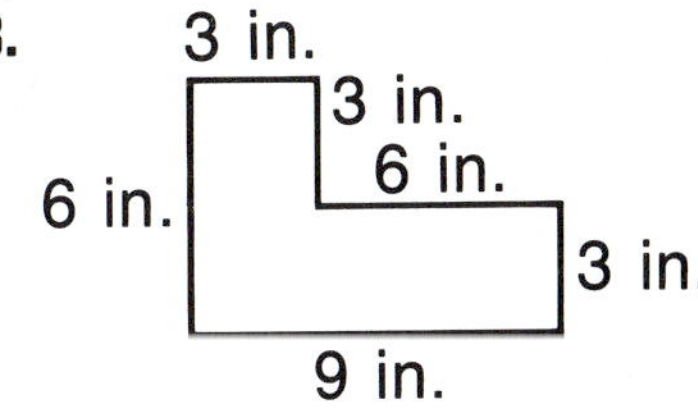

Find the area of each figure. pages 358–359

4. 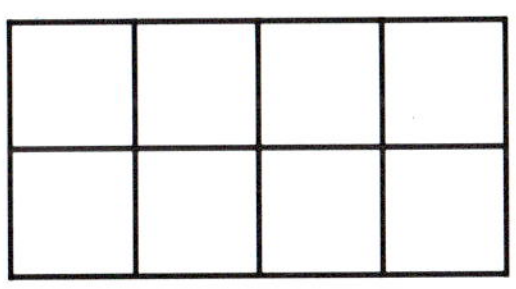

5. 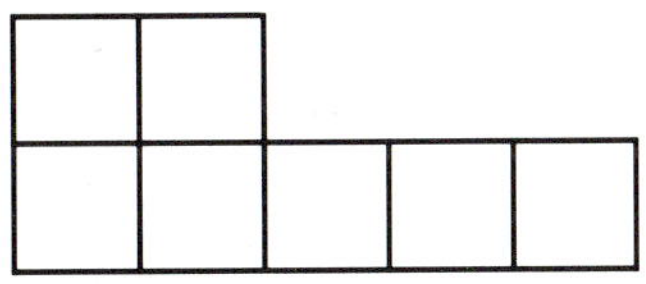

6. 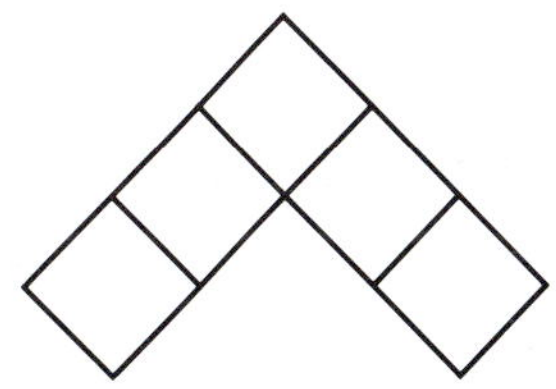

Find the volume of each figure. pages 360–361

7. 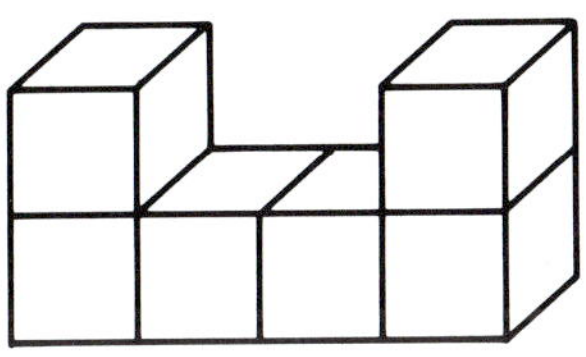

8. 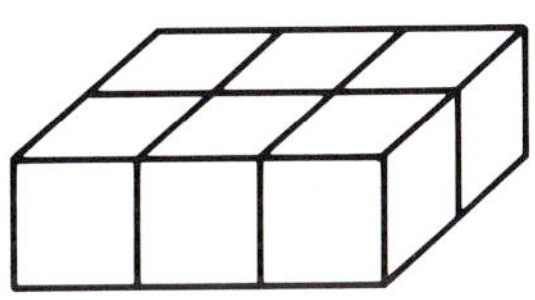

9. 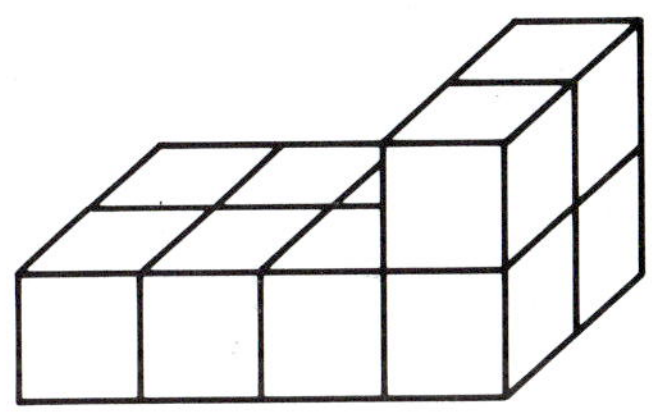

Set B

Write a decimal for the blue part. pages 370–371

1. 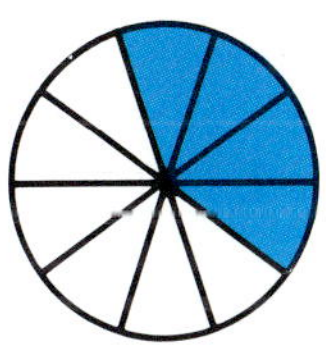

2. 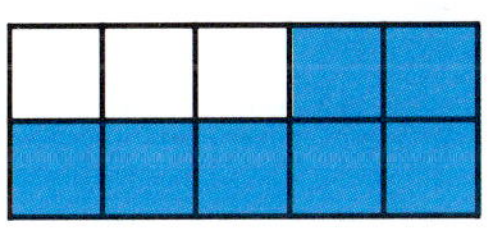

3. 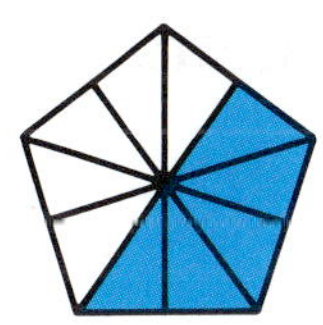

4. 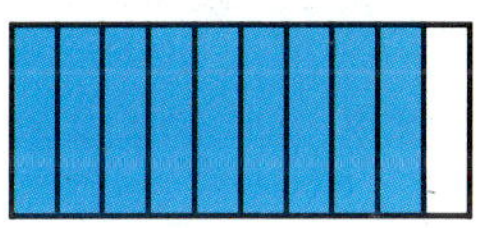

Write a decimal for the blue part. pages 372–373

5. 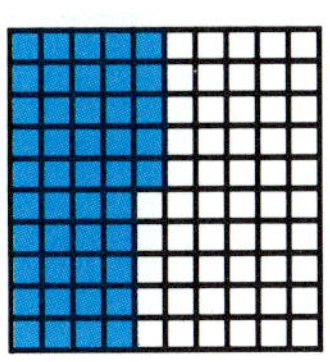

6. 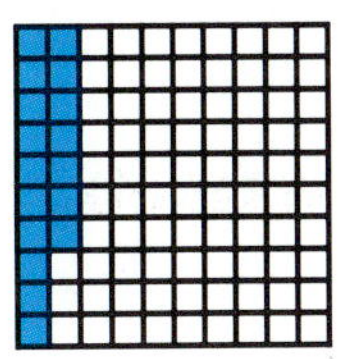

7. 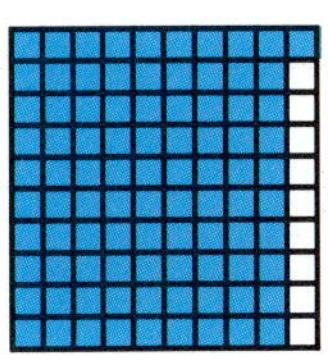

Write as a decimal. pages 374–377

8. five tenths
9. $\frac{8}{10}$
10. three tenths
11. $\frac{1}{10}$
12. two hundredths
13. $\frac{75}{100}$
14. sixty hundredths
15. $\frac{32}{100}$

Extra Practice

Set A

Write each as a decimal. pages 380–381

1. four and seven tenths
2. one and fifty-six hundredths
3. nine and nine hundredths
4. two and eighty-one hundredths
5. six and fourteen hundredths
6. three and three tenths

7. $7\frac{3}{10}$ 8. $4\frac{17}{100}$ 9. $5\frac{77}{100}$ 10. $3\frac{43}{100}$ 11. $6\frac{2}{10}$

12. $3\frac{6}{10}$ 13. $9\frac{24}{100}$ 14. $5\frac{9}{10}$ 15. $\frac{2}{10}$ 16. $2\frac{15}{100}$

Set B

Find each sum. pages 386–387

1. 4.2 + 3.5
2. 6.1 + 1.8
3. 4.8 + 4.7
4. 5.2 + 0.7
5. 0.6 + 0.8
6. 2.9 + 5.6
7. 3.5 + 4.5
8. 0.6 + 8.5
9. 1.8 + 0.7
10. 2.7 + 7.1
11. 0.8 + 0.2
12. 0.3 + 1.7
13. 2.6 + 3.1
14. 0.2 + 0.5
15. 0.9 + 0.8
16. 3.7 + 1.1 = □
17. 2.3 + 4.4 = □
18. 5.5 + 4.4 = □

Set C

Find each difference. pages 388–389

1. 0.7 − 0.4
2. 1.6 − 0.3
3. 1.1 − 0.9
4. 8.2 − 5.1
5. 4.1 − 2.9
6. 1.4 − 0.2
7. 3.7 − 1.5
8. 8.7 − 4.5
9. 6.2 − 5.8
10. 5.5 − 2.6
11. 3.2 − 3.1
12. 0.9 − 0.2
13. 4.3 − 3.9
14. 9.8 − 8.1
15. 5.8 − 1.9
16. 7.2 − 6.1 = □
17. 1.8 − 1.5 = □
18. 9.2 − 0.8 = □

Extra Practice

Set A

Use patterns to find each product mentally. pages 396–397

1. 30×2
2. 10×3
3. 50×7
4. 200×8
5. 400×6
6. $4 \times 70 = \square$
7. $5 \times 900 = \square$
8. $8 \times 500 = \square$

Multiply. pages 400–405

9. 31×3
10. 75×1
11. 44×2
12. 80×0
13. 12×4
14. 24×4
15. 19×5
16. 37×2
17. 29×3
18. 45×2
19. $0 \times 94 = \square$
20. $7 \times 64 = \square$
21. $8 \times 90 = \square$

Set B

Estimate each product. pages 406–407

1. 58×7
2. 93×8
3. 231×6
4. 487×9
5. 685×4

Estimate. Then find each product. pages 408–409

6. 613×3
7. 182×4
8. 305×8
9. 333×4
10. 413×7
11. $6 \times 231 = \square$
12. $9 \times 711 = \square$
13. $0 \times 824 = \square$

Set C

Divide. Use mental math. pages 416–417

1. $4\overline{)280}$
2. $7\overline{)350}$
3. $9\overline{)180}$
4. $3\overline{)900}$
5. $6\overline{)4,200}$
6. $8\overline{)2,400}$
7. $5\overline{)100}$
8. $2\overline{)800}$
9. $4\overline{)3,600}$
10. $9\overline{)630}$

Find each quotient. pages 418–423

11. $5\overline{)72}$
12. $7\overline{)93}$
13. $8\overline{)89}$
14. $4\overline{)76}$
15. $3\overline{)82}$
16. $9\overline{)48}$
17. $6\overline{)99}$
18. $2\overline{)65}$
19. $5\overline{)90}$
20. $7\overline{)94}$

Extra Practice

Set A

Use the picture to help you to solve each problem pages 9–11

1. Which prize needs the most tickets?
2. Flo wants a balloon and a teddy bear. How many tickets does she need?
3. How many tickets are needed to get one of each prize?
4. Pete can only get a balloon. He wants a hat. How many more tickets does he need?
5. Chun has 8 tickets. Can she get two different prizes? Which ones?

Look at the picture below. Solve each problem. pages 20–21

6. Which player has the longest name?
7. Which player has the least number?
8. Whose uniforms have even numbers?
9. Is the sum of the uniform numbers odd or even?

Set A

Solve. Use a calculator where appropriate. pages 43–45, 56–57

1. Estimate how many pieces of paper are needed to cover your desk. Experiment to check. Try to use the fewest number of pieces.
2. Place your pencil along the left side of your paper. If you roll it across the paper, how many times will it roll completely around? Estimate the number. Then experiment to check.
3. Estimate how many steps from your desk to the pencil sharpener in your classroom. Experiment to check.
4. Draw 4 lines on a piece of paper to make 8 equal sections. How many different ways can you draw the lines?
5. Henry wants to buy an apple that costs $.79. He gives the clerk a dollar bill. How much change will he get back?
6. The book costs $5.95. Andrea has 5 dollars, 3 quarters, and 6 dimes. Can she buy the book? Explain.

Set B

Solve. Use a calculator where appropriate. pages 75–77, 90–91

1. John had 63 baseball cards. Gina sold him a pack of 15 cards for $.40. How many cards does John have now?
2. At the roadside stand, corn costs 30¢ an ear, and tomatoes are 75¢ each. How much will you pay for two tomatoes?
3. Loc rode his bike 4 blocks to Sandy's house. Then they rode their bikes 5 blocks straight to school. How far from school does Sandy live?
4. Alison picked 14 pints of strawberries and sold 8. Joe picked 16 pints and sold 12. How many pints did they sell?
5. Robert scored 95, 81, 78 and 87 on his tests. Which two test scores have the greatest total?
6. On one test Andrea's score had the digits 8 and 5. On another test her score had the digits 6 and 7. What is the greatest sum the two tests could have?

Extra Practice

Set A

Solve. Use a calculator where appropriate. pages 115–117, 128–129

1. Sue has 3 pencils in her desk. Lou has twice as many pencils as Eva. Who has the most pencils?
2. Look back at **1.** If Lois has more pencils than Sue, who has the most pencils? How many more pencils does she have?
3. Carl borrowed 5 books from the library. Nina borrowed the same number, and Tao borrowed 2 more than Nina. How many books did all three borrow?
4. Fifteen students said they would buy their lunch in the cafeteria. The rest of the class brought their lunch. How many brought their lunch?
5. The school ordered 575 books for the book fair. They sold 216 books the first day. How many books were left?
6. During a period of three days the school sold 216, 197, and 143 books. In all how many books did they sell?

Set B

Solve. Use a calculator where appropriate. pages 151–153, 162–163

1. Ana, Ben, Cain, and Don must each play each other person once in a game. How many games are needed to do this?
2. You have 1 nickel, 1 dime, and 1 quarter. How many different amounts of money can you make using any of your coins?
3. Kim is using crayons to print the alphabet on his paper. He uses red for A, green for B, yellow for C, blue for D, and starts the pattern again. What color will he use for T?
4. Leonard practiced the piano for 15 minutes on Monday, 20 on Tuesday, 25 on Wednesday, and so on through Friday. How many minutes did he practice during the week?
5. Jenny sang for 15 minutes. There are 30 songs in the songbook. Chris sang for 25 minutes. Who sang longer?
6. Antonio read 256 pages in his book last week and 137 pages this week. How many more pages did he read last week than this week?

Extra Practice

Set A

Find and complete each pattern. pages 185–187, 194–195

1. □, ○, □, ○○, □, ○○○, ___, ___
2. 99, 88, 77, 66, ___, ___, ___
3. 0, 4, 8, 12, ___, ___, ___
4. AA, AB, AC, AD, ___, ___, ___
5. ◐, ◓, ◑, ◒, ___, ___, ___
6. 10, 1, 9, 2, 8, 3, ___, ___, ___
7. 10 + 1, 9 + 2, 8 + 3, ___, ___
8. X, Y, Z, U, V, W, ___, ___, ___

Solve. Use a calculator where appropriate.

9. You have \$50. Can you buy items that cost \$22, \$18, and \$15? Explain.
10. The 3rd grade class has 17 rulers. The 4th grade class has 22 scissors. Which class has more rulers?

Set B

Solve. Use a calculator where appropriate. pages 211–213, 220–221

1. The cog railway to the top of the mountain leaves every two hours starting at 7:45 A.M. Seylan and her family want to leave at 2:45 P.M. Can they? Explain.
2. Anita is making a bracelet by stringing 1 blue, 2 white, 1 green, and 2 yellow beads in that order. Then she repeats the pattern. What color is the fifteenth bead?
3. The mountain goats in the zoo are fed three times a day, four hours apart. The first feeding is 8:15 A.M. When are the others?
4. Animal figures cost \$1.25 for one, \$1.20 each for two, \$1.15 each for three, and so on. If the pattern continues, how much will each of 6 figures cost?
5. Martin wants to make license plates with 2-digit numbers for his friends' bicycles. He can use the digits 4, 5, and 6. How many license plates can he make? What are they?
6. Pat saw 12 squirrels in one day. **What if** she saw the same number of squirrels each day for a week? How many squirrels would she see.

Extra Practice

Set A

Solve. Use a calculator where appropriate. pages 243–245, 258–259

1. Luis and Rita went to the store with $1.50. They paid $.79 for 3 pens and $.59 for 5 crayons. How much money do they have left?

2. Mrs. Guthrie bought 3 pounds of grapes for $2.67 and 2 pounds of bananas for $.98. How much change did she receive from a $5.00 bill?

3. Six apples cost $2.00. How much will 12 apples for a class picnic cost?

4. Ten marbles cost $1.00. How many marbles can Becky buy if she has $3.00 to spend?

5. Trading cards come in packs of either 5, 6, or 7 cards. Harry bought 3 packs and has 16 cards. How many cards were in each pack?

6. Look at **5**. How many packs must Harry buy to make sure he gets at least 20 cards?

Set B

Solve. Use a calculator where appropriate pages 283–285, 290–291

Mu Lan wrote down the distance and the time that her family spent driving on a trip. Use this table to solve **1–4**.

	Miles	Hours
Monday	235	5
Tuesday	260	5
Wednesday	192	4
Thursday	336	6

1. On which day did they drive the most miles?

2. On which day did they spend the least time driving?

3. How many miles did they drive in the four days?

4. How many hours were spent driving?

5. Frank tossed 5 cubes and got only 5's and 6's. His total was 27. How many 5's and how many 6's did he toss?

6. A red and a green number cube, each with the digits 1 through 6, are tossed. What are the different ways the sum 8 can be shown on the cubes?

Set A

Solve. Use a calculator where appropriate. pages 304–305, 315–317

1. Mitchell plants a shrub in each corner of a square lot. He plants 5 more shrubs on each side of the lot. How many shrubs did he plant?
2. Erik scored 3 touchdowns in the first game and 4 more in the second. A touchdown is worth 6 points. How many points did Erik score in both games?
3. At Rinker's Orchard, one crate holds the apples from 2 trees. There are 14 apple trees along the driveway. Will the apples fit in 6 crates? Explain.
4. Mr. Payat has 12 cows on his farm. There are 2 times as many black and white cows as brown ones. How many cows are brown?
5. A loaf of bread costs $.76. How much would ½ a loaf of bread cost?
6. Pete uses 9 pounds of flour to make one batch of roll mix. How much flour should he use for ⅓ of a batch?

Set B

Solve. Use a calculator where appropriate. pages 340–341, 349–351

1. Leona is making a quilt block. She has 24 pieces of cloth in the shape of triangles and squares. She has 3 times as many triangles as squares. How many of each shape does she have?
2. At Anna's Quilt Shop there is a display of 16 full-size and baby-size quilts. There are 4 more full-size quilts than baby size quilts. How many of each size are there?
3. Frank looked at the top, front, and side of a figure. Each view was the same size square. What was the figure?
4. The movie is shown every 2 hours. The first show starts at 12:30. What times does the fourth show start?

Extra Practice

Set A

Solve. Use a calculator where appropriate. pages 378–379, 383–385

1. Alama has a square piece of paper. She wants to cut it into 9 squares of equal size. Draw a picture of what it might look like.

2. Leo is 3 inches taller than Ken. Matt is 2 inches shorter than Leo. Which boy is the shortest?

3. Four cars are traveling in the same lane of a highway. The red car is ahead of the blue car. The white car is behind the tan one. The white car is ahead of the red one. Which car is last?

4. Beth walked 1,500 meters. Joan walked 300 m less than Beth but 200 m more than Hanna. Lois walked 500 m more than Joan. Who walked the farthest?

5. Nadia received a score of 8.6 on the balance beam, 7.6 on the floor exercise, and 8.0 on the rings. List her events in order starting with the best score.

6. David started weeding the garden at 3:00 P.M. He spent 30 minutes weeding. He spent 1 hour and 15 minutes mowing the lawn. Then he went swimming for 30 minutes. What time was it when he finished swimming?

Set B

Solve. Use a calculator where appropriate. pages 411–413, 424–425

1. The pine tree is 6 meters tall. The maple is 4 meters taller than the pine. The spruce is 2 meters shorter than the pine. What is the difference in height between the tallest and shortest tree?

2. Sean, Doug, and Laura are the only children in their family. Sean is not the tallest child in his family. Doug is not the shortest child. Laura is between her brothers in height. Who is tallest?

3. Judy and Kate read 12 books. Judy read 2 more books than Kate. How many books did Kate read?

4. Together Luis and Rodney had 14 baseball cards. Luis had 4 fewer baseball cards than Rodney. How many baseball cards did each boy have?

Skill Hints

HINT Strategies and patterns can help you add or subtract.

Making 10 can help you add 8 + 3.

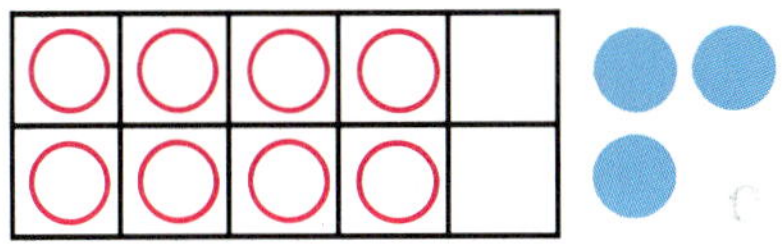

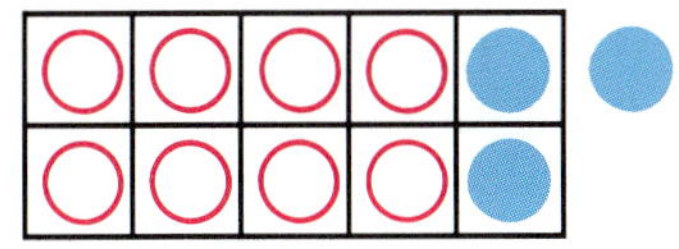

Make 10 from 8 and 3.

Find each sum. Use mental math strategies to help you.

1. $\begin{array}{r}5\\+9\\\hline\end{array}$	2. $\begin{array}{r}4\\+3\\\hline\end{array}$	3. $\begin{array}{r}6\\+6\\\hline\end{array}$	4. $\begin{array}{r}9\\+7\\\hline\end{array}$	5. $\begin{array}{r}8\\+7\\\hline\end{array}$	6. $\begin{array}{r}7\\+8\\\hline\end{array}$	7. $\begin{array}{r}9\\+8\\\hline\end{array}$
8. $\begin{array}{r}9\\+9\\\hline\end{array}$	9. $\begin{array}{r}7\\+6\\\hline\end{array}$	10. $\begin{array}{r}6\\+7\\\hline\end{array}$	11. $\begin{array}{r}7\\+5\\\hline\end{array}$	12. $\begin{array}{r}0\\+5\\\hline\end{array}$	13. $\begin{array}{r}7\\+7\\\hline\end{array}$	14. $\begin{array}{r}4\\+7\\\hline\end{array}$
15. $\begin{array}{r}7\\2\\+4\\\hline\end{array}$	16. $\begin{array}{r}3\\1\\+6\\\hline\end{array}$	17. $\begin{array}{r}6\\3\\+5\\\hline\end{array}$	18. $\begin{array}{r}4\\5\\+3\\\hline\end{array}$	19. $\begin{array}{r}2\\2\\+7\\\hline\end{array}$	20. $\begin{array}{r}5\\1\\+2\\\hline\end{array}$	21. $\begin{array}{r}5\\4\\+5\\\hline\end{array}$

Find each missing addend. Then write each difference.

22. 9 + □ = 17	23. 3 + □ = 11	24. 8 + □ = 16	25. 4 + □ = 13
17 − 9 = □	11 − 3 = □	16 8 = □	13 − 4 − □

Find each difference. Use mental math strategies.

26. $\begin{array}{r}8\\-8\\\hline\end{array}$	27. $\begin{array}{r}15\\-7\\\hline\end{array}$	28. $\begin{array}{r}12\\-9\\\hline\end{array}$	29. $\begin{array}{r}5\\-1\\\hline\end{array}$	30. $\begin{array}{r}3\\-0\\\hline\end{array}$	31. $\begin{array}{r}13\\-4\\\hline\end{array}$	32. $\begin{array}{r}7\\-7\\\hline\end{array}$
33. $\begin{array}{r}17\\-6\\\hline\end{array}$	34. $\begin{array}{r}8\\-0\\\hline\end{array}$	35. $\begin{array}{r}14\\-8\\\hline\end{array}$	36. $\begin{array}{r}9\\-9\\\hline\end{array}$	37. $\begin{array}{r}14\\-5\\\hline\end{array}$	38. $\begin{array}{r}18\\-9\\\hline\end{array}$	39. $\begin{array}{r}10\\-8\\\hline\end{array}$
40. $\begin{array}{r}6\\-6\\\hline\end{array}$	41. $\begin{array}{r}16\\-8\\\hline\end{array}$	42. $\begin{array}{r}13\\-9\\\hline\end{array}$	43. $\begin{array}{r}8\\-6\\\hline\end{array}$	44. $\begin{array}{r}12\\-4\\\hline\end{array}$	45. $\begin{array}{r}9\\-0\\\hline\end{array}$	46. $\begin{array}{r}12\\-9\\\hline\end{array}$

Skill Hints

HINT

Rounding numbers tells **about** how many instead of **exactly** how many. 23 is about 20. 179 is about 200.

When a number is halfway between, round to the greater number.

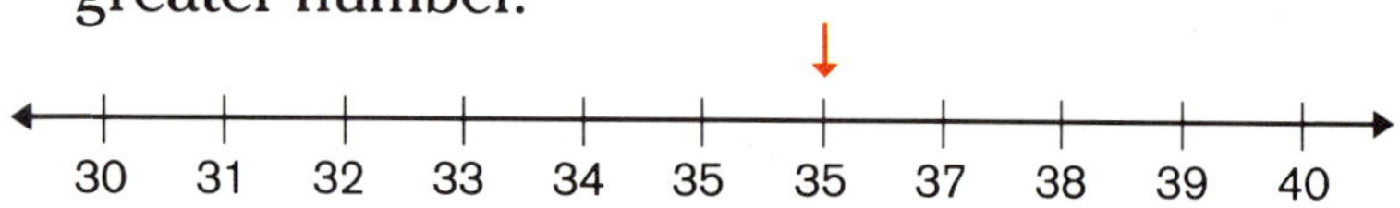

35 is halfway between 30 and 40. 35 rounded to the nearest ten is 40.

Round to the nearest ten. You may draw a number line to help you.

1. 34 **2.** 19 **3.** 25 **4.** 63 **5.** 38 **6.** 51

7. 95 **8.** 86 **9.** 37 **10.** 21 **11.** 45 **12.** 56

13. 42 **14.** 5 **15.** 15 **16.** 77 **17.** 69 **18.** 83

Round to the nearest hundred. Use the number line to help you.

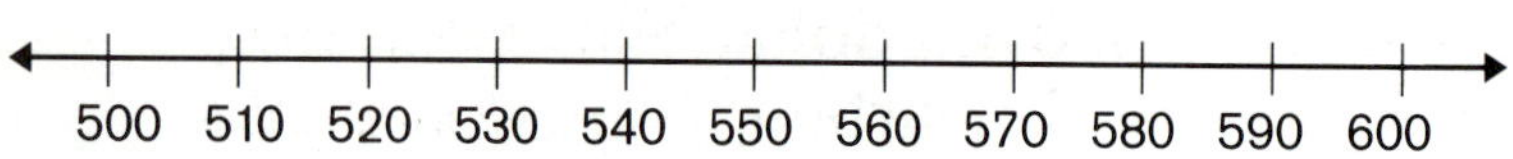

19. 583 **20.** 550 **21.** 518 **22.** 503 **23.** 595 **24.** 534

25. 150 **26.** 290 **27.** 750 **28.** 674 **29.** 432 **30.** 19

31. 350 **32.** 609 **33.** 888 **34.** 215 **35.** 798 **36.** 469

37. Use the digits 2, 6, and 8. Write a number that rounds to 300. to 600. to 900.

38. Use the digits 1, 5, and 7. Write a number that rounds to 200. to 500. to 700.

Skill Hints

HINT To count money, start with the coin or bill of greatest value. End with the coin or bill of least value.

10, 11 dollars 25, 35, 36 cents
11 dollars and 36 cents = $11.36

Write each value.

1. 

2. 

3. 4 pennies, 2 quarters, 1 nickel, 1 five dollar bill
4. 2 five dollar bills, 4 quarters, 2 dimes, 1 nickel
5. 2 one-dollar bills, 3 five-dollar bills, 5 nickels, 8 pennies
6. 3 ten-dollar bills, 4 pennies, 2 quarters, 2 one-dollar bills
7. 3 quarters, 8 one-dollar bills, 1 ten-dollar bill
8. 1 nickel, 1 quarter, 2 pennies, 1 one-dollar bill, 1 five-dollar bill

Name the fewest bills and coins needed to make each amount.

9. $5.38
10. $7.16
11. $8.25
12. $12.47
13. $15.35
14. $13.78
15. $2.26
16. $3.98
17. $10.64
18. $11.15

Skill Hints

HINT To estimate a sum, add the values of the front digits. Use zeros for the other places.

$$\begin{array}{r|c|l} & 7 & 3 \\ + & 4 & 1 \\ \hline & 11 & 0 \end{array} \quad \text{estimated sum} \qquad \begin{array}{r|c|l} & 2 & 48 \\ + & 1 & 20 \\ \hline & 3 & 00 \end{array} \quad \text{estimated sum}$$

Estimate each sum.

1. 23 + 16	**2.** 55 + 48	**3.** 14 + 41	**4.** 95 + 21	**5.** 62 + 48	**6.** 88 + 94
7. 436 + 109	**8.** 528 + 214	**9.** 325 + 673	**10.** 222 + 222	**11.** 430 + 304	**12.** 862 + 111
13. 294 + 643	**14.** 387 + 409	**15.** 197 + 109	**16.** 500 + 299	**17.** 103 + 946	**18.** 599 + 498

19. 30 + 56 = □ **20.** 67 + 82 = □ **21.** 85 + 82 = □

22. 473 + 128 = □ **23.** 641 + 294 = □ **24.** 342 + 520 = □

Add. Check your work.

25. 508 + 403	**26.** 394 + 128	**27.** 234 + 678	**28.** 880 + 109	**29.** 211 + 747	**30.** 463 + 326
31. 8,427 + 248	**32.** 4,956 + 2,039	**33.** 2,731 + 3,642	**34.** 7,928 + 47	**35.** 1,308 + 5,295	

36. 3,147 + 149 = □ **37.** 9,279 + 430 = □ **38.** 14 + 5,892 = □

39. 1,370 + 6,432 = □ **40.** 4,332 + 3,997 = □ **41.** 7,403 + 2,048 = □

Skill Hints

HINT To estimate a difference, round to the nearest ten or hundred and subtract.

$$\begin{array}{r} 49 \\ -23 \\ \hline \end{array} \xrightarrow{\text{rounds to}} \begin{array}{r} 50 \\ -20 \\ \hline 30 \end{array} \text{ estimated difference}$$

$$\begin{array}{r} 674 \\ -429 \\ \hline \end{array} \xrightarrow{\text{rounds to}} \begin{array}{r} 700 \\ -400 \\ \hline 300 \end{array} \text{ estimated difference}$$

Estimate each difference.

1. 79 − 42
2. 83 − 69
3. 43 − 17
4. 98 − 71
5. 55 − 37
6. 68 − 59
7. 821 − 307
8. 376 − 219
9. 791 − 630
10. 591 − 368
11. 615 − 482
12. 908 − 176
13. 690 − 247
14. 897 − 448
15. 339 − 115
16. 999 − 635
17. 485 − 257
18. 725 − 448
19. 73 − 21 = □
20. 48 − 15 = □
21. 99 − 47 = □
22. 432 − 187 = □
23. 915 − 462 = □
24. 581 − 443 = □

Estimate first. Then subtract.

25. 637 − 218
26. 949 − 522
27. 453 − 127
28. 782 − 368
29. 215 − 173
30. 873 − 551
31. 845 − 409
32. 586 − 394
33. 904 − 683
34. 231 − 180
35. 724 − 628
36. 397 − 276
37. 828 − 56 = □
38. 200 − 49 = □
39. 675 − 380 = □
40. 401 − 280 = □
41. 527 − 195 = □
42. 814 − 646 = □

Skill Hints

HINT When telling time, remember that the hours between 12 midnight and 12 noon are A.M. The hours between 12 noon and 12 midnight are P.M.

7:15 A.M.
Eat breakfast.
quarter after 7

8:30 P.M.
Go to sleep.
half past eight

Match the times

1.	2:30 in the morning	**a.**	6:15 P.M.
2.	half past three in the afternoon	**b.**	2:30 A.M.
3.	eleven o'clock in the morning	**c.**	3:30 P.M.
4.	quarter after six in the evening	**d.**	11:00 A.M.

Write each time in two ways.

5. 

6. 

7. 

8. 

9. 

10. 

11. 

12. 

13. 

14. 

15. 

16. 

Skill Hints

HINT The same information may be graphed in different ways.

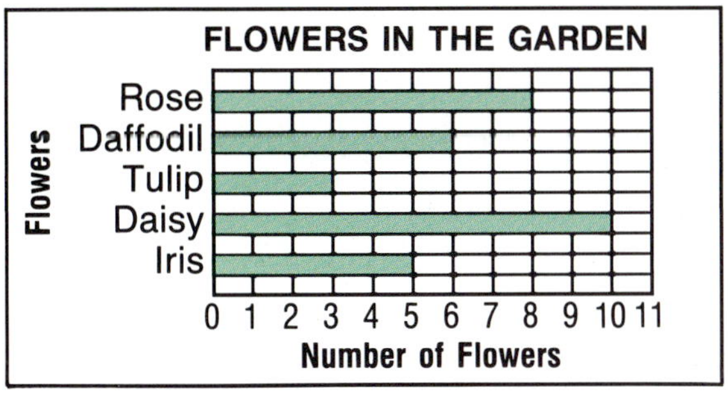

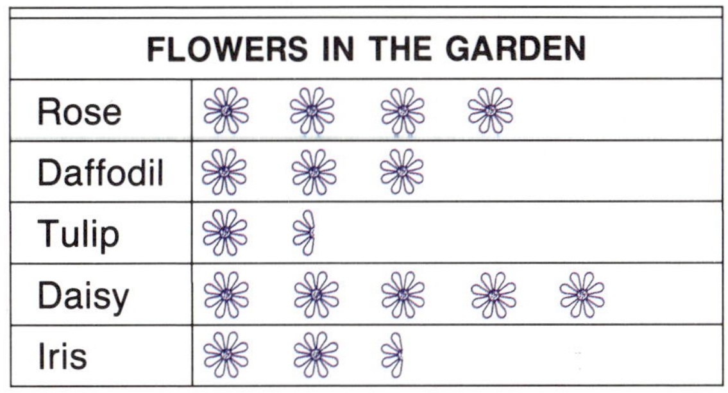

Each stands for 2 flowers.

The **bar graph** and the **pictograph** show the same information.

Use the graphs above to answer these questions.

1. How are the two graphs alike?
2. How are the two graphs different?
3. What does the symbol mean?
4. What do the bars on the graph show?
5. Which graph shows that there are 10 daisies in the garden?
6. Which graph shows that there are 3 tulips in the garden?
7. How many more daisies are there than tulips?
8. What does mean?
9. How many flowers are in the garden in all?
10. How many roses and tulips are in the garden?
11. Show the same information in another way. Put it in a chart, make tallies, or draw a picture.

Skill Hints

HINT When using 2 as a factor think of a related addition double. When using 4 as a factor think of a multiplication fact with 2 and double it.

$2 \times 3 = \square$

Think $3 + 3 = 6$

$2 \times 3 = 6$

$4 \times 3 = \square$

Think $2 \times 3 = 6$ 6 doubled is 12

$4 \times 3 = 12$

Multiply

1. 0×2
2. 3×2
3. 5×2
4. 0×4
5. 6×4
6. 5×4
7. 1×2
8. 9×2
9. 9×3
10. 8×2
11. 8×4
12. 7×5
13. 0×5
14. 1×3
15. 4×1
16. 3×0
17. 3×3
18. 2×4
19. 3×5
20. 6×3
21. 7×2

Find each missing number.

22. $2 \times \square = 0$
23. $3 \times 2 = \square$
24. $5 \times \square = 10$
25. $\square \times 4 = 8$
26. $2 \times 2 = \square$
27. $3 \times 8 = \square$
28. $\square \times 9 = 45$
29. $3 \times 5 = \square$
30. $5 \times 4 = \square$
31. $\square \times 7 = 28$
32. $3 \times \square = 12$
33. $5 \times 5 = \square$
34. $5 \times \square = 5$
35. $4 \times \square = 16$
36. $3 \times 7 = \square$
37. $\square \times 9 = 36$
38. $5 \times 8 = \square$
39. $4 \times 0 = \square$
40. $\square \times 6 = 6$
41. $5 \times 6 = \square$

Skill Hints

HINT You can use a table to find multiplication facts.

column ↓

X	0	1	2	3	4	5	6	7	8	9
0	0	0	0	0	0	0	0	0	0	0
1	0	1	2	3	4	5	6	7	8	9
2	0	2	4	6	8	10	12	14	16	18
3	0	3	6	9	12	15	18	21	24	27
4	0	4	8	12	16	20	24	28	32	36
5	0	5	10	15	20	25	30	35	40	45
row → 6	0	6	12	18	24	30	36	42	48	54
7	0	7	14	21	28	35	42	49	56	63
8	0	8	16	24	32	40	48	56	64	72
9	0	9	18	27	36	45	54	63	72	81

$6 \times 3 = \square$

Find the 6 row.
Find the 3 column.
The product is where the row and column meet.

$6 \times 3 = 18$

Multiply.

1. 9×9 **2.** 1×9 **3.** 8×8 **4.** 3×6 **5.** 7×7 **6.** 0×6 **7.** 3×9

8. 8×7 **9.** 5×8 **10.** 4×9 **11.** 6×6 **12.** 5×7 **13.** 7×8 **14.** 5×9

15. 9×8 **16.** 7×6 **17.** 7×9 **18.** 3×8 **19.** 8×6 **20.** 9×7 **21.** 4×6

22. 6×9 **23.** 9×0 **24.** 2×7 **25.** 6×8 **26.** 2×9 **27.** 6×7 **28.** 9×6

Find each missing number.

29. $6 \times \square = 6$ **30.** $7 \times 4 = \square$ **31.** $\square \times 5 = 30$ **32.** $\square \times 7 = 0$

33. $9 \times 0 = \square$ **34.** $8 \times \square = 32$ **35.** $6 \times 2 = \square$ **36.** $7 \times \square = 21$

Skill Hints

HINT You can divide to find how many equal groups or how many in each group.

8	÷	4	=	□
in all		groups		in each group

8	÷	4	=	□
in all		in each group		groups

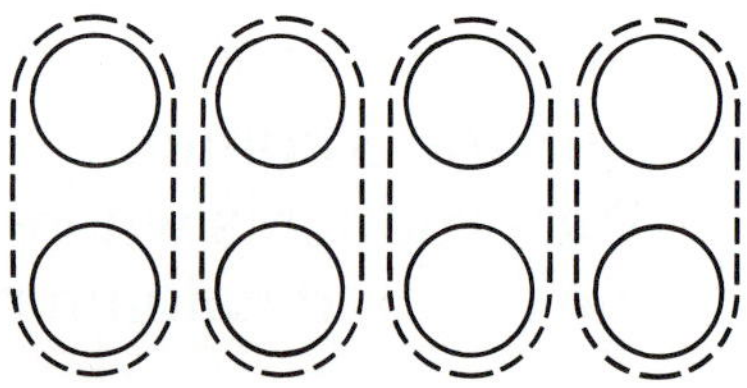

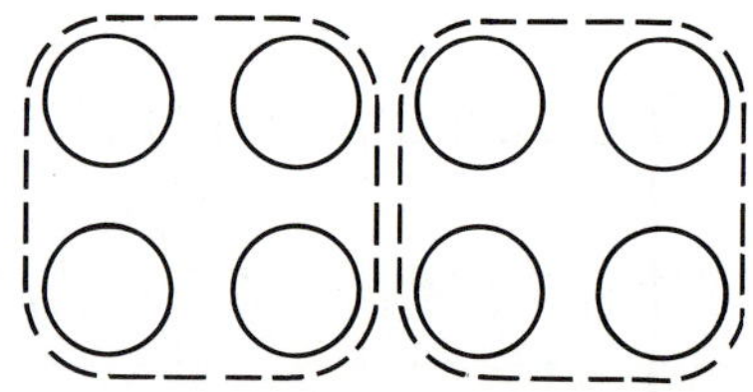

Write a division fact for each.

1. 18 circles
   6 equal groups
   How many are in each group?

2. 10 circles
   2 equal groups
   How many are in each group?

3. 28 circles
   4 in each group
   How many groups are there?

4. 21 circles
   3 in each group
   How many groups are there?

Find each quotient.

5. $9 \div 3 = \square$ **6.** $6 \div 2 = \square$ **7.** $10 \div 5 = \square$ **8.** $42 \div 6 = \square$

9. $7 \div 7 = \square$ **10.** $28 \div 4 = \square$ **11.** $0 \div 4 = \square$ **12.** $36 \div 6 = \square$

13. $4\overline{)12}$ **14.** $5\overline{)45}$ **15.** $1\overline{)8}$ **16.** $3\overline{)27}$ **17.** $7\overline{)56}$

18. $6\overline{)0}$ **19.** $2\overline{)14}$ **20.** $5\overline{)25}$ **21.** $9\overline{)72}$ **22.** $2\overline{)4}$

Compare. Use $>$, $<$, or $=$ for ⬬.

23. $18 \div 3$ ⬬ $12 \div 3$ **24.** $15 \div 3$ ⬬ $15 \div 5$

25. $24 \div 3$ ⬬ $8 \div 8$ **26.** $0 \div 9$ ⬬ $21 \div 7$

Skill Hints

HINT When you divide, you can think of a related multiplication fact.

$28 \div 7 = \square$

Think 7 times what number equals 28?

$7 \times \square = 28$

The missing factor is 4. $7 \times 4 = 28$

So $28 \div 7 = 4$ $\qquad 7\overline{)28}$ with quotient 4

Find each missing factor. Then find each quotient.

1. $6 \times \square = 0$ $\quad 0 \div 6 = \square$
2. $4 \times \square = 32$ $\quad 32 \div 4 = \square$
3. $2 \times \square = 18$ $\quad 18 \div 2 = \square$
4. $9 \times \square = 45$ $\quad 45 \div 9 = \square$
5. $7 \times \square = 49$ $\quad 49 \div 7 = \square$
6. $1 \times \square = 8$ $\quad 8 \div 1 = \square$
7. $9 \times \square = 54$ $\quad 54 \div 9 = \square$
8. $3 \times \square = 24$ $\quad 24 \div 3 = \square$

Divide.

9. $3\overline{)18}$
10. $7\overline{)7}$
11. $4\overline{)20}$
12. $5\overline{)25}$
13. $2\overline{)10}$
14. $9\overline{)63}$
15. $7\overline{)35}$
16. $6\overline{)24}$

Complete each fact family.

17. $6 \times 8 = \square$ $\quad 8 \times 6 = \square$ $\quad 48 \div 6 = \square$ $\quad 48 \div 8 = \square$
18. $8 \times 9 = \square$ $\quad 9 \times 8 = \square$ $\quad 72 \div 8 = \square$ $\quad 72 \div 9 = \square$
19. $3 \times 5 = \square$ $\quad 5 \times 3 = \square$ $\quad 15 \div 3 = \square$ $\quad 15 \div 5 = \square$
20. $4 \times 9 = \square$ $\quad 9 \times 4 = \square$ $\quad 36 \div 4 = \square$ $\quad 36 \div 9 = \square$

Skill Hints

HINT A fraction names part of a group or a region.

$\frac{\text{1 part shaded}}{\text{2 equal parts}}$

$\frac{1}{2}$ of the square is shaded.

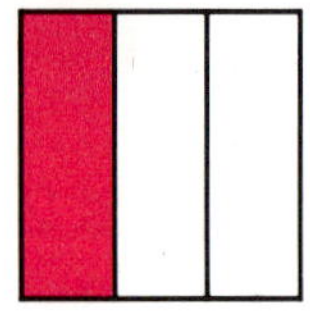

$\frac{\text{1 part shaded}}{\text{3 equal parts}}$

$\frac{1}{3}$ of the group is shaded.

Write a fraction for the shaded part.

1. 

2. 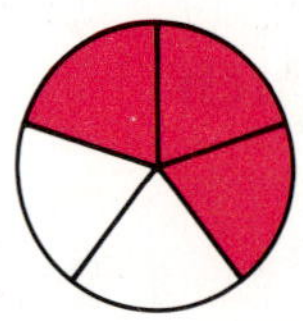

3. 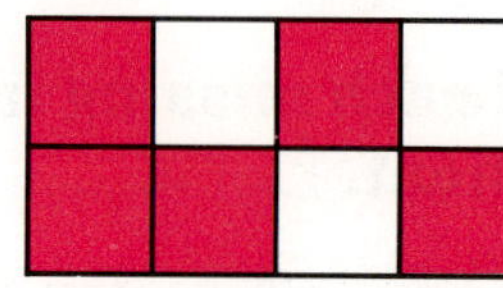

4. 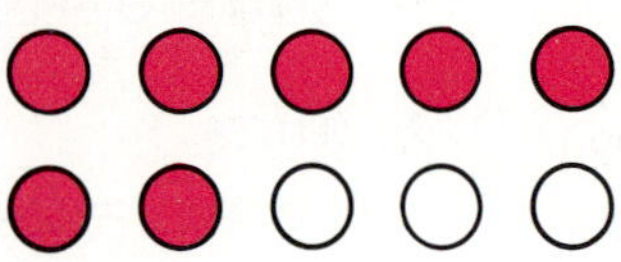

5. 

6. 

HINT To find a fractional part of a group, divide the group by the number of equal parts in all.

$\frac{1}{5}$ of $15 = \square$

Divide 15 by 5.

$15 \div 5 = 3$

$\frac{1}{5}$ of $15 = 3$

B C D E F G H I J—RRD—96 95 94 93 92 91 90

Find the fractional part of each.

7. $\frac{1}{2}$ of 18
8. $\frac{1}{4}$ of 12
9. $\frac{1}{7}$ of 28
10. $\frac{1}{9}$ of 45
11. $\frac{1}{3}$ of 30
12. $\frac{1}{6}$ of 54
13. $\frac{1}{5}$ of 40
14. $\frac{1}{8}$ of 72
15. $\frac{1}{2}$ of 20
16. $\frac{1}{10}$ of 60
17. $\frac{1}{4}$ of 36
18. $\frac{1}{7}$ of 63

Skill Hints

HINT To estimate measures, compare two objects whose measure you know.

A fish tank is about 10 mL or 10 L.

Think A teaspoon is about 5 mL and
a large container of milk is about 4 L.
10 L is the better estimate for the fish tank.

Choose the better measure for each of the following.

		a.	b.
1.	the distance from Texas to New York	a. meters	b. kilometers
2.	the length of a car	a. centimeters	b. meters
3.	the amount of water in your body	a milliliters	b. liters
4.	the amount of water in a swimming pool	a. milliliters	b. liters
5.	the weight of a slice of bread	a. grams	b. kilograms
6.	the weight of your math book	a. grams	b. kilograms

Choose the better estimate for each.

		a.	b.
7.	the weight of an apple	a. 5 oz	b. 50 oz
8.	the length of a dollar bill	a. 1 in.	b. 6 in
9.	the amount of juice in a 3-in. tall glass	a. 6 oz	b. 6c
10.	the temperature of a good swimming day	a. 93°F	b. 60°F
11.	the height of a flag pole	a. 2 ft.	b. 20 ft.

Skill Hints

HINT To find the perimeter of a figure, add the lengths of all the sides

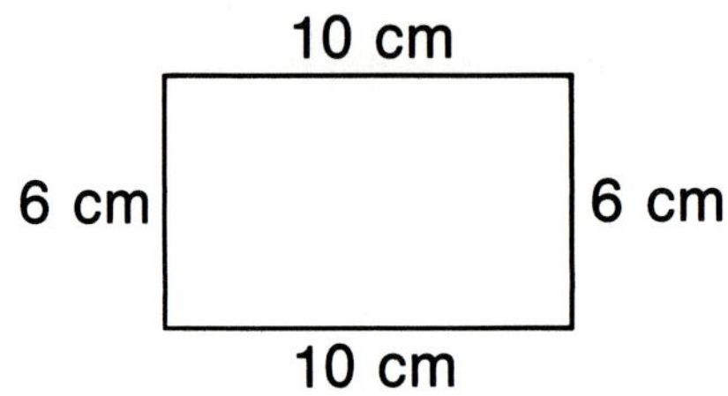

Perimeter = 10 + 6 + 10 + 6
= 32 cm

Find each perimeter.

1. 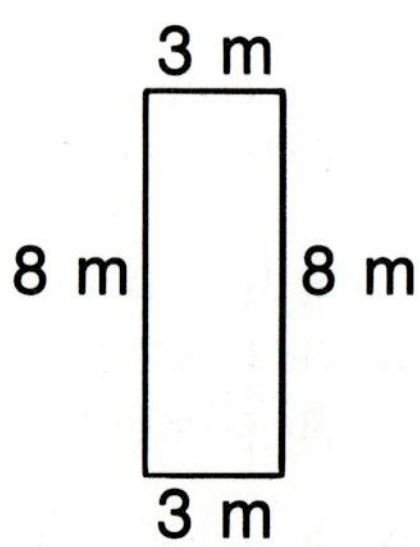

22m

2. 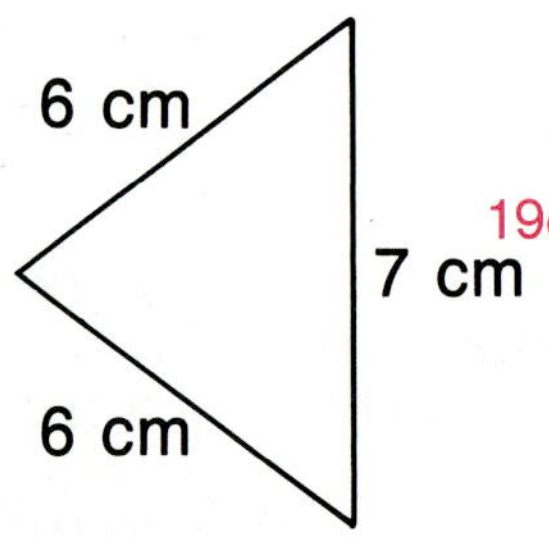

19cm

3. 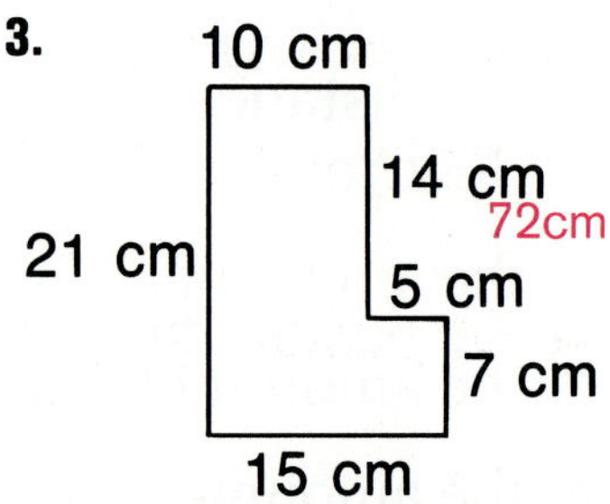

72cm

4. a triangle with each side 12m long 36m

5. a square with each side 32cm long. 128cm

HINT To find the area of a figure, count the square units which cover it.

1	2	3
4	5	6

The area of this figure is 6 square units

Find the area of each.

1. 9 square units

2. 6 square units

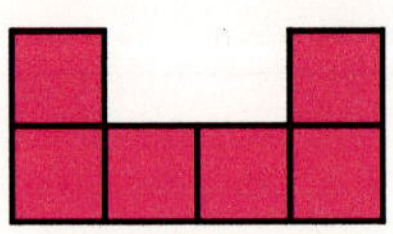

3. 8 square units

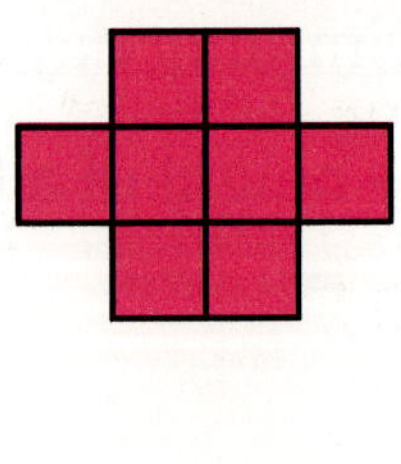

Skill Hints

HINT Tenths are written one place to the right of the decimal point.
Hundredths are written two places to the right of the decimal point.

$\frac{3}{10} = 0.3$ ↑ tenths place

$\frac{4}{100} = 0.04$ ↑ hundredths place

Write each as a decimal.

1. nine tenths
2. forty-three hundredths
3. seven tenths
4. one hundredths
5. 56 hundredths
6. five hundredths
7. five tenths
8. one and one tenth
9. $\frac{2}{10}$
10. $3\frac{7}{10}$
11. $\frac{26}{100}$
12. $\frac{4}{10}$
13. $2\frac{9}{100}$
14. $\frac{7}{100}$
15. $5\frac{27}{100}$
16. $\frac{39}{100}$
17. $\frac{87}{100}$
18. $8\frac{3}{10}$
19. $\frac{8}{10}$
20. $\frac{99}{100}$

HINT To add or subtract decimals, line up the decimal points. Place the decimal point in the answer under the points in the problem.

↓ line up decimals

$$\begin{array}{r} 0.3 \\ +0.2 \\ \hline 0.5 \end{array}$$

↑ decimal point in answer also lines up

Add or subtract.

1. $\begin{array}{r} 0.1 \\ +0.5 \\ \hline \end{array}$
2. $\begin{array}{r} 2.1 \\ +0.3 \\ \hline \end{array}$
3. $\begin{array}{r} 1.1 \\ -0.8 \\ \hline \end{array}$
4. $\begin{array}{r} 1.0 \\ -0.5 \\ \hline \end{array}$
5. $\begin{array}{r} 1.3 \\ +2.4 \\ \hline \end{array}$

Skill Hints

HINT Estimate first when multiplying. It will tell you if the product you find is reasonable.

$$\begin{array}{r} 49 \\ \times\ 2 \\ \hline 98 \end{array} \quad \xrightarrow{\text{Rounds to}} \quad \begin{array}{r} 50 \\ \times\ 2 \\ \hline 100 \end{array}$$

estimated product.
The actual product will be less than 100.

98 is a reasonable answer since the estimate is 100.

Estimate the product first. Then multiply

1. 38×4	**2.** 19×7	**3.** 52×3	**4.** 81×5	**5.** 94×8
6. 121×3	**7.** 125×4	**8.** 374×2	**9.** 181×5	**10.** 249×3

HINT The remainder in division must be less than the divisor. Change your quotient to a higher number if remainder is not less than divisor.

$$\begin{array}{r} 11 \\ 8\overline{)95} \\ 8\times \\ \hline 15 \\ -18 \\ \hline 7<8 \end{array}$$

Think $95 \div 8$ is about 10

$95 \div 8 = 11R7$

check

Divide.

1. $2\overline{)19}$	**2.** $3\overline{)29}$	**3.** $4\overline{)19}$	**4.** $7\overline{)69}$
5. $5\overline{)72}$	**6.** $4\overline{)58}$	**7.** $8\overline{)87}$	**8.** $6\overline{)82}$

Glossary

A.M. Used to show the time between midnight and noon. p. 140

addends The numbers that are added. p. 4
Example: 7 + 8 = 15
The addends are 7 and 8.

addition An operation on two or more numbers to find the sum. p. 7
Example: 4 + 2 + 3 = 9
The sum is 9.

angle Two rays with a common endpoint. p. 346 *Example:* 

area The number of square units needed to cover a region. p. 358

bar graph A graph with bars of different lengths to show information. p. 154

calendar A device that shows the 12 months of the year in order. p. 148

capacity The amount of liquid a container can hold. p. 306

centimeter (cm) A standard unit in the metric system to measure length. p. 300

chance A possibility or probability. p. 288

circle A closed figure. All the points of a circle are the same distance from a point called the center. p. 342

cone A space figure with one circular flat surface and one curved surface that meets in a point. p. 338
Example: 

congruent figures Figures that have the same size and shape. p. 352

corner A point where more than two faces of a space figure meet. Also, a point where two sides of a plane figure meet. p. 338

cube A space figure with six square faces. p. 338 *Example:* 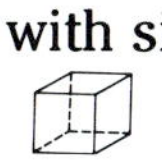

cup (c) A unit of volume in the customary system. p. 322

customary system A measurement system that measures length in inches, feet, yards, and miles; capacity in cups, pints, quarts, and gallons; weight in ounces, pounds, and tons; and temperature in degrees Fahrenheit. *See* Table of Measures. p. 478

cylinder A space figure with two faces that are circles. p. 338
Example: 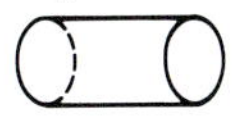

data Information that is gathered. p. 158

decimal A number with places to the right of a decimal point. p. 370
Examples: 0.4, 1.8

decimal point The dot used to separate dollars from cents and ones from tenths. p. 50 *Examples:* \$1.54, 1.3

decimeter (dm) A unit of length in the metric system equal to 10 centimeters. p. 300

degree Celsius (°C) A unit for measuring temperature in the metric system. p. 312

degree Fahrenheit (°F) A unit for measuring temperature in the customary system. p. 328

difference The answer in subtraction. p. 12
Example: 9 − 4 = 5
The difference is 5.

digit Any of the symbols used to write numbers: 0, 1, 2, 3, 4, 5, 6, 7, 8, and 9. p. 28

dividend The number to be divided. p. 234
Example: $6\overline{)36}$ or $\frac{36}{6}$.
The dividend is 36.

divisible A number is divisible by another number if the remainder is 0 after dividing. p. 428

division An operation on two numbers that results in a quotient. p. 230
Example: 18 ÷ 2 = 9

divisor The number by which another number is to be divided. p. 234

Example: $7\overline{)28}$ or $\frac{28}{7}$
The divisor is 7.

Glossary

dollar sign ($) A symbol used to mean dollars. p. 54 *Example:* $2.35

doubles Facts in which the same number is added to itself or multiplied by itself. p. 2

edge The segment where two faces of a space figure meet. p. 338
Example:

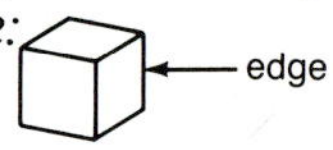

elapsed time The amount of time that has passed. p. 144

endpoint A point at the end of a line segment or ray. p. 345

equal (=) A symbol that shows one value is the same as another. p. 2

equivalent fractions Fractions that name the same number. p. 272
Examples: $\frac{1}{2}$ and $\frac{2}{4}$

estimate To give an approximate rather than an exact answer. p. 80

even number A number that has 0,2,4,6, or 8 in the ones place. p. 20

exchange Give one base-ten block for base-ten blocks of equal value or the reverse. p. 64

experiment: To carry out a plan in order to test a prediction. p. 44

face A flat surface of a space figure. p. 338
Example:

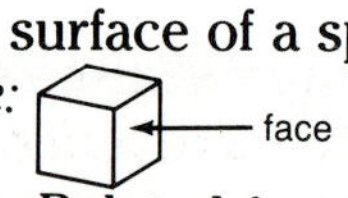

fact family Related facts using the same numbers. p. 18
Example: $2 + 3 = 5$ $5 - 3 = 2$
$3 + 2 = 5$ $5 - 2 = 3$.

factors The numbers that are multiplied to give a product. p. 175
Example: $3 \times 5 = 15$
The factors are 3 and 5.

flowchart A diagram that shows a step-by-step way to solve a problem. p. 13

foot (ft) A unit of length in the customary system equal to 12 inches. p. 320

fraction A number that names part of a whole or a group. p. 266
Examples: $\frac{1}{2}, \frac{2}{3}, \frac{6}{8}$ are fractions.

front digit The digit in the greatest place, used for estimation. p. 80

gallon (gal) A unit of volume in the customary system equal to 4 quarts. p. 322

gram (g) A unit in the metric system used to measure the weight of light objects. p. 308

graph A drawing used to show information. p. 154

greater than (>) The symbol used to compare two numbers when the greater number is written first. p. 46
Examples: $7 > 3$, $9 > 6$

grouping property of addition The way in which numbers are grouped does not change the sum. p. 6
Example: $2 + (4 + 5) = (2 + 4) + 5$

grouping property of multiplication The way in which numbers are grouped does not change the product. p. 219
Example: $2 \times (3 \times 5) = (2 \times 3) \times 5$

hour (h) A unit of time equal to 60 minutes. p. 140

hundreds Groups of 10 tens. p. 28

hundredths One or more of one hundred equal parts of a whole. p. 372

inch (in.) A standard unit in the customary system to measure length. p. 318

kilogram (kg) A unit in the metric system used to measure the weight of heavy objects. p. 310

kilometer (km) A unit in the metric system used to measure long distances. p. 302

less than (<) The symbol used to compare two numbers when the lesser number is written first. p. 46
Examples: $3 < 7$, $6 < 9$

like fractions Fractions that have the same denominator. p. 276

line A collection of points along a straight path that goes on and on in opposite directions. A line has no endpoints. p. 344

line of symmetry A line on which a figure can be folded so that both sides match. p. 354
Example: 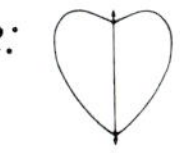

line segment A part of a line having two endpoints. p. 344

liter (L) A unit in the metric system used to measure amounts of liquid. p. 306

meter (m) A unit in the metric system equal to 100 centimeters. p. 302

metric system A measurement system that measures length in millimeters, centimeters, meters, and kilometers; capacity in milliliters and liters; mass in grams and kilograms; and temperature in degrees Celsius. *See* Table of Measures. p. 478

mile (m) A unit of length in the customary system equal to 5,280 feet. p. 320

milliliter (mL) A unit in the metric system used to measure small amounts of liquid. p. 306

minute (min) A unit used to measure a short period of time. p. 142

missing addend The number that is added to the lesser of two numbers to equal the greater number. p. 12

mixed number A number that has a whole number and a fraction. p. 280
Example: $3\frac{4}{5}$

multiplication An operation on two or more numbers, called factors, to find a product. p. 175
Example: $4 \times 5 = 20$
The product is 20.

multiplication fact The multiplication of two factors that are whole numbers from 0 through 9. p. 256

multiplication table A table that organizes multiplication facts. p. 216

multiplication sentence A multiplication statement written in horizontal form. p. 175
Example: $3 \times 4 = 12$

number line A line that shows numbers in order. p. 30
Example: 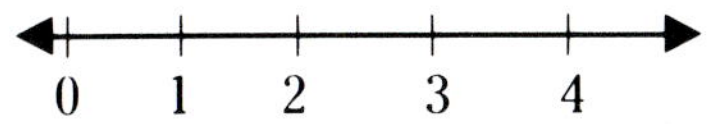

odd number A whole number that has 1, 3, 5, 7, or 9 in the ones place. p. 20

order property of addition The order in which numbers are added does not change the sum. p. 4
Example: $9 + 3 = 3 + 9$

order property of multiplication The order in which numbers are multiplied does not change the product. p. 176
Example: $3 \times 2 = 2 \times 3$

ordered pair A pair of numbers that give the location of a place on a map or a graph. p. 160

ordinal number A number used to tell order or position. p. 32 *Examples:* first, fifth

ounce (oz) A unit of weight in the customary system. p. 324

outcome A possible result in an experiment. p. 288

P.M. Used to show the time between noon and midnight. p. 140

pattern A repeated sequence. p. 186

perimeter The distance around a figure. p. 356

Glossary

pictograph A graph that shows number information by using picture symbols. p. 156

pint (pt) A unit of volume in the customary system equal to 2 cups. p. 322

place value The value of a digit determined by its position in a number. p. 28
Example: In 562, the digit 5 means 5 hundreds, the digit 6 means 6 tens, the digit 2 means 2 ones.

plane figure A geometric figure whose points are all in one plane. p. 342

point The location of an ordered pair of numbers. p. 160

pound (lb) A unit of weight in the customary system equal to 16 ounces. p. 326

predict To make a reasonable statement about what might happen. p. 158–159

probability The chance that an event will occur. p. 288

product The answer in multiplication. p. 175 *Example:* $4 \times 8 = 32$
The product is 32.

property of one for division Any number divided by one is that number. Any number except 0 divided by itself is 1. p. 142
Example: $6 \div 1 = 6$,
Example: $3 \div 3 = 1$

property of one for multiplication The product of any number and 1 is that number. p. 183
Example: $6 \times 1 = 6$ and $1 \times 6 = 6$

pyramid A space figure whose base is a polygon and whose faces are triangles with a common corner. p. 338

quart (qt) A unit of volume in the customary system equal to 4 cups, or 2 pints. p. 322

quarter hour A unit of time equal to 15 minutes. p. 140

quotient The answer in division. p. 414
Example: $24 \div 3 = 8$ or $\begin{array}{r} 8 \\ 3\overline{)24} \end{array}$
The quotient is 8

ray A part of a line that has one endpoint and goes on and on in one direction. p. 344

rectangle A plane figure with four right angles. p. 342

rectanglular prism A space figure whose faces are all rectangles. p. 338
Example: 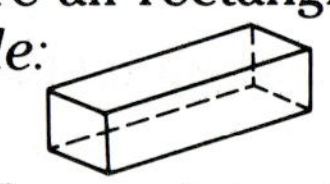

regroup To use 1 ten to form 10 ones, 1 hundred to form 10 tens, 12 ones to form 1 ten 2 ones, and so on. p. 66

remainder the number that is left over after dividing. p. 414
Example: $42 \div 8 = 5$ R 2
The remainder is 2.

repeated addition To add the same number again and again. p. 175

right angle An angle that has the shape of a square corner. p. 346
Example: 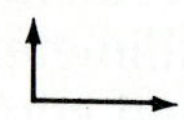

rounding Expressing a number to the nearest tne, nearest hundred, and so on. p. 34
Example: 43 rounded to the nearest ten is 40.

space figure A geometric figure whose points are in more than one plane. p. 338

sphere A space figure shaped like a round ball. p. 338
Example: 

square A rectangle with four equal sides. p. 342

square number The product of two equal factors. p. 202.
Example: $5 \times 5 = 25$
25 is a square number.

subtraction An operation on two numbers to find the difference. p. 12
Example: $15 - 3 = 12$
The difference is 12.

sum The answer in addition. p. 2
Example: 8 + 7 = 15
The sum is 15.

symmetry A figure has symmetry if it can be folded along a line so that the two parts match exactly. p. 354

tally A mark made to keep score or to count. p. 158

tens Groups of 10 ones. p. 106

tenths One or more of ten equal parts of a whole. p. 371

triangle A figure with three sides and three corners. p. 342

unlike fractions Fractions that have different denominators. p. 278

volume The number of cubic units that fit inside a space figure. p. 360

yard (yd) A unit of length in the customary system equal to 3 feet, or 36 inches. p. 320

zero property of addition The sum of any number and 0 is that number. p. 6
Example: 3 + 0 = 3

zero property of division Zero divided by any number except 0 is 0. You cannot divide a number by 0. p. 246
Example: 0 ÷ 12 = 0

zero property of multiplication The product of any number and 0 is 0. p. 183
Example: 5 × 0 = 0 and 0 × 5 = 0

zero property of subtraction When zero is subtracted from any number, the difference is that number. When a number is subtracted from itself, the difference is 0. p. 14
Examples: 7 − 0 = 7, 12 − 12 = 0

Computer Terms

BACK (BK) A Logo command that moves the turtle back.
Example: BK 3 moves back 3 turtle steps.

CLEARSCREEN (CS) Clears the screen. In some versions, also homes the turtle.

DRAW or **CG** In some versions, clears the screen and homes the turtle.

FORWARD (FD) A Logo command that moves the turtle forward.
Example: FD 10 moves forward 10 turtle steps

HOME A Logo command and that positions the turtle in the middle of the screen.

LEFT (LT) A Logo command that turns the turtle left.
Example: LT 90 turns the turtle a right angle to the left.

Logo A computer language.

path The movement of the turtle when drawing in Logo.

PENDOWN (PD) Allows the turtle to draw again.

PENUP (PU) Allows the turtle to move without drawing.

PRINT (PR) A Logo command that shows information on the screen.

RANDOM A Logo function that returns a random number.
Example: RANDOM 5 returns a number from 0 to 4.

REPEAT A Logo command that repeats commands.
Example: REPEAT 4 [FD 10] moves the turtle forward 4 times.

RIGHT (RT) A Logo command that turns the turtle right.
Example: RT 90 turns the turtle a right angle to the right.

turtle step How far the turtle moves using the command FORWARD 1 or BACK 1.

TABLE OF MEASURES

Metric		Customary	
Length		**Length**	
1 centimeter (cm)	= 10 millimeters (mm)	1 foot (ft)	= 12 inches (in.)
1 meter (m)	= 100 centimeters	1 yard (yd)	= 36 inches, or 3 feet
1 kilometer (km)	= 1,000 meters	1 mile (mi)	= 5,280 feet, or 1,760 yards
Mass		**Weight**	
1 kilogram (kg)	= 1,000 grams (g)	1 pound (lb)	= 16 ounces (oz)
Capacity		**Capacity**	
1 liter (L)	= 1,000 milliliters (mL)	1 pint (pt)	= 2 cups (c)
		1 quart (qt)	= 2 pints
		1 gallon (gal)	= 4 quarts

Time		Money	
1 minute (min)	= 60 seconds (s)	1 nickel	= 5 cents (¢)
1 hour (h)	= 60 minutes	1 dime	= 10 cents, or 2 nickels
1 day (d)	= 24 hours	1 quarter	= 2 dimes and 1 nickel
1 week (wk)	= 7 days	1 half-dollar	= 2 quarters
1 month (mo)	= 28 to 31 days, or about 4 weeks	1 dollar ($)	= 4 quarters
1 year (yr)	= 12 months, or 52 weeks, or 365 days		

SYMBOLS

=	is equal to	10¢	ten cents
>	is greater than	$1.60	one dollar and sixty cents
<	is less than	6:45	six forty-five o'clock
. . .	and so on	°C	degree Celsius
		°F	degree Fahrenheit

Index

Index

Index

Credits

CONTRIBUTING ARTISTS: Anthony Accardo: 72-73, 84-85, 206; Ernest Albanese: 2, 3, 7, 8-9, 24, 25, 38-39, 94, 135; Robert Alley: 86-87; Bryn Barnard: xii, 376-377; Laurie Bernero: 110, 122; Alex Bloch: 4, 26, 28, 29, 30-31, 32-33, 34, 35, 36, 37, 39, 40-41, 46-47, 48, 51, 52, 53, 55, 60, 186, 202, 208, 219, 223, 244, 283; Denise Brunkus: 102; Susan Johnston Carlston: 315, 316; Penny Carter: 378-379, 392; Carolyn Croll: 2-3; Ray Dallasta: 10, 128-129, 140, 220-221, 304, 310-311, 312-313, 314, 318, 416-417, 425; Susan David: 192; Daniel DeValle: 378; Nancy Didion: 4, 22, 23, 215; John Dispenza: 211; Chris Drosse: 100; Mac Evans: 70, 96; Simon Galkin: 73, 104, 105, 130, 339; Patrick Gnan: 15, 144-145, 250-251, 352-353, 404-405; Meryl Henderson: 273, 303, 306, 307, 322, 323, 326, 328, 354, 355, 462; Robin Hotchkiss: 225, 240; Paul Izzo: 414; Marilyn Janovitz: 64, 65, 104, 172, 175, 342, 396-397; Sandford Kossin: 384; Ron Lehew: 254-255; Morissa Lipstein: 2, 207, 209; Steve Macanga: ix; Cyd Moore: 18, 19; Jared Moore: 415; James Needham: 151, 156, 178, 182-183, 188; John Nex: 8, 60, 120, 124, 134, 208, 224, 262, 300, 318, 319, 320, 321, 405; Jan Palmer: 16, 17, 88, 161, 168; Jan Pyk: 80, 82, 176-177, 197, 238, 268, 269, 270, 278-279, 280; Katie Quann: 190; Larry Raymond: 204; Bob Roper: 160; Sandy Ruda: 37, 157; Bob Shein: 246-247; Jerry Smath: 9, 11, 126, 127; Patrick Soper: 68-69, 316-317; Susan Spellman: 34-35, 66-67; Ken Spengler: 406-407; Peggy Tagel: 148-149; Joe Tonna: 76, 77, 185; Joe Veno: 338; Isabell Weberbauer: 415; Mel Williges: 370-371; Christopher Ziemann: 415.

Line and Graph Art: Burmar

PHOTOGRAPHS: All photographs by Silver Burdett & Ginn (SB&G) unless otherwise noted.

All sports equipment, Courtesy of Fitzgerald's, Morristown, NJ.

FRONT AND BACK COVER: William Edward Smith/The Stock Market.

TABLE OF CONTENTS: iii: *t.* Frank Oberle/After Image; *m* Dan DeWilde for SB&G. iv: *t.* David Ball/Allstock; *m.* Mickey Gibson/Animals Animals; *b.* Jeff Rotman/Peter Arnold, Inc.; v: *t.* V. Clevenger/H. Armstrong Roberts; *m.* Comstock. vi: *t.* Lawrence Migdale/Photo Researchers, Inc.; *m.* R.S. Virdee/Grant Heilman Photography. vii: *t.* Comstock; *m.* The Image Bank. viii: Gary Braasch.

INTRODUCTORY CHAPTER: ix: *t.r.* Michael Philip Manheim/The Stock Market; *m.m.* Cheryl Griffin for SB&G; *b.r.* Werner H. Muller/Peter Arnold, Inc.

CHAPTER OPENER 1 xii-1: Bryn Barnard.

CHAPTER 1 11: *b.* Barbara Burnes/Photo Researchers, Inc. 12: *t.* Dan DeWilde for SB&G. 14: Andrew Rakoczy/Photo Researchers, Inc.

CHAPTER OPENER 2 All SB&G

CHAPTER 2 38-39: Dan DeWilde for SB&G. 44: Courtesy of Willowbrook Mall, Wayne, NJ. 50: Victoria Beller-Smith for SB&G.

CHAPTER OPENER 3 62-63: Dan DeWilde for SB&G. 62: Bob Daemmrich/Stock, Boston. 63: *l.* Frank Oberle/After Image; *r.* Mitchell Reibel/Sportschrome East/West.

CHAPTER 3 70: Kamyar Samoul. 76-77: Murray & Associates/After Image.

CHAPTER OPENER 4 98-99: Dan DeWilde for SB&G.

CHAPTER 4 106: Reibel-Lovine/Sportschrome, Inc.

CHAPTER OPENER 5 139: *l.* David Ball/Allstock; *r.* Ken Kerbs for SB&G. 142: *t.r.* Roy Morsch/The Stock Market. 152: *b.l.* John Elk III/Stock, Boston. 152-153: Steven R. Krous/Stock, Boston. 153: *r.* J. & L. Weber/Peter Arnold, Inc.

CHAPTER OPENER 6 170-171: Mickey Gibson/Animals Animals. 171: *l.* Boyd Norton/Comstock; *r.* Jeff Rotman/Peter Arnold, Inc.

CHAPTER 6 178: *m.r.* Clyde H. Smith/The Stock Shop. 185: Jim Pickerell, TSW, Click/Chicago. 187: *m.r.* Peter Rosendale, TSW, Click/Chicago; *b.r.* Art Stein/Photo Researchers, Inc. 192: Karl & Steven/Photo Researchers, Inc. 194-195: *t.* Kevin Schafer/Peter Arnold, Inc; *m.* Manfred Daneggler/Peter Arnold, Inc.; *b.* Johnny Johnson/Animals Animals.

CHAPTER OPENER 7 200-201: V. Clevenger/H. Armstrong Roberts.

CHAPTER 7 212-213: John Buttenrant/Photo Researchers, Inc. 214: Roger Ressmeyer/Starlight. 216: *t.r.* Kamyar Samoul.

CHAPTER OPENER 8 228-229: Comstock.

CHAPTER 8 236: J. Gerard Smith for SB&G. 248-249: *b.* Dwight R. Kuhn/Bruce Coleman. 248: *t.* Werner H. Muller/Peter Arnold, Inc.; *b.m.* Robert Maier/Animals Animals. 250: *m.r.* Rothwell/After Image. 252: *t.r.* *L.L.*T. Rhodes/Animals Animals; *m.r.* Gregory G. Dimijian/Photo Researchers, Inc.

CHAPTER OPENER 9 264-265: Lawrence Migdale/Photo Researchers, Inc.

CHAPTER 9 268: *t.r.* The Bettmann Archive. 286: John Running for SB&G.

CHAPTER OPENER 10 298-299: R.S. Virdee/Grant Heilman Photography. 299: *l.* Runk-Schoenberger/Grant Hielman Photography; *m.* Gregory Dimijian/Photo Researchers, Inc.; *r.* Tim Fitzharris/Allstock.

CHAPTER 10 300: *t.* Ray Yaros/Imagery. 302. Leonard Lee Rue/Stock, Boston. 303: TOM STACK & ASSOCIATES. 306: Rick Stewart/AllSport USA. 309: *t.m.* J. Gerard Smith for SB&G. 315: Four by Five. 324-325: Dan DeWilde for SB&G. 328-329: Grant Heilman/Grant Heilman Photography.

CHAPTER OPENER 11 336-337: Lawrence Migdale/Stock, Boston. 336: Comstock.

CHAPTER 11 338: Doll house & Furniture Courtesy THE DOLL HOUSE FACTORY, Lebanon, NJ. 342: *t.r.* Art Resource, 346-347: Richard Weiss/Peter Arnold, Inc. 349: F.B. Grunzweig/Photo Researchers, Inc. 354: Kyoko Ilaga. 355: Glenn Short/After Image. 360: IMAGERY for SB&G

CHAPTER OPENER 12 368-369: The Image Bank.

CHAPTER 12 370-373: Jerry Jacks. 374-375: IMAGERY. 376: The Adler Planetarium. 380: Henley & Savage/The Stock Market. 383: *t.* Cheryl Griffin; *m.r.* Ken Kerbs for SB&G. 389: *b.r.* Dan DeWilde for SB&G.

CHAPTER OPENER 13 394-395: Gary Braasch.

CHAPTER 13 400; *t.r.* Mark Boulton/Photo Researchers, Inc.; *m.r.* Geo. H.H. Huey/Animals Animals. 402: *t.r.* William Johnson/Stock, Boston. 404: *t.r.* Douglas Faulkner/Photo Researchers, Inc. 408: *t.r.* Gregory G. Dimijian/Photo Researchers, Inc. 408-409; *b.* C.W. Perkins/Animals Animals. 409: Terry G. Murphy/Animals Animals. 411: Tom McHugh/Photo Researchers, Inc.; *b.l.* Alan Levenson/After Image. 412-413: Paul Dix/After Image. 418: *t.r., m.* Stephen Dalton/Photo Researchers, Inc.; *b.* John Gerlach/Animals Animals. 422: m.r. Gary Braasch/Wheeler Pictures. 423: *m.l.* Calvin Larsen/Photo Researchers Inc.